T0245933

MILITARY, NAVAL AND CIVIL
AIRSHIPS

MILITARY, NAVAL AND CIVIL
AIRSHIPS

THE HISTORY AND DEVELOPMENT OF THE DIRIGIBLE AIRSHIP IN PEACE AND WAR

DANIEL GEORGE RIDLEY-KITTS MBE

The
History
Press

To all the National Servicemen who served in Germany 1956 to 1958
during the Cold War.
RAF Celle, Sylt, Ahlhorn
'Boys doing a man's job'

Cover illustrations: *front:* North Sea class airship NS9 at Longside RNAS
station, 1918; *back:* Britain's first military airship *Nulli Secundus* in flight at
Farnborough, 1908. (Both © Airship Heritage Trust Collection
(www.airshipsonline.com))

First published 2012 as *Military, Naval and Civil Airships since 1783*
This paperback edition published 2023

The History Press
97 St George's Place, Cheltenham,
Gloucestershire, GL50 3QB
www.thehistorypress.co.uk

© Daniel Ridley-Kitts, 2012, 2023

British Library Cataloguing in Publication Data.
A catalogue record for this book is available from the British Library.

ISBN 978 1 80399 527 4

Typesetting and origination by The History Press
Printed and bound in Great Britain by TJ Books Limited, Padstow, Cornwall

Trees for LYfe

Contents

Acknowledgements 7

Introduction: The Conquest of the Air 9

1   The Airship: An Explanation 13
2   The Development of the Dirigible 1783–1914 23
3   Early British Efforts 1878–1914 41
4   The First British Rigid: *Mayfly* 59
5   Interim 1912–14: The Astra-Torres, Parseval & Forlanini Airships 69
6   The Second Navy Rigid: *R9* 83
7   War at Sea: Sea Scouts vs the U-boat 91
8   War at Sea: Airships with the Grand Fleet 103
9   Air Defence of Great Britain: Opening Moves 117
10 Air Defence of Great Britain: Nemesis 127
11 *Zeppeline Gegen England und Weiter* 1915–18 145
12 The *R23 & 23X* Classes 157
13 Short Brothers' Wooden-Framed Y Type: *R31 & R32* 165
14 *R34* Across the Atlantic 175
15 The Last of the Aerial Cruisers: *R36 & Shenandoah* 185
16 The Vickers *R80*, Airship Sheds & Mooring Masts 203
17 The Admiralty A Type: *R38/ZR2* 213
18 The Airship vs the Flying Boat 227
19 Airship Development in the Interwar Years 235
20 From the Second World War to the Present Day 263

Appendix 1   The *R34*'s Transatlantic Flight 277
Appendix 2   Airship Development in Great Britain 1900–30 283
Appendix 3   The *LZ-127 Graf Zeppelin* 287
Appendix 4   The First Zeppelin Raid on the British Isles 289

Bibliography 294
Index 295

Acknowledgements

I first became interested in the history and development of airships in the mid-1950s, when, as a young national serviceman, I was posted to RAF Ahlhorn in Lower Saxony, Germany.

I soon learnt that this air station had been the headquarters of the Imperial German Naval Airship Service in the First World War. In the evening, after servicing the de Havilland Venom fighters of our squadron, we would visit the local bars in the village where pictures of naval airships adorned the walls. Here we learnt something of the activities that had taken place on the *flugplatz* (airfield) during the Great War from a couple of old boys who had worked on the base at the time.

Out on the field there were still the outline traces of the massive sheds that had existed there long ago, and I discovered that my own billet had in fact been the headquarters building used by the redoubtable Fregattenkapitän Peter Strasser – leader of naval airships throughout the war period until his death aboard the *L70*, in the last raid on the British Isles in August 1918, off the Norfolk coast.

So my first acknowledgement must go to the two old German gentlemen who first fired my interest in the subject of airships.

Over the intervening years as I put the book together, I have had help and encouragement from various friends, including my long-suffering wife, Christine, who very sensibly has never read a page of my writings but has patiently encouraged me to 'get it printed'.

More recently, as the book approached completion, various people have assisted me in a number of different ways. I am indebted to my neighbour, Kapil Joshi, who has come to my aid in solving problems with my computer that arose from time to time, and to Peter Lindsley of CBM Publishing Services, Cromer, who expertly reproduced the line drawings contained within the text.

I am also grateful to the Airship Heritage Trust of Cardington, Bedfordshire, who have kindly allowed me access to their photographic archive for the photos used in the book, with special mention being made to Dr Edwin Mowforth and Derek Millis for their assistance and encouragement.

Finally, I should like to thank The History Press, for the original commission and this new edition.

INTRODUCTION

The Conquest of the Air

The dream of human flight is a recurring image in the history of mankind, with man's earliest attempts to fly based, not unnaturally, on an imitation of bird flight. The history of manned flight from Icarus onward is littered with mad tower jumpers launching themselves into the void, with flapping wings strapped to their backs. Later came the more reasoned and considered works of Leonardo da Vinci, and other early pioneers, who struggled to understand the secrets of flying with all their essays based on the principle of heavier-than-air flight.

There is no obvious parallel in nature that corresponds to the lighter-than-air principle that airships rely on to 'float' in the air, but the monk and natural philosopher Roger Bacon (1214–94) speculated that the atmosphere above the earth possessed an upper surface on which 'an aerial vessel could be made to float on this sea of air'. In his work *Secrets of Art and Nature*, written in 1250, he describes such a craft as being: 'a large hollow globe fashioned from copper, wrought very thin in order to save weight, and to be filled with "Ethereal air" or "Liquid fire", then to be launched from an elevated point into the atmosphere, where it will float as a boat on water.'

Francisco de Mendoza (d. 1626) also considered the employment of what he termed 'elementary fire' to maintain such a vessel in the air, where it could be propelled by oars. Better known are the writings of Francesco de Lana (1631–87), who proposed a 'flying boat' that would float in the atmosphere by virtue of four large copper globes from which the air had been evacuated; the resulting displacement of air being sufficient to provide the necessary lift.

De Lana had in fact conducted experiments to determine the weight of air, the results of which – despite the primitive apparatus at his disposal – were very close to the modern accepted measurements. Unfortunately, the scheme was completely impractical, as de Lana had failed to appreciate that the atmospheric pressure acting on the evacuated globes, which he thought would help to consolidate the spheres, would in fact crush them flat. De Lana did, however, foresee the use of flying machines in warfare, with the bombing of fortresses and ships at sea by the dropping of 'fire balls' and the rapid transportation of troops, together with describing their employment for commerce.

The invention of the hot air balloon is generally ascribed to the Montgolfier brothers in 1783, with the first manned flight made by Pilâtre de Rozier on 15 October of that year, in Paris. This event was followed, in December 1783, by the first manned ascent of a hydrogen balloon under the direction of the physicist J.A.C. Charles, which was to become the more practical and convenient form of craft for the purpose of aerial navigation. Indeed, the superiority of the hydrogen balloon was quickly established over its hot air competitor in terms of convenience of operation and its ability to conduct flights of greater duration.

The rapid pace of the development of the hot air and the hydrogen balloon following the Montgolfiers' first flight was extraordinary: within just over a year Blanchard and Jeffries had completed the first crossing of the English Channel, and ascents were being made in London and other European capitals during this period.

Throughout the nineteenth century, as intrepid balloonists became ever more familiar with the ocean of the air, undertaking long voyages and making high-altitude ascents to study the composition of the upper atmosphere, they were still at the mercy of the winds that determined the direction of travel. The question of effective control, together with a method of propulsion, exercised the mind of the early aeronauts, and various wildly impractical schemes were suggested, including the use of oars and sails, or even being towed by trained birds.

The first scientifically based solution for controlled dirigible flight was put forward only days after the first ascent of the hydrogen balloon in 1783, when the brilliant French army officer and savant, Jean Baptiste Meusnier, proposed the design of an elongated ellipsoidal balloon. Driven through the air by three propellers and operated by manual power, this was the first application of the airscrew in aviation. Although it was not built, Meusnier's design incorporated almost all the features to be found in modern airships.

In 1852, Henri Giffard built a steam-powered dirigible with which he was able to achieve a degree of control, although he was unable to navigate a circular course and return to his starting place. Twenty years on, in 1872, Austrian Paul Haenlein flew an airship propelled by a Lenoir gas engine, the fuel for which was supplied from the airship's envelope, again with some success.

Later, in 1883, the famed balloonists the Tissander brothers applied electric power in the search for an effective and powerful prime mover, although their results were of limited success. This effort was quickly followed, in 1884, by the more successful experiments under the direction of Renard and Krebs of the French army aeronautical establishment at Chalais Meudon, again employing electrical power with the airship *La France*. In a series of trials the airship demonstrated controlled flight, achieved a speed of 14mph and made several flights over Paris. However, the disproportionate weight of the electric batteries it carried, and its limited range and carrying capacity, militated against further development.

Throughout the later years of the nineteenth century ballooning became a popular sport of the wealthy upper classes, particularly in Europe where many flights of long duration were accomplished, flights from England to Russia or other parts of the Continent being not uncommon. It was not until the development of the internal combustion engine in the late 1890s, however, that the airship finally became a practical craft.

One such wealthy aeronaut was a young Brazilian heir to a coffee fortune, Alberto Santos-Dumont. In 1898, tiring of the thrills of ballooning, Santos-Dumont turned his attention to the building of a series of small and, as it turned out, highly efficient pressure airships propelled by petrol engines. His highly publicised aerial activities over Paris – including winning the prestigious Deutsch de la Meurthe prize in November 1901 for the first flight from St Cloud, around the Eiffel Tower and returning to the starting point – did much to stimulate interest in the future development of the airship.

Concurrently, Count Ferdinand von Zeppelin was conducting experimental flights with his giant rigid-framed dirigible on Lake Constance in southern Germany. Whilst the French, led by the Lebaudy brothers, were developing another form of airship structure: the semi-rigid type.

From 1900 to 1914 several hundred airships of all designs were produced in Europe and the United States; some for sporting purposes and some for commercial passenger carrying. However, in the period in question – one of rising international tension – many airships were designed with a military purpose in mind, as the airship was seen to be a potential weapon of war.

In Germany Count Zeppelin, who had conceived his airship for war purposes in a patriotic desire to provide an air fleet to protect the nation, was frustrated by an initial lack of interest from the military authorities. In May 1909, with no military contract in the offing, Alfred Colsman, the general manager of the fledgling Zeppelin concern, suggested that a separate company should be formed to operate a passenger-carrying service. This would initially operate pleasure flights of short duration, and later the company would initiate regular scheduled services between major German cities.

The count was not himself in favour of such commercial activities, but under Colsman's persuasive and reasoned arguments he finally agreed to the scheme, which would allow for the further development and perfection of his invention. The company, *Deutsche Luftschiffahrts-Aktiengesellschaft* (DELAG), was thus founded in November 1909 with a capital of 3 million Marks. They ordered their first airship, *Deutschland*, while Colsman negotiated with the burgomasters of large towns throughout Germany to provide 'aerial harbours' to house the envisaged fleet of airships.

The proposed regular service between cities failed to materialise, but DELAG did operate seven passenger airships between June 1910 and July 1914, carrying 34,000 passengers and crew on 1,500 flights without any fatalities – although three airships were destroyed in accidents. The success of these early commercial undertakings encouraged the military and naval authorities to develop the airship for war purposes, with the army operating eight Zeppelin and a score of other airships in service at the outbreak of the war in 1914.

The parallel development of the heavier-than-air flying machine during this period created two groups of adherents: those who favoured the airship were convinced that, due to its greater lifting capacity, it would ultimately carry the main aerial commerce of the world; whilst the visionary pioneers in their frail 'flying machines' were equally certain that the success of air transport lay with the aeroplane.

At the start of the Great War the Zeppelin was regarded as a potent weapon, which, thanks to popular writers of the time, was imbued with tremendous powers of destruction and initially appeared to be immune from attack by the primitive aeroplanes. As the war progressed the aeroplane gained the upper hand, however. By the war's end, with the development of fast 'scout' aircraft that could rapidly climb to the Zeppelin's altitude, and with the aid of phosphorous ammunition, the fate of the airship for military use was largely sealed.

In the 1920s the use of airships for commercial purposes was considered by several European countries, specifically for establishing transcontinental and trans-oceanic air links. France and Italy made a series of experimental flights between their colonies, but without major success, whilst Great Britain embarked on plans for an ambitious airship service to link the empire with the home country.

After spending millions on the project between 1924 and 1930, the enterprise ended in failure, and from then on British engineers concentrated their efforts on the development of the aeroplane. This change of emphasis, fortunately for the country, culminated in the development of the Spitfire and the Hurricane; planes that were key to winning the Battle of Britain in 1940.

During the 1920s and 30s, America produced two huge naval dirigibles designed to work with the fleet, but both, after a relatively short life, were lost at sea. This, however, did not cause the United States to abandon the airship, as during the Second World War the US navy operated some 200 small blimps for ocean patrol and convoy duty.

Germany, restricted as it was by the terms of the Versailles Treaty of 1919, was forbidden to construct large airships, but nevertheless later built a large rigid for the US navy in payment of war reparations. Then, in 1928, the Zeppelin Company (that had largely managed to keep in business by manufacturing aluminium kitchenware) built the most famous airship of all time: the *Graf Zeppelin*.

Between 1928 and 1937 this famed airship traversed 1,053,000 miles in 590 flights, which included a round-the-world flight; numerous ocean crossings of both the North and South Atlantic on passenger service; and scientific flights into the icy Arctic regions and the extreme heat of the Middle East. All these journeys were accomplished without serious mishap or loss of life, and did more to encourage the view that there was a place for the airship in long-distance passenger travel.

A larger successor to the *Graf Zeppelin*, the 806ft-long *Hindenburg* of 1936 was, for a while, hailed as the first of a fleet of giant passenger airships that would span the globe. Such ill-founded optimism was short lived, however, and her career ended in a spectacular fiery denouement at Lakehurst, New Jersey in May 1937, effectively drawing a line under the concept of long-distance airship travel.

Following the war, the airship faded from the aeronautical scene, until small pressure airships were used in increasing numbers as aerial camera platforms. These activities promoted a spate of airship development around the world, including a revival of the Zeppelin Company, who re-entered the field with the highly sophisticated Zeppelin NT (or New Technology).

Today, military interest has been revived, with several US government programmes dedicated to deploying high-altitude surveillance airships to monitor military operations in war zones, or even as launching platforms for missiles. These and other even more unusual uses for the airship and its derivative designs indicate that, far from being an anachronistic machine from a distant age, the airship still appears to have the ability to surprise us.

1

The Airship: An Explanation

The principle by which an airship maintains itself in the air is, of course, displacement: whereby a measured volume of air is replaced by an equal volume of a lighter gas, such as helium or hydrogen, which imparts a static lift to such a craft.

In the case of a hydrogen-filled airship, 1,000cu ft of air at standard pressure and temperature weighs 80lb, while an equal volume of hydrogen will weigh 65.5lb, the difference in weight between these two gases giving a resultant lifting force of 14.5lb per 1,000cu ft of hydrogen. In reality, the complex equation required to calculate the lift of an airship includes such variables as ambient air temperature, gas pressure, humidity, barometric pressure and gas purity (that is to say, to what degree the hydrogen has been contaminated with common air due to osmosis through the cell walls). These calculations assume a standard pressure and temperature at sea level with the cells 95 per cent full and the craft fully laden with crew, petrol, water ballast and its useful load of cargo or weapons.

During the First World War, hydrogen was the sole lifting agent – an explosive element first identified by the English physicist Cavendish in 1742. Hydrogen and later coal gas (a cheaper alternative that, although affording less lift, was more readily available in Victorian times from town gas supplies) were used by the early balloonists for their aerial voyages. Whilst they provided the necessary lifting force, they were both extremely dangerous to use in practice, a fault which, in the end, militated against their employment for both civilian and military use.

The use of helium – an inert and, therefore, safe gas – was introduced in 1922 by the Americans for use in their airships, following a series of fatal fires involving hydrogen-filled airships in the United States and Europe. The Americans had a monopoly of this rare gas; first detected in the spectrum of the sun before being discovered on earth in the 1900s in Texan oil wells, from where it was produced by a complex and expensive refining process.

Helium was not only some ten times more expensive to produce than hydrogen in those early days, but also possessed only 85 per cent of the lifting capacity compared with an equal volume of hydrogen. Consequently, a helium-filled airship would have to be proportionally larger than a hydrogen-filled craft in order to lift the same load.

The efficiency of an airship is measured by a comparison between the 'gross lift', or the total lifting power, and the 'useful or disposable lift'. This represents the difference between the fixed weights of the structure, framework, gas cells, outer cover, engines and equipment, and the useful payload in terms of crew, passengers, cargo, fuel and water ballast. In most early airships the useful load was in the order of 25 per cent of the total lift, although in the 'height climber' class produced by the Zeppelin Company towards the end of the Great War, efficiencies in excess of 60 per cent were achieved.

These savings were only accomplished at the cost of lightening the structure to such an extent that structural failure was an ever-present danger, which could and did occur when these lightly built ships were subjected to high-speed manoeuvres in the denser air at low altitude.

Airships fall into three categories that relate to their structure and the method of construction employed. The first type is the pressure airship, or blimp. This is, in effect, a motorised balloon – from which they were first developed effectively in the mid-nineteenth century – initially using steam, then electricity and finally the internal combustion engine as prime mover.

In its simplest form the pressure airship consists of an elongated balloon of streamline form, in order to present a surface of least resistance to the airflow, and containing the lifting agent which would have been hydrogen before and during the First World War period. Below the envelope would be suspended a car containing the crew, controls, engines, ballast and bombs etc, with the fuel tanks either suspended above the car or situated within the envelope. Steerable control surfaces were fitted to the rear portion of the envelope in order to achieve directional control in flight.

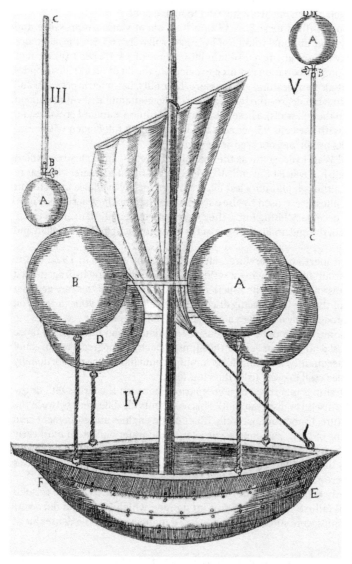

1. Francesco de Lana's aerial ship proposal, 1680.

In a pressure airship, the shape of the gas-tight envelope is maintained by the internal gas pressure being at a higher value than that of the supporting air, so maintaining the integrity of the structure under varying conditions of air pressure relative to altitude. Also, the temperature differential caused by meteorological conditions will affect the density and volume of the lifting agent, and therefore its lifting capacity.

As previously stated, the displacement of the heavier air by the lighter hydrogen or helium provides the static lift that enables the airship to float in the air. Around 1,000cu ft of hydrogen will lift 14.5lb at standard pressure and temperature. However, barometric and atmospheric conditions can adversely affect the lift equation with great rapidity. For example, the shadow of a cloud falling on the nose of an airship can cause an instant cooling effect on the gas, contracting the volume and lessening its lift, causing the nose to pitch down. Conversely, with a ship emerging from a cloud into warm sunshine, an instant heating effect will expand the gas: increasing the lifting capacity and causing the nose to rise.

A colder gas temperature regime within the envelope, in relation to the ambient air temperature, affords a greater percentage of lift than a warmer, more rarefied gas regime. On the other hand, if the gas contained in the cells is warmer than the outside air, this will result in less lift being created from an equal volume of gas. Paradoxically, more lift can be temporarily generated in a heavy airship by driving the craft down rapidly from a higher altitude: so compressing the gas molecules and producing heat, which leads to a temporary increase in lift. This method was often employed by airships when returning to base following a long patrol, having expended much of their lifting gas and ballast, in an 'exhausted' condition, and in danger of making a heavy landing.

In pressure airships it was (and still is) necessary to inflate the envelope to a higher pressure than that of the surrounding air in order to maintain the required degree of rigidity against the airflow – this pressure being in the order of 5lb/sq ft. This positive pressure is maintained by the introduction within the envelope of air-filled ballonets, usually two or more in number, occupying around a fifth to a quarter of the total volume of the envelope.

The ballonets are maintained at a positive pressure by an aluminium tube positioned in the wake of the propeller slipstream, allowing air to enter the ballonet via a one-way valve known as a crabpot. Any excess air is valved off through automatic air valves as the hydrogen expands in response to changes in pressure when the ship rises into less dense air. The ballonets are also employed to adjust the longitudinal trim of the craft, by partially filling or emptying the fore and aft compartments about the airship's centre of gravity (CG) to tilt the ship up or down, relative to the line of flight.

In order to prevent the fabric of the bow being pressed in under the pressure of the airflow, it was common practice to reinforce the bow-section cone with stiffening rods, usually made from bamboo radiating from a central aluminium nose cone and secured to the envelope by fabric pockets.

In pressure airships the fixed weights – that is to say the distribution of engines, control car, fins etc – are supported from the envelope by a system of rigging wires, carried either internally or externally, and designed to spread the load evenly while maintaining the aerodynamic shape of the envelope and the trim of the ship. During the First World War, the British predominated in the development of this type of craft, building and operating over 200 very efficient and handy airships that were credited with some notable performances.

At the present time, the pressure airship remains almost the sole surviving type of dirigible airship that is still operating.

The second type of airship for consideration is the semi-rigid, which, as its name implies, is possessed of some elements of a rigid structure in the form of a metal keel. This keel is either attached to, or suspended from, the envelope, which, as well as acting as a walkway within the ship, also assists in absorbing the stresses acting on the structure imposed by aerodynamic loading in flight.

In some examples of this type of ship, a substantial tubular metal nose cone is fitted at the bow, while a cruciform girder work is incorporated in the rear keel structure to carry the external control surfaces. The gas was contained within the fabric envelope and this typically included a system of ballonets, as found in the pressure airships, to maintain the external shape. Semi-rigids were generally of a larger size than pressure airships, and the gas space was usually divided transversely into a series of separate compartments by fabric curtains to prevent gas from surging to either end of the envelope and upsetting the trim of the craft.

The first successful semi-rigids were those built in France in the 1900s, most importantly the creations of two brothers, Paul and Pierre Lebaudy, sugar refiners who, from their factory at Moisson, Seine en Oise, supplied a series of ships to the French army.

During and after the war period, it was the Italians who were to develop this particular type of craft, becoming the undisputed experts in this type of construction, which culminated in the remarkable flight in 1926 of the polar dirigible *Norge*. This noteable airship was captained by its designer, Umberto Nobile, and carrying the famed Norwegian explorer Roald Amundsen. The ship flew from Spitsbergen to Teller in Alaska, directly over the North Pole, in a continuous flight of seventy hours forty minutes, covering 3,180 nautical miles – a distance equivalent to the *R 34*'s transatlantic flight by an airship of one-third its capacity.

The third category of airship is the rigid, characterised by a metal or wooden framework forming a system of braced ring girder bulkheads, or transverse frames, interspersed with lighter intermediate transverse frames enclosing the separate gas cells, and connected by longitudinal latticed girders. The purpose of the girders is to absorb the powerful bending moments imposed on this long and relatively fragile craft. The completed framework is then covered by a stout, tightly doped fabric outer cover, in order to protect the fragile gas cells from the effects of weather and abrupt temperature variation.

In all rigids, the loads are distributed as evenly as is practical throughout the framework, so as to maintain the longitudinal trim and equalise the stresses acting on the structure. The heavier weights, such as engine cars and control gondola, are suspended from the main transverse frames, whilst petrol tanks, water ballast bags and other movable weights are arranged along the internal keel. These, in turn, are either attached directly to the frame structure or suspended from the radial wiring of the transverse frame bulkheads.

The lifting force developed within the gas cells is transmitted through the wire mesh netting enclosing the cells, through a system of lift wires attached to catenary wiring (similar in principle to the support cable in a suspension bridge) and connected to the transverse frames and longitudinal girders to spread the lift evenly throughout the entire structure. Another set of diagonal shear wiring holds the cells in place within the framework to minimise chafing against the girder work, while at each mainframe radial diagonal shear wiring forms a bulkhead to prevent cell surging.

Running from bow to stern, passing through the centre of each gas cell and anchored to the radial shear wiring bulkheads at each frame, runs the axial cable – designed to absorb and spread the static and aerodynamic loading exerted on the framework. Should a gas cell become deflated, the axial cable will compensate, to a degree, for the hogging effect on the adjacent structural members.

The control car was typically attached directly, or suspended below the framework at a position near the bow, containing at the front the steering position from where the commander could pilot the ship; directing the steering and height coxswains and controlling the engine-room telegraphs, ballast and gas valve controls etc. Behind this area was the navigation room, wireless room and a companionway giving access into the body of the ship.

TYPES OF AIRSHIP STRUCTURE

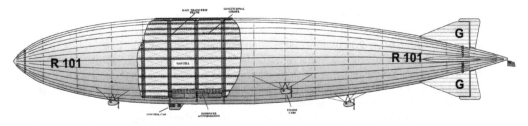

RIGID CONSTRUCTION - R101 1930

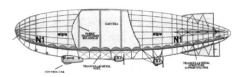

SEMI - RIGID CONSTRUCTION - "NORGE" 1926

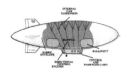

PRESSURE AIRSHIP "SKYSHIP" 1986

The navigation of an airship was akin to that of a seagoing ship, with the use of dead reckoning and astral navigation, supplemented by an early form of wireless fixes from shore stations and ships, being the main aids to these early voyagers of the air.

Before the days of controlled air routes and modern navigational aids, the commander of an airship was required to make minute-to-minute decisions regarding course, speed and altitude, forever keeping an eye on the developing weather situation in order to guide his craft safely through the sky. The steering coxswain would be facing forward with his eye fixed on his compass course, while the elevator coxswain, the most experienced helmsman, would be standing sideways to the direction of flight watching the inclinometer and was responsible for the trim, attitude and altitude of the airship.

The engine cars were mounted along the sides of the ship in small streamlined nacelles, each with its own engineer who responded to the instructions received from the control car, via the engine room telegraphs, to control the speed of the engines.

At the stern of the ship, horizontal and vertical stabilisers were fitted. Originally these were of a flat externally braced non-aerodynamic section. These were later superseded by cantilever aerofoil-shaped fins, which were attached to the rear of the ship with strong cruciform girders traversing the face of the main ring structure – to provide a strong attachment point for the rudder and elevator controls.

This part of the structure needed to be of great strength due to the powerful aerodynamic forces and bending moments acting on the rudders and elevators in flight; many airships over the years experienced structural failure of the fins in flight due to inadequate stressing of the structure of these surfaces.

In rigid airships, ballonets were not employed as the gas pressure of the cells was equivalent to the external air pressure relative to the height. In early rigids, the cells were open at the bottom to allow for expansion of the excess gas, this being allowed to disperse through the outer cover. Later, automatic gas valves were fitted to the bottom of the cells and connected to fabric exhaust trunking. This ran between the gas cells to discharge through hoods on the upper surface of the hull; a safer option than discharging gas in the vicinity of hot engine exhausts.

Another means of discharging gas was by a series of 'manoeuvring valves', whereby measured quantities of gas could be released from the control car during flight or when landing.

Airships required large handling crews to bring them in and out of their massive sheds, sometimes in the order 200 men would be required to bring out or hold down a large rigid in a brisk breeze. While the Germans relied on large crews for ground handling, the British in particular utilised the mooring mast, pioneering this method towards the end of the Great War. Most British rigids were fitted with mooring cones which helped to alleviate the problems of handling large rigids on the ground.

The Americans later favoured the use of the 'stub mast', where the ship was moored on the ground with a car attached to the aft gondola, so that the airship could vane in the wind on a circular track. This arrangement was superior from the point of view of accessibility and servicing, and is the method still in use.

The process of the preparation of an airship for a voyage during the First World War period was a complex and labour-intensive operation, and a process always constrained by the vagaries of the weather. An airship was able to operate in only the most favourable conditions, sometimes unable to either enter or leave its shed due to strong crosswinds or beset by adverse

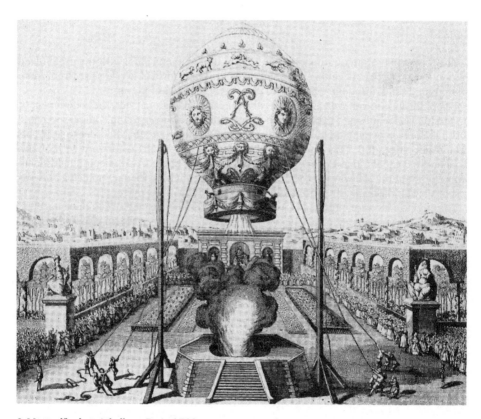

2. Montgolfier hot air balloon, Paris, 1783.

meteorological conditions en route that could hinder its progress or, if severe enough, endanger the craft.

In the case of a rigid, at the commencement of a voyage – once all the cells had been filled with hydrogen, fuel tanks filled and water ballast, oil, war load, provisions and crew had boarded – it was customary for the ship to be 'weighed-off' in the shed. That is, to put the ship into a state of neutral buoyancy with the cells 95 per cent full, so that in this state the ship was just floating and could be lifted easily by the ground crew.

Next, the engines were started and run up without the propeller clutches being engaged. If the conditions of wind and weather were judged to be favourable, the ship would then be walked out by the ground crew on to the field, downwind of the shed and clear of all obstructions, where it was then turned into the wind.

With all preparations complete, the order 'let go' (in Great Britain) or 'up ship' (in the United States) would be given, whereupon the ground crew would drop all the handling lines and push the ship up by hand. As they did so, a simultaneous discharge of water ballast fore and aft would cause the airship to lift away with a force equivalent to the weight of water dropped. When the ship had risen to a suitable height, and before the wind could cause it to 'free balloon', the engine clutches would be let in and the airship would move forward under the impetus of the propellers.

At the outset of the voyage, the captain would allow the airship to rise statically to a height where the cells were 100 per cent full. Here the craft was once more in equilibrium – this being the pressure height for the full load being carried – and he would thereafter endeavour to keep his ship below this height in order to conserve hydrogen. This height was usually around 8,000ft for a Zeppelin in the First World War.

Typically, as the voyage progressed, the airship would become lighter due to the consumption of fuel and would tend to rise progressively to a higher altitude. The captain, in order to keep his chosen pressure height, could valve gas to keep below this altitude height or, should the ship be allowed to rise beyond its original pressure height, it would continue to rise until it was once more in equilibrium at a higher altitude.

When an airship rises, the gas within the envelope expands in response to the natural physical laws, becoming less dense while occupying a greater cubic volume within the cells. If an airship passes beyond the pressure height, gas will be expended through the automatic valves resulting in a loss of lift, while, conversely, if a ship is descending into denser air the gas will be compressed, occupying less cubic volume in the cells.

In addition to the static lift provided by the hydrogen, dynamic lift can also be generated – dynamically – by inclining the nose upward at an angle of attack in relation to the direction of flight to produce aerodynamic lift over the hull. This effect can also be used by adopting a nose-down attitude in order to stop the ship rising when in a light condition.

Modern airships undertaking long trips invariably take off in a 'heavy' or overloaded condition, and utilise the aerodynamic effect to generate sufficient lift to become airborne until sufficient fuel has been used to allow the static lift to fully take over the lift function. During flight, the relationship between the weight and lift equation will alter, with the consumption of fuel causing the ship to become lighter and rise to a higher altitude, as previously mentioned. Other variables that effect lift are barometric pressure, temperature and humidity.

Where an airship is flying in trim in a region of high pressure and low temperature, it will develop greater lift if the sun heats the outer cover; the dense gas is heated through 'superheating', thus causing the ship to become 'light'. Conversely, if flying in a regime where warm air and low barometric pressure dominate, the density of the gas will fall, resulting in the ship rapidly becoming heavy. If, however, the airship continues to fly in a region of warm air, this will eventually warm the gas and increase the lift.

During a voyage, a constant watch would be required on the varying factors that govern an airship in flight, with continuous adjustments being made to the trim of the ship and its

3. Blanchard and Jefferies' cross-Channel flight, 1785.

attitude, being flown nose up or down by several degrees in order to dynamically maintain a chosen altitude. Management of the lifting gas is of primary importance, as once valved it cannot be replaced, with controlled quantities being released through the manoeuvring valves as required should the ship become light, or the dropping of water ballast if the craft is heavy.

A large rigid would be capable of picking up a ton of rainwater in a few minutes, requiring an equivalent quantity of water ballast to be dropped to compensate. Similar quick action would be required to cope with changing barometric conditions that would instantly and adversely affect the lift and trim. The commander of a Zeppelin attacking the English coast would need to balance these factors as he climbed to his attack altitude of around 13,000ft.

One of the areas where the airship had a distinct advantage over the aeroplane was in its ability to climb with great rapidity. A Zeppelin, if necessary, could climb at a rate of 2,000ft per minute; a rate far higher than that which could be achieved by contemporary aircraft. In practice, however, a rate of climb of 1,000ft per minute was more than adequate to escape a pursuing aeroplane, and put considerably less strain on the structure.

A Zeppelin airship on a raid could spend between eighteen and twenty hours in the air, the majority of that time flying above 13,000ft in the freezing and often anoxic conditions of the substratosphere which had a debilitating effect on the efficiency of both crews and engines alike.

Changes in wind strength and direction at varying altitudes frequently caused Zeppelin navigators to make errors in landfall over the British Isles of up to 100 miles, often leaving them lost and wandering over the darkened countryside trying to identify targets, such were the primitive nature of navigational methods employed.

It is obvious from the foregoing that airship operation and navigation is of a most complex and difficult nature, quite unlike that of heavier-than-air navigation, and requiring constant attention to the changing conditions in which an airship flies. Modern airships have at their disposal the most modern navigational, radio and satellite aids, which ensure that the captain can be certain of his position in the sky within a metre. However, the captain is still required to possess a highly developed understanding of meteorology, being able to recognise the developing weather conditions and apply their skills to the best advantage, guiding the craft safely through the skies.

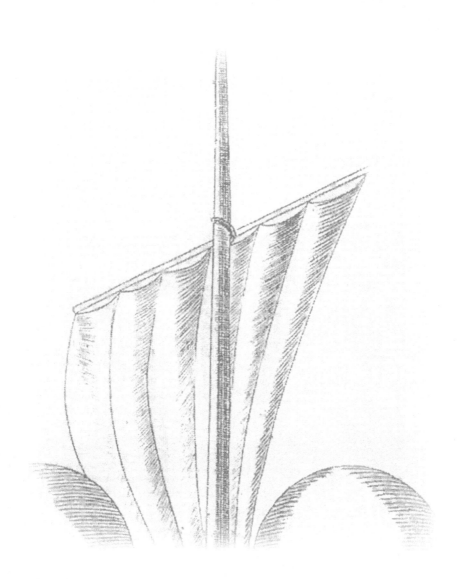

2

The Development of the Dirigible
1783–1914

In recent years the airship, after a long period in the aeronautical wilderness, has again become familiar to a new generation due to its increasing use as an aerial camera platform at major outdoor sporting events. Cruising effortlessly over the heads of the crowds with purring engines, its silver envelope glinting in the sunshine as it manoeuvres gracefully over the stadium or racetrack, it has now become an accepted part of the sporting scene.

A dwindling group of people are old enough to remember the great silver rigid airships of the 1920s and '30s, and are familiar with their adventurous voyages in those distant days. For the majority of younger people, however, their perception of what these sky giants achieved has been coloured by the terrifying filmed spectacle of the *Hindenburg* being consumed by fire at Lakehurst, New Jersey, at the end of a transatlantic crossing from Frankfurt am Main on an evening in May 1937.

The black-and-white film and dubbed radio commentary have an immediacy that is both shocking and compelling to watch, involving the viewer in a way that, despite the passage of time, few other filmed images of catastrophe can equal. The *Hindenburg* disaster drew a line under any further development of the rigid airship for commercial purposes. A disaster in part abetted by the US government's refusal to sell to the German Nazi government supplies of the inert, non-explosive helium gas.

This rare element was at that time found only in gas wells within the United States, for which the ship was originally designed, and therefore this decision ensured that the only alternative was to fill the cells with 7 million cu ft of explosive hydrogen gas as the lifting agent. With the coming of the Second World War, only the United States employed airships, or blimps, for patrol and escort duties with the US navy. In the course of which, while performing this valuable work of convoy protection and reconnaissance, they flew many hundreds of thousands of miles.

This included, in 1944, a transatlantic crossing by two squadrons of six airships from the US Atlantic coast to North Africa, from where they played their part in anti-submarine duties in the western Mediterranean Sea.

At the war's end they continued to be built, in ever-increasing sizes, for use along the US Atlantic seaboard and in the high Arctic regions of Canada, as part of the US DEW Line radar defence system. These airships operated as flying radar pickets to detect incoming Russian ICBMs during the Cold War period, each with a huge 50ft diameter radar scanner revolving within the 1.5 million cu ft envelope.

The ZPGW class were the last of these craft built in the 1960s. Using new materials that were being developed for the space race, equipped with vastly more powerful engines and the most modern navigational and radio aids, these craft were technologically far in advance of anything that had gone before.

These airships performed their duties in the worst imaginable weather conditions: arctic blizzards and storms in which, just a few years earlier, it would have seemed impossible for such craft to survive. Their endurance exceeded that of any other type of aircraft, with missions of over 100 hours becoming routine. An endurance record for any type of aircraft was set in August 1954, when the *ZPGW-2* maintained a continuous patrol of the high Arctic latitudes for 205 hours – or over eight days – in the air.

By 1966 the last of the radar picket ships had been decommissioned, their duties having been taken over by high-flying jet propelled Awacs aircraft, and increasingly by surveillance satellites. On the civilian side, by the end of the war transport aircraft had advanced to such a degree in speed and load-carrying capacity that it was obvious the aircraft had superseded the airship as an economic and reliable alternative to long-distance air travel.

As Dr Eckner (who had collaborated with Count Zeppelin to develop the rigid airship in Germany before and after the First World War) declared in the late 1940s, reflecting on the increasing frequency of Atlantic crossings by speedy Constellations, Skymasters and Stratocruisers, 'A good thing has been replaced by a better one'.

The origin of the dirigible, or steerable airship, lies in the spherical hydrogen balloon of the early nineteenth century, and the attempts to find some way to control the direction of its flight path rather than being at the mercy of the wind.

In France, as early as 1783, and not long after the invention of the balloon itself, Jean Baptiste Meusnier, then a lieutenant in the Corps of Engineers of the French army, produced a proposal for an exceptionally advanced design incorporating many features that were to be used in later successful airships. The balloon was of an elongated ellipsoidal shape some 260ft long, as Meusnier appreciated the need to reduce the frontal area relative to the line of motion to reduce resistance.

Meusiner's Manually Powered Dirigible - 1783
Académie des Sciences, Paris, France

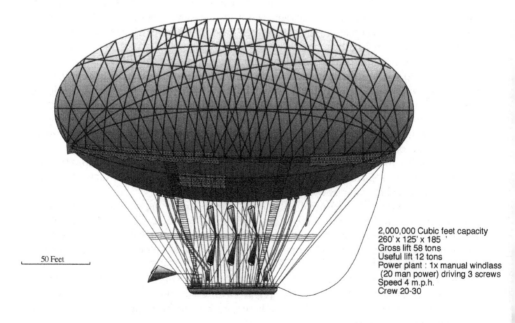

50 Feet

2,000,000 Cubic feet capacity
260' x 125' x 185 '
Gross lift 58 tons
Useful lift 12 tons
Power plant : 1x manual windlass
 (20 man power) driving 3 screws
Speed 4 m.p.h.
Crew 20-30

4. Count Ferdinand von Zeppelin, 1838–1917.

The design also proposed the use of a ballonet, within the envelope, to be filled by a hand-driven pump in order to vary the weight of the airship and so control ascent and descent. Although it was not the original intention, the ballonet would also help to preserve the shape of the envelope under varying atmospheric conditions. Suspended below the envelope was a long, boat-shaped car containing a hand-driven crankshaft driving three pusher airscrews mounted above the car – this being the first recorded instance of the application of the propeller in aviation – and a steering vane at the rear of the car to control direction.

Meusnier also designed a shed to house the ship, neither of which was ever built on the grounds of expense. Its sheer size of 2 million cu ft was beyond the constructional capabilities of the day, but proof that this was a remarkably far-sighted design. The concept was the work of an extremely gifted engineer and was a practical proposition, as demonstrated in the 1990s when a small-scale replica was successfully flown at Cardington, even managing to make progress against the wind under the power of the hand-driven propeller.

Meusnier rose to the rank of general in the French army, gaining great distinction in the field of military science before being killed in the fighting at Mainz at the age of 39 during the early days of the French Revolutionary Wars; a loss both to his country and to aeronautical science.

The most significant figure to emerge in the field of aviation during the nineteenth century was Sir George Cayley, a Yorkshire squire; a man of brilliant intellect who published the principles of mechanical flight and was the first to fully understand the significance of the forces acting on an aerofoil.

Cayley also produced plans for steam-driven elongated balloons, propelled either by airscrews or 'wafting wings', but took these plans no further, concentrating his efforts on experiments with heavier-than-air craft. Over the years he also designed and built several

Giffard Steam Powered Dirigible - 1852
Hippodrome, Paris

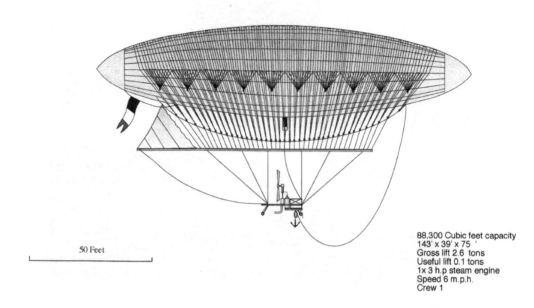

50 Feet

88,300 Cubic feet capacity
143' x 39' x 75 '
Gross lift 2.6 tons
Useful lift 0.1 tons
1x 3 h.p steam engine
Speed 6 m.p.h.
Crew 1

gliders, and in 1853 at his estate in Yorkshire, one of his full-scale machines, piloted by his coachman, achieved the first successful manned, controlled flight in history.

Meanwhile, other experimenters throughout the nineteenth century continued to grapple with the problems of dirigibility, but still the lack of a suitable prime mover defeated their efforts. The first practical airship was the work of another Frenchman, Henri Giffard, who in 1852 constructed a 143ft-long craft. Pointed at both ends and driven by a 3hp steam engine; weighing, with its boiler, 350lb and suspended 20ft below the envelope to reduce the obvious danger of fire; driving an 11ft diameter propeller at 110rpm to give a still air speed of 6mph.

On 14 September 1852 Giffard lifted off from the Hippodrome in Paris and, in near-perfect weather conditions, flew 17 miles to Trappes, south of Paris in around three hours.

Whilst Giffard was able to effect a degree of control with the triangular sail that performed the function of a rudder in directing his flight, he was, however, unable on this or later flights to either navigate in a circle or return to his starting point due to the low power of his engine.

The lack of a compensating ballonet, whilst not adversely affecting the rigidity of the envelope during these trials, did subsequently cause his second and larger airship to collapse a few years later. Giffard, nevertheless, deserves to be recognised for his practical approach to the problem, and to the high degree of mechanical and technical skill demonstrated in the construction of his airship.

Further applications of steam power, followed by electric, gas and finally petrol engines, were employed over the coming years in the endeavour to overcome the problem of a suitable source of motive power.

In 1860 a Belgian, Étienne Lenoir, patented a gas engine fuelled by coal gas. This innovative development represented the first example of the successful use of the internal combustion engine. This invention was quickly taken up by an Austrian engineer Paul Haenlein (1835–1905), who employed a four-cylinder Lenoir gas engine producing 5hp in his airship built at Brunn in 1872.

The airship's varnished silk envelope was filled with coal gas, which supplied the engine at a rate of 250cu ft of gas per hour, turning the large four-bladed propeller at 40rpm. Internal pressure was maintained by a ballonet supplied with air from a mechanical pump, fitted to compensate for the coal gas burned in flight. During the first trial at Brunn in December 1872, perceptible control and a degree of acceleration were evident, with the craft achieving a speed of 11mph.

Despite the initial success of this and further trials, the early Lenoir engines suffered from a poor power-to-weight ratio and low power, which militated against its further use in the development of the airship.

Haenlein eventually abandoned his work, frustrated by the lack of a suitably powerful engine that could enable his airship to make headway against a wind of any force.

Again, it was two Frenchmen who took the next step forward when, in 1882, the Tissander brothers built a 92ft-long airship. At 37,000cu ft capacity, it was driven by a Siemens electric motor of 1.5hp driving a primitive two-bladed airscrew, which drew its power from the twenty-four bichromate of potash cells weighing 500lb, contained in the minute car suspended by netting below the envelope.

The first trial took place at Auteuil on 8 October 1883 when, despite a fair breeze, a measure of control was achieved with the aid of the sail rudder. A second trial on 26 September produced somewhat better results, although due to the low power of the motor the craft was unable to

Haenlein Gas Powered Dirigible -1872
Hanlein Works, Brunn, Austria

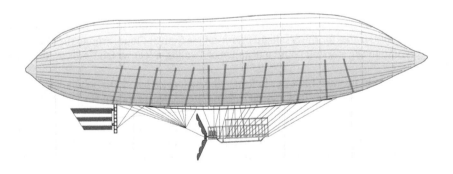

25 Feet

72,000 Cubic feet capacity
150' x 30' x 44'
Gross lift 2.1 tons
Disposable load 0.30 tons
1x 5 h.p Lenoir 4 cylinder gas engine
Speed 11 m.p.h.
Range 10 miles
Crew 1

make any headway in a wind of any magnitude, and the brothers concluded their work having expended considerable sums of money in the course of their experiments.

Their pioneering work had, however, been brought to the notice of the French government and the military balloon establishment at Chalais Meudon. It was here, in 1884, the distinguished engineers Charles Renard and Captain A.C. Krebs designed and constructed the airship *La France*, which was to prove capable of steady, navigable flight, and was able to describe circular flights against the wind and return to its departure point.

La France, with a capacity of 66,000cu ft, was 165ft long with a diameter of 28ft, the envelope being made from Chinese varnished silk and of excellent streamlined form. Below the envelope an enclosed car, 108ft in length, contained the Graham electric motor, developing 8.5–9hp for a weight of 210lb, together with the chromium-chloride batteries that provided the current, to drive a cloth-covered tractor airscrew of 23ft diameter. Steering in the horizontal plane was effected by a large rectangular rudder at the stern, while vertical movement and attitude were controlled by a sliding weight mounted within the body of the car assisted by an 'elevating rudder'.

The first trial of *La France* took place at Chalais Meudon on 9 April 1884. The flight proving a limited success: making a circular flight of some 5 miles in twenty-three minutes at a speed of around 12.5mph in still air, then returning to its starting point – the first time this had been accomplished. Six other flights were made in 1884–85, including two over Paris, where it made a great impression in government circles who were quick to realise its military potential.

However, despite this modest success in aerial navigation it was recognised that, due to the weight of the batteries for a relatively low power output, some other lightweight power source was required. It was thus to the newly developed petrol engine that attention was directed.

Alberto Santos-Dumont, the son of a wealthy Brazilian coffee planter, was intrigued by the problem of motorised flight as soon as he arrived in Paris in 1891 – then the centre of scientific research into the art of flying – where he was immediately accepted into influential society. Alberto found himself in contact with those who had access to the latest ideas in the development of modern science, and surrounded by an atmosphere of progress in all fields of transportation – he was inspired.

Santos-Dumont built his first petrol-powered dirigible in 1898, and over the next few years constructed a total of fourteen airships of steadily improving performance. His dirigible *No. 6*, powered by a petrol motor of a nominal 12hp, succeeded in winning the Deutsch de la Meurthe prize of 125,000 francs for the first flight over a course from St Cloud, around the Eiffel Tower and returning to the starting point; a distance of 7 miles, a feat which he accomplished in thirty minutes.

Santos-Dumont, through his feats of airmanship performed before the population of one of Europe's most sophisticated cities, helped to stimulate an acceptance of the possibilities that aerial navigation presented.

At around the same time, in 1895, Austrian timber merchant David Schwarz produced plans for a most advanced rigid-framed airship. It was built in St Petersburg for the Russian government but, although completed, never flew due to political complications; the Russians constantly regarding the unfortunate Schwarz as a spy. Schwarz was forced to leave Russia in secrecy, returning to Germany where he built a second airship of similar design for the Prussian Airship Division in Berlin.

Unfortunately, Schwarz died before construction was complete. The work was, however, taken over by his extremely able and practical widow, who ensured her husband's work was completed. The airship itself was of a remarkable and unique design, consisting of a bottle-shaped craft of cylindrical form with a conical nose and a shallow rounded stern, an internal framework constructed of tubular aluminium members and girders was covered by a skin of sheet aluminium '008" in thickness'.

5. Schwarz metal-clad dirigible, Templehof, Berlin, 1897.

This all-metal rigid airship, one of only three to be built, was 156ft in length by 40ft in diameter with a gas capacity of 131,000cu ft. The filling of the envelope was accomplished by introducing a series of paper bags within the framework to be filled with hydrogen gas, which at the same time expelled an equal volume of air contained within the metal hull. A two-cylinder Daimler petrol motor of 12hp output provided the motive power, driving three airscrews through fabric transmission belts.

Two wing screws mounted above the car were designed to drive the craft through the air, while a third larger diameter propeller mounted on the centre line, able to move through 90° vertically and horizontally, was supposed to control direction. In the event, the effectiveness or otherwise of this method of steering was not to be put to the test.

The first trial and only attempted flight of this far-sighted design took place at the Tempelhof Feld near Berlin on 4 November 1897 in gusty, overcast and, one would think, unsuitable weather conditions for the first flight of an untried craft. With the engines running, the ground crew released the ropes and the airship rose to approximately 100ft. Maintaining its position against the brisk breeze for some minutes, the airship was beginning to make progress against the wind when, suddenly, the left-hand driving belt transmitting the power jumped off its sheaves.

The airship continued to rise to 1,300ft as the pilot, a soldier named Ernst Jagels, struggled to maintain control, adjusting the angle of the central propeller to counter the wind. Just as he was turning into the wind the right-hand driving belt also slipped off, leaving the airship powerless. Jagels valved gas to bring the craft down near to the Tempelhof Feld but the descent was too rapid, causing it to crash heavily on the ground, completely wrecking the airship.

Fortunately, no fire followed and Jagels escaped unharmed, but without further funds Schwarz's widow had to abandon this innovative project.

Meanwhile, in 1899 at Manzel on Lake Constance in southern Germany, Count Ferdinand von Zeppelin had begun the construction in a floating hangar of a giant 420ft-long rigid airship, which, after many trials and tribulations with both it and subsequent craft, proved

Renard & Krebs Electric Powered Dirigible "La France" - 1884
French Military Balloon Establishment, Chalais-Meudon

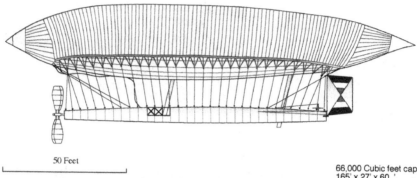

50 Feet

66,000 Cubic feet capacity
165' x 27' x 60 '
Gross lift 2.0 tons
Useful lift 0.1 tons
1x 9 h.p.Gramme electric motor
(Chromium chloride battery)
Speed 13 m.p.h.
Crew 3

"La France" was designed by Captain Charles Renard
and Captain Krebs and built at the Military Aeronautical
establishment at Chalais Meudon, where it accomplished
a series of successful flights during 1884 -85 proving to
be capable of steady controlled flight and demonstrated
it's ability to make cicular flights returning to it's original
starting point.

to be the precursor to a long line of successful rigid dirigibles. The Zeppelin design contained within its structure the essential spark of genius that, to a large extent, was to elude builders in other countries.

The Zeppelin Company were to go on to build 130 rigid airships over the next forty years, giving them an unparalleled wealth of experience in the construction and handling of large rigid airships. Zeppelin airships were to operate successfully over the coming years in every climatic zone and adverse weather condition, making the Germans the premier exponents of this type of aircraft.

Count Zeppelin was born in what is now Baden-Württemberg in 1838, the son of an aristocratic family for whom military and state service were as a duty an accepted part of their lives. In 1858 the young Zeppelin joined the 8th Infantry Regiment as a second lieutenant, but soon put his name forward to study at the University of Tubingen for courses in mathematics, chemistry, political economy and history. He was eventually posted to an engineer unit where he was able to study the latest developments in modern military techniques and mechanical advances, and where he further developed his ability as an engineer.

In 1863, he was sent to America where he was attached to the Union Army during the Civil War as a military observer to study the practices of modern warfare, during which time he had the opportunity to make his first ascent at St Paul in a Union observation balloon. Following this experience he returned to Europe to take an active part in the Franco-Prussian War of 1870, serving with distinction in leading his cavalry regiment in attacks on French positions. By the war's end he had been promoted to major.

Zeppelin made steady progress within the Imperial Army, commanding several regiments and attracting the attention of his superiors, including the kaiser and von Moltke, for his zeal and efficiency. However, he also incurred the dislike of certain sections of the Prussian military establishment for his outspoken criticism of Prussian domination of the army. That was to inhibit his progress and ultimately to force his retirement from his chosen career.

In 1884 he was promoted full colonel and became aware through the press and other sources of the success achieved in France by Renard and Krebs with the dirigible *La France*. Count Zeppelin was imbued with the desire for Germany to be in the forefront of such developments. Within three years he was writing to the Chancellor deploring the state of aeronautical development in his own country compared to France, and began to consider how best to remedy this situation.

In 1892 his military career came to an end at the age of 52, having been passed over for promotion in the Imperial Army where his enemies had achieved their end to remove him. On leaving the army with the rank of general, the count took up the post of military advisor at the court of the King of Württemberg, where he was welcomed and appreciated by his fellow countrymen.

Here the count, now free from the restrictions of the Imperial Army, turned his attention to the problem of dirigible flight, which he had been considering over many years. He had taken his inspiration from the work of Renard and Krebs, and latterly the Alsatian Spiess' rigid design, together with Ganswint's proposal for a giant metal airship. Using these designs he brought his not inconsiderable skill as an engineer to bear on the problem.

His great contribution to dirigible flight was that, as early as the 1870s in his notebooks, he proposed the construction of an airship on a large scale in order to achieve a high and therefore useful lifting capacity. This early design consisted of a number of separate gas cells enclosed within a rigid framework – both to protect the gas from the effects of superheating by the sun's rays, by providing a stout outer covering, and to minimise the loss of lift should one or more cells become deflated. Additionally, he recognised that the rigid structure would be better able to distribute the loading and lifting forces than a pressure airship envelope and thereby maintain the structural integrity of the craft.

Schwarz Metalclad Airship No 2 - 1897
Prussian Airship Detachment, Tempelhof, Berlin

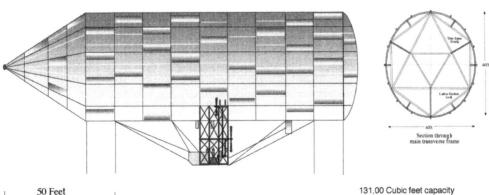

50 Feet

Section through
main transverse frame

This metal framed rigid airship's hull containing a
single gas cell, was covered with aluminium sheeting
0.008"in thickness.
Schwarz No.2 crashed on it's initial trial flight at
Tempelhof 3rd November 1897

131,00 Cubic feet capacity
154' x 46' x 60 '
Gross lift 3.8 tons
Useful lift 0.6 tons
1x 12 h.p.Daimler four cylinder
Speed 17 m.p.h.
Crew 1

Within a year his first proposal existed in model form as the *Aerial Train*, which consisted of a series of rigid-framed cylindrical gas compartments joined together like the carriages of a train. The whole structure was to be driven through the air by petrol engines, contained in gondolas under the foremost 'carriage' attached to an articulated keel structure. This was, in essence, the prototype of the successful Zeppelin airships that followed. Although this particular design was rejected by one of the many official commissions that proliferated within Imperial Germany at the time.

The structure was, however, based on the principle of the 'Schewdler Copula', which consisted of wire-braced tensioned rings joined by braced connecting girders. These were to prove to be the key to the rigid airship constructional success.

During 1898 Zeppelin formed The Joint Stock Company for the Development of Airship Flight with many prominent investors, including Carl Berg the aluminium tycoon, and in conjunction with the talented engineer Theodor Kober, who had previously worked for the Riedinger Balloon works of Augsburg. Together the three men endeavoured to develop the count's idea into a more practical craft than the cumbersome and impractical *Aerial Train*.

In these endeavours they were soon joined by a young engineer, Ludwig Durr, who was later to become the designer of all the wartime Zeppelins and subsequently the *Graf Zeppelin* and the *Hindenburg*.

Construction of the count's first full-size dirigible commenced in the following year in the floating hangar on Lake Constance. The *LZ-1* must be adjudged to have been a considerable feat of engineering using, as it did, aluminium as the main structural component on a scale not attempted before, and all the more remarkable in being a structure that could fly.

The *LZ-1* took to the air over the waters of Lake Constance on 2 July 1900, the finished product being 420ft in length by 40ft in diameter with a gas capacity of 400,000cu ft of hydrogen contained in 16 cells, supporting an all-up weight of just under 12 tons. The parallel-sided, conical-ended craft had two gondolas fixed under the hull, each containing a four-cylinder Daimler petrol motor with an output of 16hp. Lateral control was effected by two small rudders at the stern, whilst attitude and vertical control relied on a 500lb sliding weight suspended on a cable beneath the ship.

The airship was towed out of its floating hangar by the steamboat *Buchorn* and, once clear of all obstructions, vaned into the wind floating like a kite where the tow rope was released. On a near perfect day the *LZ-1* rose to about 1,300ft, from where it flew for some twenty

Zeppelin "Aerial Train" Proposal - 1895
Count von Zeppelin, Manzell, Wurtenburg

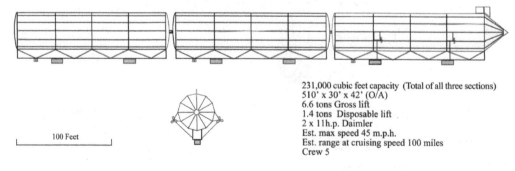

100 Feet

231,000 cubic feet capacity (Total of all three sections)
510' x 30' x 42' (O/A)
6.6 tons Gross lift
1.4 tons Disposable lift
2 x 11h.p. Daimler
Est. max speed 45 m.p.h.
Est. range at cruising speed 100 miles
Crew 5

Zeppelin LZ1 - 1900
Zeppelin "a" Type
A/G Forderung-derMotorluft-schiffahrt, Manzell, Bodensee

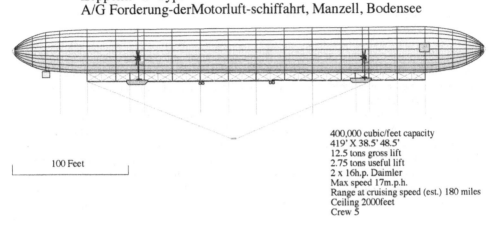

100 Feet

400,000 cubic/feet capacity
419' X 38.5' 48.5'
12.5 tons gross lift
2.75 tons useful lift
2 x 16h.p. Daimler
Max speed 17m.p.h.
Range at cruising speed (est.) 180 miles
Ceiling 2000feet
Crew 5

minutes above the calm waters of the lake. With a crew of four, including the count himself, the airship covered some 4 miles against a headwind of 16mph. The *LZ-1* successfully achieved a degree of control despite the relatively low power of the engines, but problems with the sliding weight necessitated a descent on to the lake from where the ship was towed back to the hangar.

A second flight, of thirty minutes' duration, took place on 17 October 1900. Experiments in control were undertaken, but again these tests had to be cut short due to engine trouble and there was some degree of structural failure of girders in the framework, requiring yet another, fortunately safe descent on to the water.

After some repairs and strengthening of the fractured girders the *LZ-1* made its third and final flight of some twenty minutes, achieving a speed of 17mph during which it again exhibited its ability to answer to the helm. This time the flight was terminated because of water in the petrol, but the ship again returned safely to its shed.

Despite the relative success of the count's experiments so far, he felt impelled the following year, due to the lack of a requirement for either the military or civil to use his invention in a practical way, to disband the joint stock company while personally assuming its liabilities. The count had the *LZ-1* dismantled, while the great shed was beached and closed up. He also dismissed all the workers apart from Kober, whom he retained to work on a more advanced design he had conceived.

Zeppelin was a devoted patriot despite the treatment he had received from the army high command and was convinced that his airships could be of use in the defence of the Fatherland. Accordingly, he was anxious to interest the army in its development, but felt frustrated in his efforts due to the disinterest and obstruction shown by his former adversaries within the army high command.

Imperial headquarters, for its part, had been alerted to recent developments in France with the first Lebaudy semi-rigid, *Jaune*, being accepted semi-officially into the French army during 1904. A second unit of the French air fleet, the *Patrie*, was acquired during 1906.

Spurred on by these developments the German government's response was to order Major von Gross, the officer commanding the army balloon establishment outside Berlin, to commence the design of a large semi-rigid to match the French efforts. Work began in 1906 under conditions of the greatest secrecy, and was to result in the construction of a series of airships distinguished by an envelope of elipsoidal form with a triangular-section, articulated tubular steel keel.

This structure evenly distributed the shear and compressive stresses that resulted from the weight of the cars and other loads, whilst allowing the envelope to retain its shape to the best aerodynamic advantage. The Gross airships were shown to be sturdy and well-engineered craft, the development of which benefited from the availability of almost unlimited resources provided by the German government.

The *Gross I* of 1908 was superior in every way to the contemporary British Royal Balloon Factory product, *Nulli Secundus*; with the British airship being capable of only 13mph and only able to stay aloft for four hours. Conversely, the efficient *Gross II* made a record flight of thirteen hours in September 1908, and later the larger *Gross IV* of 1913 was accepted for service by the navy for use in the Baltic where it performed useful work during the war. The *Gross II* was frequently moored out to sea anchors, demonstrating its handiness to be speedily dispatched on patrol, and was one of the few airships to successfully make an attack on a British submarine in 1915.

The earlier Gross airships, although also used by the army air battalion in a ground support role with the army, were not possessed of any distinctive transferable constructional features and, ultimately, contributed little to the development of the airship itself.

At the same time, at the instigation of the kaiser, a committee was formed, the *Moorluftschiff-Studien-Geselschaft*, to investigate the work of other promising experimental airship proposals. One of the designs selected by this committee as being worthy of further encouragement was that of Major August von Parseval. His first ship, built during 1906, was of 88,000cu ft capacity, powered by two 50hp Mercedes motors and of an advanced aerodynamic form that contributed to its relatively high speed.

Control in the vertical plane was effected by the provision of two internal ballonets forward and aft, which allowed the nose to be depressed or elevated by filling or emptying one or the other. Simple control surfaces were also fitted to the tail section of the envelope, while an automatically compensating suspension system working on rollers allowed the car to move forward or backwards whilst maintaining a horizontal axis irrespective of the attitude of the envelope.

Between 1906 and 1923 the Parseval concern built twenty-seven pressure airships, the last of which was of 1 million cu ft in capacity. In design the Parsevals showed a considerable advance over contemporary craft, being of excellent streamlined form and introducing many advanced features such as automatic valves. At that time they so impressed the British Committee of Imperial Defence by their performance and potential that an example was purchased for the Royal Navy, with orders placed for three others from the Parseval Company.

During the period from 1909 to the outbreak of the First World War, some very interesting pressure and semi-rigid airships were produced. These included a 180ft-long semi-rigid of 200,000cu ft built in 1907 in Paris for the well-known journalist and adventurer Walter Wellman and the engineer Melvin Vaniman. This airship, the *America*, powered by two 50hp engines, was shipped to Spitsbergen the same year for an attempt on the North Pole, but was forced to return after only a few hours' flight with engine trouble.

During 1909 Wellman made a second attempt to fly to the Pole but this venture was also unsuccessful. Undeterred by these failures, Wellman and Vaniman enlarged and re-engined the airship, still known as the *America*, whose equipment now included a lifeboat.

On 15 October 1910, the re-engineered *America* departed Atlantic City bound for Europe. Yet once again the airship was plagued by engine trouble and after two days in the air, in which they covered over 1,000 miles, the airship came down in the Atlantic. From here the crew were safely rescued by a British ship, the SS *Trent*.

Following this, Wellman abandoned further interest in airships. Vaniman, however, immediately built a replica craft, the *Akron*, constructed by the Goodyear Rubber Company, and in early trials the airship handled well. On 2 July 1912 during a trial from Atlantic City, preparatory to a second transatlantic attempt, the *Akron* exploded in flight killing all her crew of five. Thus ended a series of audacious attempts on both the North Pole and the Atlantic crossing.

Another fascinating yet little known attempt on a transatlantic crossing was proposed in 1911 at Kiel, with the building of the *Suchard Transatlantisches Motor Luftschiff*, a pressure airship originally of 353,200cu ft capacity.

The main backer of this extremely well-thought-out craft, which included a gondola that doubled as a powerful motor boat in the event of the airship coming down mid-Atlantic, was the Swiss chocolate company of the same name, Suchard Chocolate.

It was proposed that the flight should start from Tenerife in the Canaries, and take advantage of the north-east trade winds to aid the ocean crossing to the American mainland, however after a number of trial flights, followed by a series of setbacks, these plans came to nothing.

The Siemens-Schuckert company during this period also produced an example of a very large semi-rigid design, which was not far short of the dimensions of the contemporary Zeppelins and which enjoyed a brief popularity with the army authorities. This particular airship was housed in what, at the time, was a unique 1,200-ton revolving shed that could be aligned into the direction of the prevailing wind to aid housing and launching.

Another semi-rigid airship deserving of mention is the *Veeh 1* of the *Deutsche Luftschiff-Werft GmbH*, built in 1911 it incorporated many advanced features, as well as being of excellent streamlined form.

Prior to the war these and many other airships were built on the Continent, each making a contribution to the development of the airship in those truly pioneering days of early aviation. In Germany this development was to find its fullest expression in the work of Count Zeppelin. Although other airships were excellent – the Parseval airships in particular in performance and design – they were eclipsed by the rigid-framed Zeppelin for war purposes.

We may now return to the vicissitudes that beset the ageing count at his home in Friedrichshafen following his first essay into airship building with the *LZ-1*, as he considered the future of his semi-proved concept.

Count Zeppelin now had the benefit of first-hand experience of the construction of a large rigid airship, and had amassed a great deal of technical and constructional data that would be of value in the future. He had also realised that the cost of further development lay beyond his own personal resources and that the need for some form of sound financial backing from industry and government was required to develop the principle further. Therefore, in order to achieve this end, Zeppelin turned to his patron the King of Württemberg, who agreed to organise a state lottery which ultimately raised 200,000 Marks to finance the building of a second airship.

In addition, he again persuaded the Ruhr aluminium magnate Carl Berg to supply the alloy for the framework, and prevailed upon Gottlieb Daimler into providing light engines of suitable power that could be developed from auto engines to suit the special operating conditions imposed in flight. The floating hangar was at once refurbished and materials began arriving at the lakeside in early 1905 allowing work to commence, with construction proceeding at a rapid pace.

The new airship, which was completed in December 1905, was of sturdier construction than its predecessor and, whilst in outward appearance and dimensions similar to *LZ-1*, she was equipped with larger and more effective control surfaces fore and aft. These elevating 'Aeroplanes' were to replace the cumbersome sliding weight of the *LZ-1*, whilst the engine power had been increased to a total of 170hp.

The *LZ-2* was readied for its first flight in November 1905, but unfortunately as the ship was being towed out of the floating hangar the airship went out of trim causing the bow to dip into the water, which in turn damaged the elevator planes and structure. At the same time, the tow rope attached to the motor boat parted and the *LZ-2* began to drift across the lake whilst all efforts to start the engines failed. It was only the smart work on the part of the motor-boat crew in chasing the wayward airship that averted disaster; by securing the tow rope again and returning the airship to its floating shed.

After repairs the second flight took place on 17 January 1906, at first all went well with the airship reaching a speed of 24mph and answering the helm in a satisfactory manner, but once more the forward engine broke down due to a problem with the cooling water and shortly after this the rear engine also failed. All efforts on the part of the crew to restart the engines failed, too, and floating above the lake as a free balloon the *LZ-2* drifted away on the breeze to be brought down 25 miles away outside the village of Kisslegg.

The damage on landing was not too serious and it was hoped to save the ship. Yet although held by a hastily assembled ground crew of soldiers and villagers later that night, with a wind of increasing force the *LZ-2* was smashed into the ground and became a total loss.

Following this disaster Zeppelin vowed he would build no further airships; however, within weeks, and persuaded by popular sympathy, he managed to raise further funds from his own resources, and a second state lottery was promised by the king. Together these methods raised sufficient money quickly enough to allow the construction of a new airship to go ahead.

The third airship, *LZ-3*, was to prove to be the successful turning point in the count's fortunes. The new craft was based on *LZ-2*, of similar dimensions and utilising the same engines, but possessed of more functional bow and stern multiple elevators and having a keel cabin installed amidships.

The new airship was built within a period of some three months, making her maiden flight on 9 October 1906, which resulted in a flight of two hours, covering 60 miles and returning to its base without incident. Count Zeppelin was jubilant over the success of this and later flights, demonstrating as they did that he had truly conquered the air for the Fatherland.

Further flights followed: one being of eight hours, another of twelve hours where the *LZ-3* sailed along the Rhine valley and on to Switzerland covering 200 miles. Manoeuvring with confidence in trouble-free flight, this performance was witnessed by many thousands of German and Swiss citizens who watched the great ship sail overhead. It was a performance which galvanised the authorities into positive action: firstly by an award of 500,000 Marks from the government Airship Commission to aid research, and secondly the placing of an order for two Zeppelins for the army.

The stringent conditions laid down for the purchase contract included a twenty-four-hour endurance flight to cover a distance of no less than 500 miles with a crew of twenty, and to include a landing on return. Hitherto, Zeppelins had of course only operated from water. Now, in addition, the ship was required to return to its original base at the conclusion of the voyage.

Zeppelin realised that the *LZ-3* did not possess the endurance necessary to comply with these arduous conditions and at once proposed the construction of a larger ship.

The *LZ-4* had an increased capacity of 530,000cu ft and was powered by two engines delivering a total of 210hp. Provision was made for a keel cabin and more efficient multiple bow and stern elevators were fitted, which, in conjunction with larger rudders, improved the lateral and vertical stability.

In July 1908, the *LZ-4* made a twelve-hour flight along the Rhine and on into Switzerland; the whole trip proving to be completely trouble free. The following month the twenty-four-hour army trial took place. Initially all went well, with the airship sailing placidly along the valley of the Rhine. However, just short of Mainz the forward engine failed, causing the *LZ-4* to make an emergency landing on the River Rhine.

Here the airship was secured to the bank and repairs were carried out on a sheltered reach of the river, from where after a few hours the flight was able to continue. By now darkness

Zeppelin LZ4 - 1908
Zeppelin "c" Type
A/G Forderung-derMotorluft-schiffahrt, Manzell, Bodensee

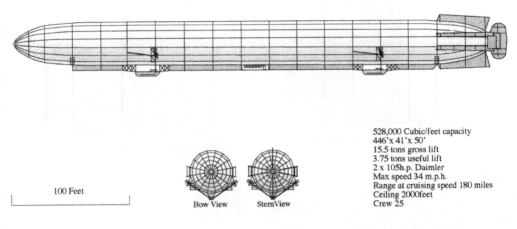

528,000 Cubic/feet capacity
446'x 41'x 50'
15.5 tons gross lift
3.75 tons useful lift
2 x 105h.p. Daimler
Max speed 34 m.p.h.
Range at cruising speed 180 miles
Ceiling 2000feet
Crew 25

100 Feet

Bow View   SternView

was coming on, and Count Zeppelin undertook the first night flight by a rigid airship, cruising confidently over the sleeping villages and towns of southern Germany.

Towards dawn, however, further engine trouble occurred necessitating a diversion to Stuttgart for repairs to be effected at the Daimler works. The airship landed at the village of Echterdingen where a landing party of soldiers was hastily assembled to take the handling lines and make the ship secure. During the afternoon, whilst waiting for the engineers from the Daimler factory to arrive, the wind freshened to such an extent that the ground crew were unable to hold the ship against its force. The luckless airship was wrenched from their hands and smashed to the ground, where it instantly caught fire and in the space of minutes it had become a mass of twisted molten metal; the first of many Zeppelins that would be destroyed by fire.

In the face of such a disaster a lesser man would have abandoned such a risky and unrewarding undertaking, but the 70-year-old count, with characteristic determination, vowed to carry on.

His optimism was not misplaced, as within hours of the news of the loss of the *LZ-4* appearing in the papers a spontaneous outpouring of popular support in what he was trying to achieve was demonstrated by the German people. From all over the nation money and promises of donations came flooding in to Friedrichshafen, and within a short time more than 6 million Marks (£5 million) had been subscribed.

With the loss of the *LZ-4* the army demanded an immediate replacement, for this purposes the old *LZ-3* was taken in hand and enlarged and equipped with more powerful engines. At the same time work began on the new *LZ-5*, which was of similar size but again with more power than its predecessor, with a capacity of 530,000cu ft and combined propulsive output increased to 220hp.

The kaiser, ever watchful of the count's growing popularity with the German people, proposed a board of trustees to manage the money Zeppelin had accumulated. The count countered this suggestion by forming the Zeppelin Foundation as a charitable body to continue research into dirigible flight, whilst separately in September 1908 establishing the *Luftschiffbau Zeppelin GmbH*.

6. Zeppelin *LZ-3* leaving the floating shed at Manzell, Bodensee, 1906.

Carl Berg's son-in-law, Alfred Colsman, was installed as managing director, which allowed the count to continue his development work with the minimum of interference from the government and army sources.

LZ-5 was completed in May 1909 and undertook a proving flight of thirty-seven hours covering 603 miles during the course of which, whilst on the way to Berlin, it encountered adverse weather conditions causing it to make a forced landing at the village of Göppingen near Ulm.

During this manoeuvre the *LZ-5* hit a tree, severely damaging the bow section and deflating the forward gas cell. Despite this serious damage, temporary repairs were effected by removing the damaged forward section and after jury rigging the outer cover over the damaged area the crippled airship was flown 95 miles back to Friedrichshafen.

In due course the *LZ-5* was fully repaired and subsequently delivered to the army as the Z-2, where it served successfully until the outbreak of the war, being further modified and lengthened whilst in service.

Despite these promising army orders Colsman, as director of the Zeppelin Airship Company, saw the need to commercialise the basis of airship production in order to fund further development. His proposal was to form a passenger airline service linking the main cities of Germany, and in November 1909 the company with a capital of 3 million Marks (£2.5 million) was founded in Frankfurt am Main.

DELAG, the *Deutsche Luftschiffahrts-Aktiengessellscaft*, was funded in part by the Hamburg-Amerika Shipping Company and from money raised by those cities involved in the undertaking. Each city, in addition, provided 'aerial harbours'; erecting airship sheds at their own expense to house the fleet of airships. Most of the larger cities invested in the undertaking with great enthusiasm and, although no actual scheduled passenger services ensued, excursion flights of several hours' duration at 100 Marks per head (£80) proved to be a popular attraction to Germans and foreigners alike over the next four years of operation.

The first airship destined for service was the *LZ-7 Deutschland*, with a capacity of 681,000cu ft and able to carry twenty passengers in comfortable Pullman-style accommodation in a cabin amidships. Unfortunately, following a successful delivery flight the airship was wrecked after crashing in the Teutoburger Wald in unfavourable weather conditions, but luckily without any loss of life.

A replacement, the *LZ-8*, or the *Ersatz Deutschland*, was put in hand. Following this in the years preceding the war came the airships *Schwaben*, *Sachsen*, *Viktoria Luise* and the *Hansa*, which between them carried over 35,000 fare-paying passengers in 1,588 flights, covering 170,000 miles without serious mishap; a remarkable achievement given the largely experimental nature of aeronautical development during that period.

While the operation did not constitute a regular scheduled service, it did provide many thousands of German people the unique opportunity to view their country and cities from the air, and imbued the general German public with a sense of national pride at their achievement of the conquest of the air.

During early 1914 DELAG were considering extending their operations to include a fully fledged passenger service, both within Germany itself and to other European countries. DELAG had previously made excursion flights to Switzerland, Austria and Denmark, and intended to

Zeppelin LZ 8 Ersatz "Deutschland" - 1910
Zeppelin "e" Type
Luftschiffbau Zeppelin GmbH, Friedrichshafen

100 Feet

Bow View

681,600 Cubic/feet capacity
485'x 45'x 54'
19.5 tons gross lift
6 tons useful lift
3 x 120h.p. Daimler four cylinder
Max speed 37 m.p.h.
Range at cruising speed 950 miles
Ceiling 2000feet
Crew 8
Passengers 20

run a regular service to Copenhagen, Stockholm and other European capitals, however the outbreak of war caused these plans to be abandoned.

By 1914 the Zeppelin airship had developed into a practical and reliable commercial vehicle with greater lifting capacity, increased speed, an adequate cruising range and improved structural design. It had proved to be a reliable and safe means of transport when handled with prudence with regard to weather conditions.

3

Early British Efforts 1878–1914

The British army began experiments with balloons for military purposes as early as 1878, and by 1884 a balloon detachment of the Royal Engineers was included in a punitive operation against dissident tribesmen in Bechuanaland. During this campaign the unit achieved a fair degree of success in its observation role, while the mere appearance of the balloon made a strong impression on the native tribesmen as a symbol of the great power of the British.

In the following year a second detachment was sent to the Sudan to assist General Wolseley's column, advancing up the River Nile in an abortive attempt to assist General Gordon then invested at Khartoum by the forces of the Mahadi. Again the balloons performed their duties with great efficiency, demonstrating the usefulness of their scouting function that only they could provide in terrain and conditions that were arduous in the extreme. Another positive side of the experience was to show that this form of aerial reconnaissance offered a degree of protection from the ever constant threat of ambush.

By 1892 a permanent school of ballooning had been set up at Aldershot, where a factory for their manufacture was established under the direction of Major James L. Templer, an officer possessed of great energy and organisational ability. Templer, amongst other pioneering work, introduced the storage of compressed hydrogen in steel flasks, an innovation that obviated the need for the complicated and bulky generating plant that until then had been necessary for the purpose. He also foresaw the need to make the balloon section more mobile and to provide for horse-drawn wagons to carry the steel flasks together with the balloon, basket and other equipment, allowing it to operate effectively as a self-sufficient unit in the field.

Templer was also responsible for the introduction of gas-tight balloon envelopes made from goldbeater's skin – this material being obtained from the intestines of oxen and produced as strips of an impermeable membrane some 30in by 20in in size, often requiring some 8,000 skins to make one gas cell. Each skin, after cleaning and preparation, were laboriously glued to a cotton fabric base to produce a gas-tight cell, where the leakage of gas was much reduced compared with the varnished silk used previously. This material continued to be used in the manufacture of gas cells for rigid airships, including the *R101*, until replaced by gelatine-based polymers in the 1930s.

One of the difficulties experienced by Templer was that he had to secure the services of the Weinling family, one of the few groups of people supposedly expert in the difficult process of preparing and joining the goldbeaters' skins together. In his dealings with them Templer found the Weinlings to be troublesome employees, being insubordinate, obstructive and generally difficult to deal with. On one occasion one of their number was even arrested by the police for a drunken affray. They were also intensely secretive regarding their craft; intent on keeping their unique skills within the family in order to protect their position at the balloon factory.

Despite all these difficulties the factory managed to produce a steady flow of reliable balloons for the army over the next few years, and eventually Templer was able to

7. Major James Templer, superintendent of the army balloon school at Aldershot, 1899.

replace the troublesome Weinlings as more tractable and reliable employees came into the workshops.

During the South African War of 1899–1902 balloons constructed at Aldershot, in what was by then the Royal Balloon Factory, saw service in the field spotting effectively for the artillery at the Battle of Magersfontein. In addition, they also served with distinction during the Siege of Ladysmith, where the Boer movements were closely watched by the advancing column, enabling them to avoid the entrenchment's that had given an unpleasant surprise to British troops earlier in the war.

During a private visit to Paris in 1902, Colonel Templer witnessed ascents by the Brazilian aviation pioneer Alberto Santos-Dumont as his diminutive airship flew over the streets of Paris. Upon his return to England Templer was sufficiently impressed by what he had seen to recommend to the War Office that the development of this type of craft should be pursued by the army for war purposes.

Despite a cut in the budget apportioned to the balloon establishment in the army estimates for 1904, sufficient funds were made available for the construction of two elongated envelopes made from goldbeaters' skin for use in experiments to produce a powered airship.

In 1906 Colonel Templer was retired from the army list due to his age and was superseded by Colonel James Capper, another officer of great resource and ability. Together with his civilian assistant, the colourful pioneer aviator Colonel Samuel Cody, an American who had become a naturalised British subject, Capper formed a team that was to design and build the first army airship.

Colonel Capper, it should be pointed out, did not solely devote his considerable engineering and administrative skills only to the area of lighter-than-air navigation, but also showed a great interest in the development of heavier-than-air flying machines.

Capper had visited the Wright brothers in early October 1908 in Dayton, Ohio representing the British government with the intention of negotiating the purchase of their flying machine or obtaining a licence for its manufacture by the British War Office. In the end these negotiations came to nothing due to the excessive secrecy displayed by the Wrights and the exorbitant sum they demanded for the purchase of their system.

Colonel Cody was an able and innovative engineer who had been conducting experiments with man-lifting kites and gliders at Laffans Plain near Farnborough for a number of years with considerable success, receiving some financial assistance to this end from the War Office in his role as civilian kiting instructor. The War Office encouraged his efforts, being interested in the man-lifting kite for possible employment in the British army as a more convenient and manageable instrument of reconnaissance than the balloon.

Cody's work produced a successful and stable system consisting of a train of small, stabilising kites that eventually lifted a larger kite, to which was attached a basket containing the observer. These 'man lifters' could be flown with every confidence to heights of around 2,000ft in perfect safety, this form of kite being used up to the outbreak of the war by both the army and the navy for artillery spotting and reconnaissance until replaced by the later kite balloons.

Cody continued to develop his kites and as a natural progression soon turned to gliders, spurred on by the rumoured success of the Wright brothers in America. Subsequently, in his additional role as civilian advisor for aeroplane development at Farnborough (where after much experiment he was to achieve success in the building and flying of the army's first heavier-than-air flying machine, *Army Aeroplane No. 1*), Cody performed some of the earliest aeroplane flights made in this country during 1907–08.

Some mention should now be made of the efforts of other civilian pioneers in the field of airship development at the turn of the century.

One of the earliest of these was Stanley Spencer who, with his brothers, ran a balloon factory at Highbury in North London. In 1902 they built a small 20,000cu ft capacity airship constructed along the lines of the Santos-Dumont craft.

Stanley Spencer Airship No1 - 1902
C.G.Spencer Balloon Works, Highbury, London

25 Feet

The first successful navigable flight in Great Britain of a manned, powered aircraft took place on 22nd Septenber 1902, when Stanley Spenser piloting his airship No1 flew fom Crystal Palace via St Pauls to Eastcote Middlesex a distance of some 28 miles in 94 minutes

Capacity 20,000 cubic feet
75' x 20' x 35'
Gross lift 0.58 tons
Disposable lift 600lbs
1 x 3hp JAP single cylinder water cooled
Speed 20 mph
Crew 1/2

With this airship, *Spencer No. 1*, driven by a single cylinder JAP petrol engine of a meagre 3hp, a number of successful flights were made. The first of these, from Crystal Palace in September 1902, was the first manned, powered flight by any form of aircraft in Britain. It was intended to fly from Crystal Palace, round to St Paul's and return to the starting point, but contrary winds combined with the low power of the engine caused the little airship to be driven to the north-west, where after a flight of some ninety minutes a landing was made at Eastcote in Middlesex.

Spencer built a series of similar but more powerful airships up to the outbreak of war, financing them in part by the use of advertising slogans on the envelopes; Bovril being one of the companies that utilised this unusual advertising medium.

Shortly before his death from malaria in 1904, Stanley Spencer was visited by Count Zeppelin, an indication of the respect he commanded amongst the group of early aviation pioneers of the period. At the time of his visit Count Zeppelin was planning his second rigid, the *LZ-2*, and discussed with Spencer the many problems facing them in developing the dirigible as a practical aerial vehicle.

A much larger semi-rigid of 230,000cu ft capacity powered by two 50hp engines was built at Alexandra Palace in 1905 by a Dr Barton, who, after demonstrating a flying scale model, received encouragement and some financial help from the War Office to build a full-size craft. The airship showed some promise in its design, but unfortunately came to grief on its maiden flight due to control problems. Diving into the ground it broke up, fortunately without any serious injury to the crew of five. A second, larger craft built on similar lines was commenced soon after the accident but was abandoned before completion due to lack of funds.

The most successful and influential of these early experimenters was undoubtedly E.T. Willows of Cardiff, starting with his diminutive *No. 1* of 12,000cu ft capacity. It first flew in September 1904 and over the next few years made a series of flights of reasonable duration, utilising its unique directional thruster propellers for lateral control.

Further craft followed, with the larger *No. 2* launched in 1909. His *No. 2* flew from Cardiff to London in August 1910, a trip which included navigating by night, a hazardous undertaking at that time, and landing at Crystal Palace at dawn.

A third airship, named the *City of Cardiff*, of 33,000cu ft capacity was completed in November 1910 and, with Willows on board, left from Wormwood Scrubs to undertake the first flight from London to Paris, a distance of 218 miles. During this voyage, after a trouble-free Channel crossing, a forced landing due to engine trouble necessitated a diversion for repairs. These were carried out at the workshops of the Clément-Bayard airship company at Levallois-Perret. After repairs were effected the little airship continued on to the French capital the next day, arriving to much acclaim and earning the distinction of being the first British airship to cross the Channel.

During 1911 Willows moved his business from Cardiff to Birmingham, from where the *Willows No. 4* was launched in June 1912. The completed ship being 110ft in length with a 24,000cu ft capacity envelope of oiled cotton, carrying a small car mounted on a long boom containing the crew of two/three and a 35hp Anzani engine driving swivelling airscrews. Simple cruciform fin and rudder planes were affixed to the rear of the envelope.

Army Airship "Nulli Secundus II"- 1908
Royal Balloon Factory, Farnborough

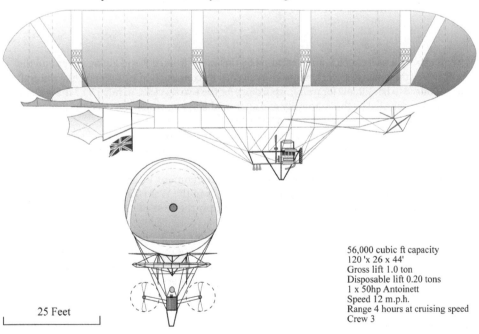

25 Feet

56,000 cubic ft capacity
120 'x 26 x 44'
Gross lift 1.0 ton
Disposable lift 0.20 tons
1 x 50hp Antoinett
Speed 12 m.p.h.
Range 4 hours at cruising speed
Crew 3

8. Britain's first military airship, *Nulli Secundus*, in flight at Farnborough, 1908.

This airship was inspected and appraised by the newly formed Royal Flying Corps and the navy, where the quality of workmanship involved in its construction was praised and the craft was adjudged to be suitable for training purposes for the two services. In July 1912 the *Willows No. 4* was purchased by the Admiralty and after modification, which included the fitting of a new envelope, she became naval airship HMA *No. 2*.

Willows built several other airships including a further order for the navy. This was to become the prototype for the early Sea Scout class of airship used for convoy protection during the war.

Prior to these developments the first official government interest began in 1907 at Farnborough, where, after receiving a small grant from the War Office of £2,000, Colonels Capper and Cody were at last able to start work on the first of the army airships, the *Nulli Secundus*.

The airship was designed as a semi-rigid, no doubt influenced by the success of the contemporary French Lebaudy airships, and was to be built under conditions of the greatest secrecy. A noticeable feature of the completed craft was the long, uncovered square section metal keel suspended some 12ft below a cylindrical goldbeaters' skin envelope of 56,000cu ft capacity. With a length of 120ft, power was provided by a 50hp Antoinette petrol motor, which doubled as the power unit for Cody's *Army Aeroplane No. 1*, and which drove two aluminium paddle propellers on each side of a small car suspended below the keel containing the crew of three.

An advanced feature of the propellers was that they represented the first use of a variable pitch mechanism in any aircraft, so that the blades could be set at the optimum angle of attack. The envelope, however, lacked an internal ballonet, a feature its designers must have been aware was essential for maintaining the envelopes rigidity, compensating for the expansion of gas as the airship rose to a moderate altitude. They did include an automatic gas valve and two hand-operated manoeuvring valves. One can only assume that the omission of this important feature must have been made on the grounds of economy.

The envelope was attached to the keel framework by four circumferential cloth bands and covered with cord netting; the netting and bands being made necessary, despite the extra weight, due to the extreme difficulty of attaching suspension wires or other attachments to the goldbeaters' skin fabric. Twin rudders were fitted aft, while an elevating plane was situated at the bow in the hope of providing vertical control.

A series of short, local flights were made at Farnborough where a speed of 16mph was obtained. Following these tests *Nulli Secundus* departed on 5 October to fly to London, landing after a three-and-a-half-hour flight at Crystal Palace. Plans for a return trip were abandoned the next day when strengthening winds threatened to damage the moored ship, the sergeant in charge instead ordered the ship to be deflated and it was returned to its base by road.

During the winter months the airship was rebuilt, emerging as *Nulli Secundus II* of a slightly increased capacity with a new, streamlined silk-covered deep keel that was attached directly to, and faired into, the underside of the envelope. Several local flights were made during July and August, where a speed of 22mph was recorded; however, both vertical and lateral control were deemed to be inadequate and the airship was dismantled at the end of 1908.

Army Airship "BABY"- 1909
Royal Balloon Factory, Farnborough

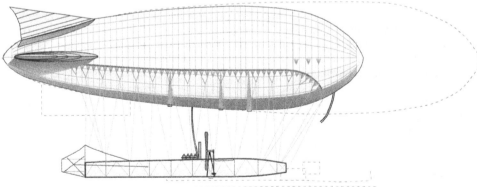

Diagram showing original configuration of 1909

Dashed outline indicates 1910 enlargement to 104 ft
and capacity to 35,000 cubic feet

50 FEET

21,000 cubic feet capacity
84' x 26' x 38'
2 x 8hp Buchet (originally)
1x 25hp REP (as modified 191
Gross lift 0.62 tons
Useful lift 400lbs
Speed 20 m.p.h.
Range 3 hours at cruising speed
Crew 3

Army Airship HMA "Gamma" - 1911
Naval Airship No 18
Royal Aircraft Factory, Farnborough

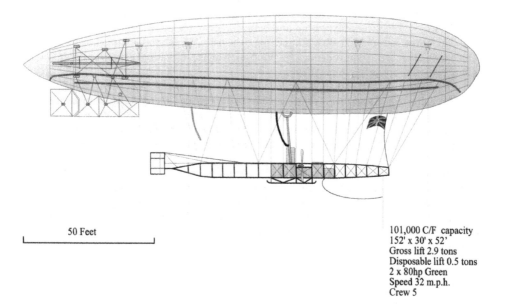

50 Feet

101,000 C/F capacity
152' x 30' x 52'
Gross lift 2.9 tons
Disposable lift 0.5 tons
2 x 80hp Green
Speed 32 m.p.h.
Crew 5

It cannot be said that *Nulli Secundus* represented any great advance in airship development, and indeed it contained some antiquated features, with its designers failing to take advantage of the considerable amount of information they possessed on current continental developments. The lack of adequate funding to the project restricted the designers' ability to fulfil their original desire: to build a British airship that was truly 'Second to None'. In reality its performance was in no way comparable to the contemporary French Lebaudy airship and could only be considered as a woefully inadequate craft possessing poor directional stability.

The next product of the Royal Balloon Factory, launched in May 1909, was a smaller 22,000cu ft airship designated the *Baby*. This was a more carefully conceived design incorporating some of the more advanced features seen in contemporary continental airships, representing a definite improvement on its predecessor.

In its original form the craft was 84ft in length by 24ft diameter, with bulbous air-filled fins at the stern. *Baby*, unlike the previous airship, was fitted with an internal ballonet. Power initially came from two wholly inadequate 8hp Buchet petrol engines, these engines having previously been installed in an early version of the Dunne tail-less biplane.

After initial testing the air-filled fins were replaced with more suitable flat section steering planes, while a 25hp REP engine was installed to provide more power. After this, and following a series of trials that highlighted poor controllability, she was further taken in hand during the winter. Emerging in the spring of 1910 in a reconstructed form and now known as *Beta*, her

ARMY AIRSHIP HMA "BETA II" - 1912
NAVAL AIRSHIP No 17
Royal Aircraft Factory, Farnborough

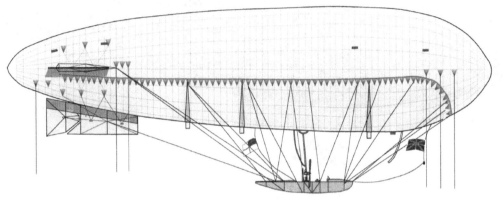

This drawing shows the final configurationof "Beta II"
The original airship "Baby" of 1909 was reconstructed in 1910
by increasing its' length to 110ft, and capacity to 35,000 c/ft, with
powere provided by a 35 h.p "Green" engine.
In 1912 a further rebuilding took place, increasing the capacity to
50,000 c/ft, and length to 116 ft.

25 Feet

50,000 c/ft capacity
116' x 28' x 40'
1 x 45 h.p. "Clerget"
Gross lift : 1.22 tons
Useful lift : 0.40 tons
Speed 32 m.p.h.
Range : 8 hours at cruising speed
Crew : 3

length had been increased to 116ft with a capacity of 35,000cu ft, giving her a gross lift of just over 1 ton together with a greatly improved all-round performance.

A further change of engine, in the form of an efficient and reliable 35hp four-cylinder, water-cooled Green engine, completed her conversion; conferring on her a normal endurance of over five hours carrying a crew of three.

This small craft could be considered the first successful British airship, on one occasion staying aloft for eight hours. She made several other flights of note, including flying over London at night in early experiments to determine the best means of defence to be employed against attack from aerial craft. *Beta* was also the first airship to be fitted with a wireless apparatus, being used in early 1911 for experiments in wireless telegraphy – which were conducted with a considerable degree of success both in sending and receiving messages at distances of over 30 miles.

During the army manoeuvres of 1910, HMA *Beta* was in continuous service together with the recently completed *Gamma*. The two airships were employed in evaluating their potential use to the army in warfare: in extended flights over Somerset, Dorset, Hampshire and Wiltshire flying reconnaissance sorties; scouting for the attacking and defending forces for a period of two months; covering, in the case of *Beta*, in excess of 1,000 miles; spending most nights moored out in the open, using the cover afforded by a screen of trees; being maintained and serviced under these basic conditions, and on one occasion having a broken crank shaft replaced within the protection of a quarry.

9. Army airship *Beta* at Farnborough, 1910.

The airships were supported in the field by the old existing balloon gas train and workshop, drawn by a steam traction engine to provide a mobile base.

During the period of the manoeuvres *Beta* had to be deflated on one occasion only, and provided an excellent example of early tactical air support to the army in the field.

In 1912 *Beta* again underwent further modification, with her envelope being split lengthwise to accommodate an additional longitudinal gore. This increased her diameter to 28ft and lengthened to 135ft overall. The swivelling propellers were retained whereas the engine was once more changed, this time to a 45hp Clerget, increasing the top speed to 35mph.

Beta was also used in early mooring mast trials; using a mast designed by Commander Masterman RN at Farnborough, known as the 'high mast' system, whereby the airship floated free of the ground attached to the mast at the nose, using weighted rollers to keep the stern of the ship in trim as it vaned in the wind.

The Germans did not employ any similar system during the war years, although finally, in the late 1930s, they adopted the American 'low' or 'stub mast' for ground handling, which proved most successful and was in fact employed originally by the early British airships.

As *Beta II* she continued to give good service, taking part in the army war games on Salisbury Plain before the war and being involved in the early parachute experiments carried out from her by General E.M. Maitland (who was later to lose his life aboard the ill-fated *R38* over the Humber).

In January 1914 the navy took over lighter-than-air operations from the army, with *Beta II* becoming HMA *No. 17* and initially used to familiarise naval personnel with airship handling.

With the outbreak of war HMA *No. 17* was for a short time sent to France. Based near Dunkirk, she made several night flights during January 1915 over the German lines and was on artillery-spotting duties before such activities were deemed to be too dangerous.

Subsequently, HMA *No. 17* was used for training at RNAS Kingsnorth before finally being deleted in mid-1916, having contributed much to the development of this type of small, handy airship in an active and varied life of seven years.

In February 1910, during a speech in the House of Commons, the Minister for War R.B. Haldane declared that:

> On the subject of aeronautics, great strides have been made with work at the aero department of the N.P.L. at Teddington increasing its staff and research programmes, while the Royal Balloon Factory have reorganised their existing management structure and construction departments. With two new dirigibles currently in hand or completing, while new facilities are being provided for the training of officers and men.

He went on to say:

> The next step is to constitute a regular 'Aeronautical Corps' such as already exists in France and Germany, while the great Naval dirigible under construction at Barrow is expected to be launched in the summer, and negotiations are in hand to purchase two further dirigibles from France, then once we have mastered the technical problems involved in operating this type of craft we will produce a fleet of these 'Flying Dreadnoughts' comparable or superior to those of other nations.

"Clement Bayard II" Army Airship -1910
British Army "Daily Mail" Airship
Clement Bayard Constructions Aeronautiques - Levallois-Perret, Paris

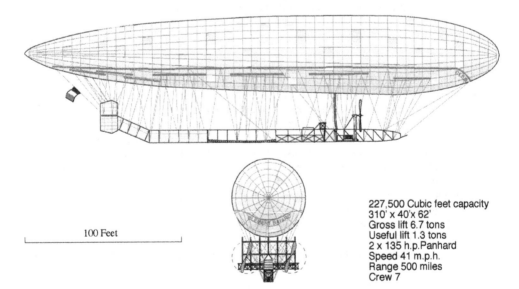

100 Feet

227,500 Cubic feet capacity
310' x 40' x 62'
Gross lift 6.7 tons
Useful lift 1.3 tons
2 x 135 h.p. Panhard
Speed 41 m.p.h.
Range 500 miles
Crew 7

At this time the French government were spending £880,000 per annum on aviation, with Germany spending even more, while Great Britain had expended only about 10 per cent of this sum on military aviation.

On 1 April 1911 the air battalion of the Royal Engineers was formed under royal warrant, initially consisting of two wings with No. 1 Company representing the airship, balloon and kiting aspect of aviation, while No. 2 Company was responsible for aeroplanes.

Over the course of the next twelve months the air battalion worked with the army, participating in manoeuvres and learning the skills of aerial reconnaissance and artillery spotting that were to become vital to the army during the conflict to come.

Both companies came under the overall command of Sir Alexander Bannerman who later became the director of the Royal Aircraft Factory. At the same time, Farnborough was undergoing a major reorganisation involving its technical staffing, together with the provision of an expanded engine research department and the building of two new, large airship sheds to complement the old balloon shed.

Farnborough now came under direct government control through the War Office, finally ending the Royal Engineers' direct involvement and responsibility for balloon and airship work. As early as 1910 the Minister for War, R.B. Haldane, had become concerned about the development of British military aviation. Realising it had reached a stage where it could no longer be left to enthusiastic amateurs like Cody and Capper, Haldane reasoned that in order to develop the aeroplane and airship for practical military use the Royal Balloon Factory must be reorganised and placed on a more scientific basis.

Under the new civilian director, Mervyn O'Gorman, the Farnborough establishment now became the Royal Aircraft Factory in March 1912, with its budget increased from the 1909 total of £6,300 to some £72,000 per annum. The whole undertaking was placed on a more scientific and organised basis, with a larger staff consisting of eight officers and 150 men.

Technical facilities were improved to provide for testing and development of both airships and aeroplanes, while concurrent with these developments the two promised large dirigibles had been ordered from France. The purchase of the first of these, the *Clément-Bayard*, was the result of agitation by factions within Parliament who were concerned by the lack of positive action to provide for the aerial defence of the nation and had persuaded the War Office to purchase an existing military dirigible, which had already seen service with the French army.

The Treasury put up the bulk of the £18,000 purchase price and several patriotic MPs supplied the remainder with the assistance of the *Daily Mail*, which paid for the construction of a 365ft-long by 75ft-wide shed to be erected at Wormwood Scrubs to house the craft.

The 310ft-long airship was flown from its base at Compiegne on 16 October 1911, making the first journey of its kind between France and Great Britain in six and half hours, flying at the respectable average speed of 38mph. The airship flew over central London landing safely at Wormwood Scrubs, to be housed in the new shed where it was the subject of much attention by the populace and a source of satisfaction to those who had put so much effort into acquiring it.

The wisdom of buying a second-hand airship that had already seen a year's usage with the French army was questioned when subsequent inspection showed the envelope to be in a much deteriorated condition. Indeed the leakage of gas was so severe as to require replacement of the entire envelope.

The airship was deflated and taken by road to Farnborough where it was stored in the old balloon shed. After further inspections of the structure, the War Office decided not to attempt to recondition the airship on the grounds of expense, and the remains were subsequently scrapped.

The second airship, the *Lebaudy*, was again purchased out of patriotic reasons. This time by the *Morning Post* newspaper as a gift to the nation, despite the fact there was at that time no shed large enough to house the 337ft-long craft – after protracted negotiations with the War Office – agreement was reached to provide a new shed at Farnborough.

The *Lebaudy* was a giant airship by the standards of the day, with a gas capacity of 353,000cu ft giving a gross lift of 10.4 tons, powered by two 135hp Panhards. Colonel Capper had flown aboard her during trials earlier in the year in France and had reported favourably on her capabilities.

The airship was flown to England on 26 October 1910, manned by a French crew of seven and two representatives of the War Office including Major Bannerman who had by then succeeded Templer as superintendent at Farnborough. They crossed the coast at Cap Gris Nez, the airship easily overhauling the escorting destroyers at a speed of 35mph.

After an uneventful flight of five and a quarter hours the airship landed at Farnborough. On being taken to the shed the major in charge of docking became concerned that the top of the airship was too high to pass under the shed entrance, but was overruled by a senior officer who insisted the ship be walked in at once.

As the *Lebaudy* entered the shed the upper portion of the envelope was torn by the roof girder work, causing the ship to collapse and become a total wreck. It would appear that the dimensions of the ship supplied before the erection of the shed were different from the craft as completed, with the diameter being some 10ft wider.

The wreckage was dragged into the shed and followed by an inquiry, during which the War Office refused to pay the purchase price to the Lebaudy Company, declaring them negligent in not advising them of the increased diameter of the envelope. It was eventually agreed that Lebaudy Frères would foot the bill to repair and make the airship airworthy.

British Army "Morning Post" Lebuady Airship -1910
Lebaudy Freres, Constructions Aeronautiques
Moisson, Seine-et-Oise, France

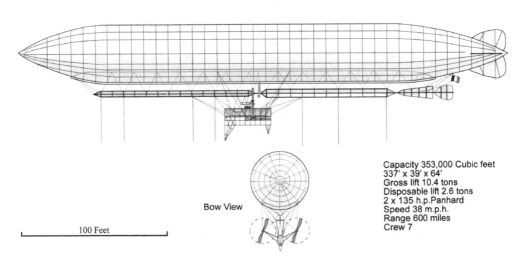

Bow View

100 Feet

Capacity 353,000 Cubic feet
337' x 39' x 64'
Gross lift 10.4 tons
Disposable lift 2.6 tons
2 x 135 h.p.Panhard
Speed 38 m.p.h.
Range 600 miles
Crew 7

10. *Clément-Bayard II*, the *Daily Mail* airship arriving at Farnborough, 1910.

During the winter months the shed roof was raised by 15ft, and a complete reconstruction of the dirigible was put in hand. A test flight of the refurbished airship took place in May 1911 in the hands of the manufacturers with an all-French crew. Soon after take-off and following a short flight, problems with directional control were experienced, most likely due to the craft being badly rigged following certain modifications to the suspension system.

The crew limped back to Farnborough where, whilst attempting to land, the airship lost control and dived into the ground, crashing on to a house and becoming a complete wreck. Amazingly, as the envelope wrapped itself around the cottage it came within feet of a smoking chimney, but fortunately there was no fire and the crew suffered only minor injuries.

So it was that within a few short months the combined efforts of the War Office, patriotic newspapers and influential groups both within and outside Parliament to bolster Britain's aerial defences had come to nothing.

Before these unfortunate events had taken place, however, the previously mentioned airship *Gamma*, which had been designed by Colonel Capper, had successfully made its maiden flight in February 1910 at Farnborough.

The new airship was a considerable advance on *Beta* both in size and lift capacity, being 152ft in length and 30ft in diameter with an initial capacity of 75,000cu ft, giving a very respectable gross lift of 2.1 tons. Later in 1912 *Gamma* was reconstructed by lengthening the envelope, increasing the capacity and lift to 101,000cu ft and to 2.9 tons, respectively.

In her final form a long, metal-framed car was suspended below a bright yellow rubberised cotton fabric envelope made by the Astra-Torres company in Paris, earning her the nickname the *Yellow Peril*. Power was supplied by a 35hp four-cylinder, water-cooled Green engine driving swivelling propellers mounted on out-riggers from the car, giving her a still air speed of 35mph.

The swivelling propellers were capable of moving through 240° about her longitudinal axis and were most effective in controlling the ship during climbing and landing. They represented a great technical achievement for the small balloon factory staff at that time.

The envelope was sub-divided by internal transverse partitions to prevent surging of gas, and had multiple ballonets to preserve envelope pressure.

During 1911 wireless experiments were carried out with *Gamma*, where messages were received at ranges of up to 30 miles distant.

The first product of the newly established Royal Aircraft Factory was the airship *Delta*, which had originally been designed in 1911 as a semi-rigid but emerged in 1912 rebuilt as a non-rigid of 175,000cu ft capacity. Contained in a rubberised fabric envelope and powered by two powerful 105hp White and Poppe petrol engines, again driving swivelling propellers, it was the fastest British airship so far with a top speed of 45mph.

A short, 28ft-long car and its crew of five were suspended some 20ft below the envelope, the suspension wires being attached to horizontal ropes sewn into the lower portion of the envelope fabric, while simple braced horizontal and vertical steering surfaces were attached at the stern.

With her increased capacity and a gross lift of 4.6 tons she was extensively used for training and wireless telegraphy experiments conducted by Captain H. Lefroy RE in 1912, during which transmissions were clearly picked up at a range of 100 miles.

Both *Gamma* and *Delta* took part in further army war games during 1912 and 1913, where they were employed both in reconnaissance and the 'bombing' of towns in 'enemy' territory

Army Airship HMA "DELTA" - 1912
Naval Airship No19
Royal Aircraft Factory, Farnborough

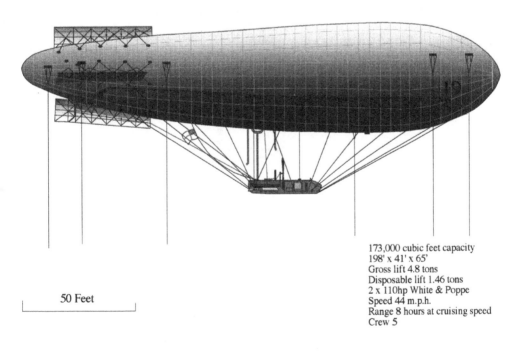

50 Feet

173,000 cubic feet capacity
198' x 41' x 65'
Gross lift 4.8 tons
Disposable lift 1.46 tons
2 x 110hp White & Poppe
Speed 44 m.p.h.
Range 8 hours at cruising speed
Crew 5

– providing valuable experience in the effectiveness of this new weapon for troops and the general staff alike.

The next airship built at Farnborough was to be the last army airship, this was *Eta*, launched in August 1913 and completed as a non-rigid of 118,000cu ft capacity incorporating twin ballonets and capable of 46mph.

Eta was noteworthy due to the introduction of the 'Eta patch' in its design as an improved anchorage system for car suspension that greatly reduced drag. In previous airships the car had been attached to the envelope by rope netting or by sewing ropes longitudinally into the envelope, from which the car was suspended by a system of wires and bridles.

In early airships the car was often of considerable length and positioned well below the envelope in order to spread the load and prevent sagging over the length of the envelope. The Eta patch allowed the car to be made smaller and attached nearer to the envelope, providing a better streamlined form and reducing wind resistance and hence drag.

The patch consisted of a steel ring through which several layers of overlapping material were rove, forming a fan-shaped patch with the ring positioned at the lower apex. The

Army Airship HMA "ETA" - 1913
Naval Airship No 20
Royal Aircraft Factory, Farnborough

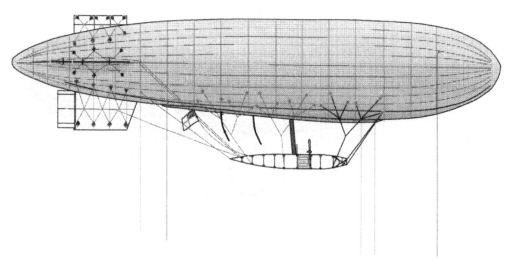

50 Feet

118,000 Cubic feet capacity
190' x 38' x 47'
Gross lift 3.26 tons
Useful lift 0.9 tons
2 x 85 h.p. Canton-Unne
Speed 42 m.p.h.
Range 10 hours at cruising speed
Crew 4-5

11. HMA *Delta* at Farnborough, 1912.

overlapping layers of fabric were glued and stitched to each other and to the envelope, forming a strong attachment point positioned in such a way as to allow for a better distribution of load on the integument of the envelope. So successful was this method that it is still used on modern airships today, where the internal suspension wires within the envelope terminate in a variant of the original Eta patch.

Eta could carry a crew of five for ten hours and gained a reputation as a fast and handy ship, on one occasion being dispatched to Odiham to perform the unusual duty of towing back to Farnborough the Willows airship *No. 2*, which was disabled due to engine trouble. She successfully accomplished this mission, proving that she had plenty of power in reserve in her twin 85hp Canton-Unne engines and swivelling propellers.

In January 1914 *Eta*, like her sisters, was taken over by the Admiralty. She became known as naval airship *No. 20*, taking part in defensive night flights over the capital during the opening stages of the war and making war patrols over the Thames Estuary. She was intended to join the BEF in France at Dunkirk, but was badly damaged in a forced landing at Redhill in Surrey in November 1914.

The following year the undamaged envelope was attached to the unused car from the Astra-Torres *No. 10*, and the combination, known as *Eta 2*, was used largely in a training role until early 1916 when she was deleted, the old *No. 10* envelope being used on the prototype Coastal *C1*.

Eta was an efficient design for the period, being lighter than contemporary foreign airships of similar size, and ultimately contributed much to the design of the mass-produced Sea Scout patrol airships of the war period.

All these early airships were from time to time extensively modified, so much so that the photographic record shows the same airship in a variety of forms. This can cause some confusion to the researcher with different cars, fins and envelopes being substituted at various times.

Further reorganisation of the aerial services took place in April 1912, when an Order in Council established the formation of the Royal Flying Corps, initially comprising two wings designated the RFC Military Wing and the RFC Naval Wing. However, within a short time the naval wing acquired the more distinctive title of the Royal Naval Air Service, underlining that service's independence of the War Office.

As previously mentioned, the Admiralty took over control of all airships in January 1914 and were destined to develop this type of craft in a most successful manner for convoy escort, mine spotting, and co-operation and reconnaissance with the fleet.

The process for the production of hydrogen in the late 1890s was a laborious, messy and not to say dangerous process involving passing sulphuric acid over iron filings and collecting the resultant gas. By the First World War efficient Silicol plants had replaced these more primitive methods and were capable of producing up to 10,000cu ft of hydrogen per hour, the gas produced being stored in gas holders, while gas flasks could be filled under pressure for more mobile use to airships operating from temporary moorings out on sites in the field.

The experimental work carried out initially by the Royal Engineers, the Royal Balloon Factory and latterly by the Royal Aircraft Factory in the years preceding the Great War was of immense importance to the future development of both aeroplanes and airships.

In the light of the prevailing technology of the time in this new area of scientific endeavour their work was of the highest technical calibre, carried out initially under conditions of the most stringent financial support and with the provision of minimal technical facilities. However, once the establishment had been put on a more organised and scientific basis and driven by the need to place Britain's aeronautical development on a par with those of the leading continental nations, resources were made available and rapid progress in the understanding of the problems and theory of flight were made, together with the development of both reliable aircraft and aero engines.

This period of development, led by the dedicated and technically brilliant group of engineers, scientists and craftsmen employed by the Royal Aircraft Factory, deserves to be recognised as one of considerable achievement in the early days of British aviation.

4

The First British Rigid: *Mayfly*

arly in 1908, in response to the success of the Zeppelin Company in Germany and other airship developments in Europe, interest began to be shown by the military and naval authorities in this new and novel form of flying machine for use in time of war.

The British government, in the form of the Committee of Imperial Defence with the enthusiastic backing of both Prime Minister Asquith and the wily First Sea Lord Admiral Fisher, now evaluated the threat posed by the advent of aerial navigation to Britain in time of war. They considered what use such craft could be put to in defence and came to the conclusion that the airship would be an indispensable asset to the Royal Navy in a scouting capacity, also possessing the advantage that it could be built at a fraction of the cost of a surface cruiser.

Accordingly, with Treasury approval granted for an estimated cost of around £35,000 to be included in the naval estimates for the coming year, the Cabinet directed the Admiralty to undertake the development of airships of a large capacity that would be suitable to perform scouting duties with the battle fleet.

At this time, the Admiralty were ahead of the Imperial German Navy in adopting the concept of utilising airships in a scouting role, although it was widely believed in official circles that the German navy were already building such craft.

The Admiralty set up a design group under the direction of Captain Murray Sueter, serving in the newly appointed post of superintendent of airship construction, and quickly followed this in July 1908 in a preliminary enquiry to Messrs Vickers Son & Maxim to 'prepare a specification for an airship with all possible despatch' in keeping with the urgency of the matter.

The original specification called for an Aerial Scout capable of 40 knots for twenty-four hours, able to rise to an altitude of 1,500ft carrying wireless and other naval equipment with a crew of twenty officers and men, the projected gross lift of the craft was to be 20 tons with a disposable load of 3 tons. Such a ship would be required to operate and keep station with the fleet at sea, scouting ahead of the column forming an extended patrol line many miles in front of the fleet, and able to report the presence of enemy vessels to the admiral by wireless.

Considering the primitive state of aeronautical development at the time, the design concept represented a bold and far-sighted vision, with the Admiralty and the Vickers engineers venturing into uncharted regions in terms of new materials and the relatively unknown forces imposed by air pressure in flight. This, together with the structural complexity of building such an experimental craft, would tax the design staff to the limit.

The task they faced may be compared in scale and difficulty to that encountered in the design and building of Concorde in the early 1960s, but the Edwardian engineer had to operate without the benefit of the existing store of technical and aerodynamic knowledge that had been so painstakingly accumulated during the intervening sixty years.

In 1908 the development of the heavier-than-air machine in Europe had only just reached the first faltering stages of practicality, with Henri Farman completing the first sustained flight of a 1km circuit at Issay near Paris in January of that year. But such was the rate of aeronautical progress in Europe that by the end of that year no one doubted that the flying machine had indeed arrived. Rapid strides were being made in engine development and most importantly in controlling a machine in the air.

The aeronautical community were further spurred on by the visit of Orville Wright to France in the autumn of 1908. Here he ably demonstrated his mastery of the air with superbly controlled flights, some of which exceeded one hour in duration at a time when European experimenters could only manage short hops.

Within the passage of a few short months even greater strides were being made in aviation, cumulating in June 1909 with Louis Blériot flying the English Channel in his diminutive, barely controllable monoplane and at the same time demonstrating that Britain was no longer an island now that the air had been conquered.

Meanwhile in Great Britain, apart from such pioneers as Cody who had been working on his *Army Aeroplane No. 1* since early 1907, the British focus was on airships. Of those so far constructed by the Royal Balloon Factory under Colonel Templer were small pressure airships whose performance and lifting capacity were far inferior to the proposed Admiralty vessel, which in turn had a capacity half as large again as the biggest Zeppelin then flying.

The contract was finally placed in May 1909 with Messrs Vickers Son & Maxim to build the experimental ship, *Mayfly*, to a design jointly devised by the Admiralty and Vickers (during

12. Naval dirigible *Mayfly* in Cavendish Dock, Barrow-in-Furness, May 1911.

the course of construction Maxim resigned his interest in the company, which after 1911 became known as Vickers Ltd).

The design of such a large vessel in the absence of any previous experience was to be based, by default, on a combination of known shipbuilding techniques, empirical principles and such available aeronautical data obtained from home and continental sources.

This project came under the general direction of a design committee composed of Captain Bacon as the Admiralty representative and Captian Murray Sueter who had played an important part in the introduction of the submarine into the Royal Navy a few years earlier. Murray Seuter, with Commander Oliver Schwann as superintendent, led a group of naval constructors together with the Vickers group headed by Sir James McKechnie, director of the naval construction works at Barrow.

Vickers were at that time a major naval shipbuilder and their experience with submarine construction work was deemed to be of benefit to the project. Analogies could be drawn between the two types of craft working, as they did, immersed completely in a supporting medium; one in water, the other in air.

Work began in conditions of the greatest secrecy and only the barest details on the progress of the vessel were made public over the next two years that were required for her construction. Vickers hoped that by providing a constructional shed at their own expense and placing it at an early date, they then would be in a good position to secure further orders in the form of a ten-year monopoly in airship construction, as they had previously obtained for submarine work.

In this their judgement proved to be in error. Although they were to build further airships by the end of the war they were not considered for further work as their shed facilities were by then too small. A further error they made was to choose to site the fixed construction shed, started in 1909 directly over the water in Cavendish Dock at Barrow.

The choice of a waterborne constructional facility was the result of the Admiralty's view that an airship was comparable to a naval ship, per se – as was evidenced by the Admiralty's insistence that the ship should carry a variety of naval gear such as anchors, capstan, etc, which was better suited to a destroyer than an aircraft. This was also influenced by Count Zeppelin's use of a floating hangar on the sheltered waters of Lake Constance for building his early airships.

The inconvenience of working directly over the water during construction and the detrimental effect in the form of corrosion to structural materials, plus the general damp environment, took its toll on the metal framework. In particular this led to the rapid deterioration of the fragile gas cells and the outer cover.

This unsatisfactory state of affairs was soon recognised in Germany, where future airships were built in dry, wooden-floored sheds on land. Vickers and the Admiralty failed to appreciate the principle, however. Indeed the only advantage of using a floating shed such as that used at Manzel was that it could be vaned into the prevailing wind to allow for ease of launching and docking.

Had a floating, revolving shed been available to *Mayfly* instead of the fixed shed, it might well have avoided its ultimate fate: being crushed against the shed wall by a crosswind while being brought in by an inexperienced ground crew, who were additionally hampered by handling the vessel partly from the land and partly on water.

Tests were conducted at the Royal Aircraft Factory at Farnborough, and at the National Physical Laboratory at Teddington in the hydrodynamic tanks, using wooden models to determine the optimum shape for the craft. The need to produce a body form affording the least resistance to the airflow was well understood, and the design team may well have been influenced by the streamlined form of the German Parseval and Gross designs (whose body forms had been determined through wind-tunnel testing at Gottingen University where the German government had the foresight to establish a research institute devoted to the problems of flight).

Naval Airship HMA No 1 "MAYFLY" (R1) - 1911
Messrs Vickers Son & Maxim, Barrow in Furness

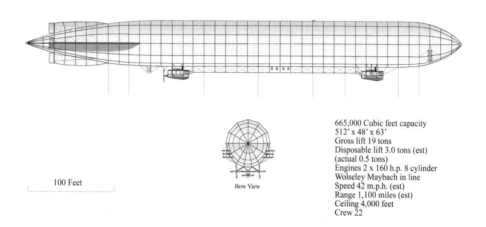

100 Feet

Bow View

665,000 Cubic feet capacity
512' x 48' x 63'
Gross lift 19 tons
Disposable lift 3.0 tons (est)
(actual 0.5 tons)
Engines 2 x 160 h.p. 8 cylinder
Wolseley Maybach in line
Speed 42 m.p.h. (est)
Range 1,100 miles (est)
Ceiling 4,000 feet
Crew 22

H.M.A. No.1 "Mayfly" (R1) - 1911 Girderwork details
Vickers Son & Maxim, Barrow in Furness

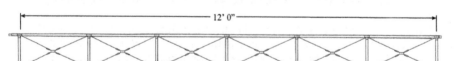

— 12' 0" —

1'0"

Section of main longitudinal "I" girder
Similarly dimensioned girders were also employed in the construction of the
intermediate transverse frames set at 12' intervals.

Whilst main transverse frames were of a triangular section to the same design set at 36' spacings

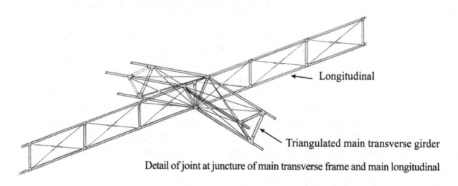

← Longitudinal

← Triangulated main transverse girder

Detail of joint at juncture of main transverse frame and main longitudinal

The design of the hull shape for *Mayfly* was in part based on the researches of a Professor Albert Zahm, an American whose studies led him to the belief that the radius of the bow and stern curvatures were determined in terms of proportions of the diameter of the parallel hull cross-section to produce a streamlined form – this theory later being superseded by more accurate scientific aeronautical data.

As constructional materials in the form of aluminium alloy became available, design work began on a trial-and-error basis. Initially a 38ft-long trial section of hull representing three bays was built in order to evaluate differing constructional methods. One bay was built of a structure comprised of combined triangular wood and aluminium girders, while a second portion was composed of a hollow wooden tube structure and the third section erected from aluminium girder work.

Of these three methods the composite triangular wood and alloy structure proved to be the strongest. However, the Admiralty team decided in favour of an aluminium framework, utilising the Zeppelin pattern of latticed girder work for the main hull.

The actual airship as it evolved was, as previously stated, of a very advanced design. It was far superior in aerodynamic form to the contemporary Zeppelin LZ-4 (this being the fourth rigid built from which any comparison could be drawn by the British constructors) and offered only 40 per cent head resistance to the airflow in comparison to a contemporary Zeppelin.

The original design also incorporating such refinements as enclosed cars, simplified control surfaces and a water recover apparatus designed to compensate for the weight of fuel consumed in flight. Water was recovered as a by-product of engine combustion to maintain the airship at or below the required pressure height.

It must be said that the concept of weight-saving was given only scant attention by the constructors, who, as shipbuilders, were more used to thinking in terms of tons rather than pounds in weight. This was of course an unacceptable state of affairs when a simple calculation could demonstrate that a total lift of an airship of that capacity could not exceed 20 tons total lift, and that, therefore, the total weight of the ship's structure, engines, fuel, crew and equipment must be kept below that figure.

In this and later British rigids the requirement to find the balance in building a ship that was sufficiently strong and capable of resisting both static and aerodynamic loads, together with the requirement of attaining a high proportion of useful load over the weight of the structure, was never to be successfully achieved.

This subtle balance in airship design was to find its most complete expression in the German Zeppelin Company and their craft, which were the result of continuous experiment and the accumulation of hard-gained experience by a company always conscious of the need to save weight without sacrificing strength.

In the event, before the construction of *Mayfly* began in March 1909 the alloy Duralumin became available from Germany. This new alloy comprised of an amalgam of aluminium mixed with small quantities of copper and iron that afforded an increase in tensile and compressive strength of 100 per cent over normal aluminium for a minute weight increase.

Initially, while working with this new alloy, some difficulty was experienced with the malleability of the metal when rolling sections, but once these problems were solved it was to prove a most satisfactory constructional material. It was not until 1914 that the German Zeppelin LZ-26 first utilised Duralumin in its structure.

The final design for *Mayfly* was for a ship 512ft long by 48ft beam with a gas capacity of 665,000cu ft contained in nineteen gas cells. The hull structure comprised of forty transverse frames braced radially by steel wire and spaced at 12ft intervals, with twelve triangular longitudinal girders joining each mainframe, running from nose to tail.

Both the transverse frames and the longitudinal girders were in the form of a tubular, wire-braced, I-shaped flat section with a depth of 12in, apart from frames 3 and 38, where triangulated girders were installed to support the 'aeroplanes'. In addition, treble longitudinal girders were fitted over five bays above the two engine cars.

Control surfaces consisted of box pattern elevators and rudders fixed to simple stabilising horizontal and vertical fins at the stern of the ship, while forward of the control car mounted on a main transverse frame were the triplane 'aeroplanes' designed to assist control in the horizontal plane in the same manner as hydroplanes are used in submarines.

The main steering rudders had a combined area of 440sq ft, producing a turning moment at a pressure of 3lb per sq ft of 1,320lb, while the auxiliary lower rudder was of 74sq ft contributed a further 222lb of force. The forward set of 'aeroplanes' were of 274sq ft whilst the stern planes were 266sq ft, giving a lifting or depressing force of 822lb and 798lb respectively.

The forward enclosed control car was boat-shaped, made from copper-sewn mahogany, containing steering and ballast controls and one 160hp Wolseley-Maybach eight-cylinder petrol motor developed from a marine unit. This drove two 15ft-diameter, four-bladed wooden airscrews fitted to brackets on the hull through reduction gearing at 500rpm. The after car contained a second Wolseley-Maybach eight-cylinder motor, in this case driving a smaller 10ft-diameter, two-bladed airscrew through direct drive at 1,500rpm.

Additionally, three smaller auxiliary rudders were positioned below the hull aft of the rear power car in the wake of the airscrew, together with the mounting of the two forward airscrews lower on the hull and closer to the longitudinal axis than in Zeppelin practice (designed to improve lateral stability and turning moment).

Between the two power cars ran a light, 7ft-deep triangular keel walkway allowing access to the interior of the ship and providing communication between the two cars. Most authorities claim that this keel structure played little or no part in providing a load-bearing role or in stiffening the hull structure in this design – as was the case in the early Zeppelins where the external keel had a key part to play in strengthening the hull. However, it seems inconceivable that such a company as Vickers, steeped in shipbuilding practice and working from first principles in airship design, would have overlooked this key element in providing rigidity to the structure.

The triangular keel structure, composed of girders of the same dimensions and strength as those used in the main transverse girders, undoubtedly did afford a significant measure of stiffening to the hull. Its removal later in the programme in order to lighten the ship would, it seems, have contributed to its loss together with other weight-saving modifications that were undertaken in June 1911.

A small, enclosed cabin was situated at the centre of the keel providing for the Marconi wireless installation and a mounting for a Zeiss telephoto lens plate camera. Above the keel ran the control wires to the elevating and steering surfaces, together with the cables for operating the Parseval-pattern manoeuvring valves to the nineteen gas cells. This was designed to allow the ship to rise at a rate, if necessary, of 3,000ft per minute while discharging gas at a rate of 4,800cu ft per minute, though what the effect on the structure while rising vertically at 32mph would impose on the framework can only be guessed at.

The later wartime Zeppelins were capable of ascending at a rate of climb of 2,000ft per minute, but in practice rarely exceeded a rate of 1,000ft per minute to avoid excessive strain on the structure.

Water ballast sacks and petrol tanks were positioned in the lower part of the hull, with the loads being evenly disposed and supported on the main transverse frames. *Mayfly*, as she unofficially became known, was intended to carry 1 ton of water ballast and 1 ton of petrol, or alternatively 2 tons of petrol and 0 tons of ballast, which could give endurance at full speed of thirty hours.

The water recovery apparatus was fitted to the engines to compensate for the weight of fuel used, and, therefore, to avoid unnecessary valving of hydrogen. When tested this showed an efficiency of over 50 per cent. It was also intended that the system could be used to trim the ship nose up or down, by pumping the recovered water through a system of tanks positioned at the bow and stern.

Seventeen gas cells were installed, but instead of following the then current British practice of making them from goldbeater's skin, the bags for *Mayfly* were instead made by the Continental Rubber Company from a fabric consisting of two layers of Egyptian cotton with a layer of a rubber compound sandwiched between to form a gas-tight seal. However, despite the manufactures' best efforts the bags, in practice, were highly permeable. Due to this experience, British airships returned to the use of goldbeaters' skins to line the fabric cells – this arrangement producing the most gas-tight medium then available.

Each cell was equipped with an automatic pressure valve in the lower portion of the bag, discharging into an exhaust trunk passing vertically through the gas cell to vent through a further automatic valve and gas hood on top of the hull. A diamond-pattern wire netting kept the gas bags in place and the lifting force of the hydrogen gas was transferred via a system of load-bearing wires placed circumferentially around the gas cells, connected to the longitudinal girders in such a way as to spread the load evenly over the whole framework.

After much experiment with different fabrics the outer cover was made from a silk-proofed material coated with one of the first aeroplane dopes to be introduced from Germany. Known as 'loco', it had excellent waterproofing and strengthening qualities. The lower part of the hull was doped a darkish yellow, while the upper portion was clear doped and treated with a silver aluminium dust to counter the effects of superheating in the gas cells.

Construction began in early 1910, but it was soon discovered in the workshops that the use of the new metal was proving to be problematical due to difficulties with rolling and heat treatment of the formed sections required for constructing the girders. After some experiment these problems were eventually overcome; a better understanding of working the metal and the annealing processes were mastered and the resulting structural members met the stress and load requirements of the design.

The mainframes were constructed first on wooden jigs, then lifted from tackles in the shed roof into a vertical position at the required frame spacing. Here the longitudinal girders were attached. Once in place, the whole structure was rotated as shear wires were fixed and tensioned to the polygons of the outer framework, after which the outer cover was applied. The next section was then built and joined to the previous portion of hull.

Work continued at a rate controlled by the supply of parts, various problems with materials, and design changes effected by new information on foreign developments as it became available to the Admiralty. By January 1911 the hull was complete. Shortly thereafter the keel together with the cabin and fore and aft gondolas were installed, fins, rudders, elevators and the forward 'aeroplanes' were attached to the hull, and finally the gas bags were installed.

During April 1911 the process of filling the nineteen gas bags began, the gas being supplied from steel bottles. The task took over a day to complete and required over 1,000 bottles to accomplish the task. During this and subsequent replenishment, several longitudinal girders and a portion of the mainframe failed under the load being applied through the shear wiring of the lifting force of the gas acting on the structure. Repairs were quickly effected, but the inherent weakness in the hull had been demonstrated in static conditions.

By 22 May 1911, the airship was complete and floating inside its shed. At this stage concerns were expressed that the ship would be far heavier than originally proposed; each of the eight-cylinder 140hp Wolseley-Maybach engines bought from the German company weighed 1,800lb each, against the projected weight of 1,300lb. This, together with the multitude of additional heavy naval equipment specified as essential by the Admiralty, which had been included regardless of weight, was a cause of great concern. Would the airship be able to lift its own weight, let alone accomplish its proposed duties?

The gas bags were by now almost two years old and were in a poor condition, leaking at a rate equal to 2 per cent of the total lift per day. The deteriorating purity of the gas resulting from air mixing with the hydrogen further reduced the useful lift to an unacceptably low level.

Mayfly was taken out of her shed for the first time on 22 May 1911, to be moored to an inclined mooring mast attached to a pontoon in Cavendish Dock where she remained for three days, successfully riding out a wind of 45mph.

During this time engine trials were conducted, giving a fuel consumption figure of 17 gallons per hour. Other tests connected with the fuel and electrical systems were also conducted by the nine officers on board. This venture, over three days with the ship in neutral buoyancy, can be considered as *Mayfly*'s only success. The ship maintained its trim and vaned in the wind, absorbing up to a 2-ton strain at the bow connection without sustaining any damage.

Following these trials it was obvious that the ship was drastically overweight, and she was returned to her shed for modification. This manoeuvre was completed with some difficulty due to the narrow width of the entrance and a brisk crosswind, the operation requiring some 200 sailors and one and a half hours to complete. *Mayfly*, perhaps unsurprisingly, incurred some minor damage in the process as she pushed against the shed wall during the process.

Once housed, shed trials were conducted. The ship was trimmed and weighed-off to become airborne, the result of which gave a gross weight of 19.5 tons and a gross lift of 19.7 tons. This gave a useful lift of a paltry 448lb, and this without petrol, stores or full crew, making the vessel still some 3 tons heavy.

In order to remedy the situation the ship was stripped of all unnecessary weight, including the water recovery apparatus and the entire keel structure, which, as noted, undoubtedly provided a degree of strength together with a strong stiffening effect to the hull.

The keel cabin was also removed along with much of the Admiralty-designated equipment. The positions of the fore control and aft engine car were then changed, although to what effect is debatable as each of the wooden boat-shaped control cars weighed 1.5 tons each, and would have been better replaced by lighter cars of Dural alloy construction.

Other changes included the direct drive 10ft aft airscrew being replaced by two of 15ft diameter. On the recommendation of the Committee of Aeronautics some consideration was also given to cutting the ship in half and inserting an additional bay of 40,000cu ft, which would have increased the gross lift by 1 ton (as was resorted to in the later *R101*). This proposal was not implemented, however, due to the limitations of the shed size.

By 22 September all alterations were complete and the ship was again airborne in the shed. *Mayfly* was now provisionally accepted by the Admiralty, although the disposable lift that followed the modifications was probably in the region of 1 ton, and thus still far from adequate.

On 24 September *Mayfly* was again taken out on to the waters of the dock for further trials at the mooring mast, although no attempt at flying the airship was made as deteriorating weather conditions forced the return of the airship to the shed. It was during this procedure, with the bow of the ship partly in the shed, she had the misfortune to be struck by a squall, which forced the hull against the shed entrance seriously damaging the structure and breaking her hull amidships. Further efforts to re-house the ship led to more damage and she became a total loss.

The original cost of *Mayfly* had been calculated at £35,000, but as with all experimental work the costs rapidly escalated. The total expenditure to the exchequer was in excess of £100,000 (not including the cost to Vickers of the shed), more than equal to the cost of a scouting cruiser it was meant to replace. It can, therefore, only be said that the Admiralty's first attempt to build a rigid airship had been a costly failure; a failure that was to call a halt to any further rigid airship development in Great Britain for the next two years.

At the subsequent court of inquiry, the official position was that no blame should be apportioned to any individual or group connected with the debacle. The inquiry further went on to stress that the removal of the keel together with the other modifications were in no way related to the loss of the vessel, with the court stressing that due to the highly experimental nature of the work in a hitherto untried area of research, that there would always be a high element of risk.

The outcome of the inquiry satisfied the Admiralty, who were anxious that the blame for the failure did not fall directly on them, and the outcome proved to be a politically expedient way for those with political and technical responsibility for the project to avoid the worst of the odium of failure.

Mayfly, or HMA *No. 1*, was in hindsight a flawed design: it was too heavy to fly irrespective of what further modifications may have been made to the structure. The hull, which was in essence strong, could have been made 50 per cent lighter by increasing the mainframe spacing of 10ft without weakening the hull, thus reducing the number of frames and allowing the keel to remain in place.

The replacement of the heavy mahogany cars with ones made of aluminium, and the substitution of the Wolseley-Maybach for lighter engines (perhaps Green's or White & Poppe) would also have helped *Mayfly* to achieve her 3-ton useful lift.

Mayfly's dramatic denouement would have been the fate any airship caught in a similar situation: a 512ft-long craft trying to enter the narrow confines of a shed entrance, positioned on the side of a dock in a strong crosswind without room to manoeuvre freely.

Although *Mayfly* never took to the air, many valuable lessons had been learned from the experiment and she provided the basis for much useful research in the area of materials, constructional techniques and as a test bed for future developments. Unfortunately, these hard-learned lessons bequeathed by the *Mayfly* project were, in the main, later to be forgotten. Due to political interference and just plain incompetence by those responsible for later designs, many of the same mistakes were to reoccur in the following Vickers designs, to their detriment.

5

Interim 1912–14: The Astra-Torres, Parseval & Forlanini Airships

Following the setback caused by the wreck of *Mayfly* and the apparent inability on the part of the British to successfully build and fly a rigid dirigible, the Admiralty were reluctant to proceed with further work in this field of aviation.

The government concluded that at the present state of development, it should for the time being abandon the airship construction programme, with the decision being formally taken in February 1912. At the same time the Navy Airship Department had also been disbanded, the officers and the design group dispersed to other posts within the navy or the dockyards. Captain Sueter himself was temporarily placed on half pay.

Between 1911 and 1914 heavier-than-air machines were making great strides in technical development, and great interest in their military potential was shown by governments on the Continent. In Germany, for instance, considerable advances continued to be made in the development of the Zeppelin airship and many notable performances were recorded.

In November 1911 (prior to the disbanding of the Navy Airship Department) it was indicative of the Admiralty's ambivalent attitude towards airship development when, only two months after the wreck of the *Mayfly*, they dispatched Admiral Sir John Jellicoe and Captain Sir Hugh Watson, the naval attaché in Berlin, to appraise and report on the latest developments in Austria and Germany.

The visit was made openly with permission sought from Admiral Tirpitz for the two high-ranking officers to be allowed to undertake a tour of the airship construction works in that country. To this request the Germans readily agreed and the British visitors were afforded every co-operation by their hosts. The two officers flew aboard the new Zeppelins *Schwaben*, then in service with DELAG, and also the Parseval *PLVI*, being most impressed by the high speed and endurance achieved by these vessels.

At the Parseval works the officers initiated preliminary enquires for the purchase of an airship from this firm for the navy. During their trip they also inspected the large Siemens-Schukert semi-rigid at Biesdorf near Berlin and the kite balloon establishment at Tempelhof.

On their return to London, the two officers reported positively on the performance of the German airships, being most impressed by their speed, radius of action, and the energy with which the German government were rapidly developing their air arm.

At the Admiralty, ambiguity on policy still prevailed. On one hand the complete cessation of all airship work was advocated, but on the other hand a faction at the Admiralty were preparing to enter into negotiations with the Zeppelin Company for the purchase of a *Schwaben*-type airship for use with the fleet.

This last course of action would have resulted in a Zeppelin airship serving in the Royal Navy before the war. These negotiations failed to materialise, however, due to the German government forbidding the sale of Zeppelin products to any foreign countries. Fortunately,

13. Astra-Torres naval airship *No. 3*, 1913.

this restriction did not apply to the Parseval concern, and subsequently one of their ships was purchased for the Royal Navy.

Despite the uncertain course of policy with regard to aeronautical matters, the accelerated pace of progress in Europe rendered a policy of inactivity undesirable, with Germany, France and Italy spending more in a year than Great Britain had spent over the past five years.

In view of these troubling developments, in June 1912 the Committee of Imperial Defence again sent two of its members, Captains Sueter and Mervyn O'Gorman, to France, Austria and Germany to determine and report further on the state of aeronautics in those countries.

The whole trip was this time being conducted under a cloak of secrecy and the two officers assumed American civilian identities, although to what extent they fooled the German secret service is doubtful.

In the three countries they found some thirty airships flying or under construction, and were most impressed and alarmed by what they saw.

In Germany they flew aboard the *Viktoria Luise*, which was operated by DELAG and employed in sightseeing flights from Hamburg, carrying up to twenty passengers and crew with an ease and facility which in itself must have been galling to the two officers who had so recently been involved in the *Mayfly* fiasco. The flight took them over Lubeck and along the Baltic coast, and they had by good fortune to be flying whilst a naval crew were on board under training.

Amongst the constructional facilities they visited was the Schütte-Lanz factory in Mannheim to see the wooden-framed *SL1*, then in an advanced state of building. The *SL1* was a remarkable craft of 730,000cu ft capacity, built to the design of Dr Schütte, a professor of naval architecture at Danzig.

The airship first flew on 17 October 1911, being 420ft in length by 59ft beam and capable of a speed of 40–45mph. The method of construction consisted of a series of wooden hoops placed diagonally to the horizontal axis of the craft, forming a geodetic structure of great strength and an excellent streamlined form. This form of construction was also used to great effect twenty-five years later by Dr Barnes Wallis in the successful Wellington bomber.

With a gross lift of 21 tons and a disposable load of 5 tons the airship was powered by two powerful Austro-Daimler petrol engines of 270hp each, contained in two gondolas attached to the framework by hollow wooden spars. Within the ship an internal keel and walkway ran the full length of the hull to allow access between the cars and to gas cells, fuel tanks etc. Control was effected by simple monoplane elevator and rudder surfaces fixed to the stern of the ship, assisted by biplane elevators at the bow.

The airship was purchased by the German army and contained many advanced features, which were later to be incorporated by the Zeppelin factory, such as the axial cable running through the centre of the gas cells from bow to stern, which allowed the stresses imposed on the structure by the lifting forces of the gas cells to be equalised.

The *SL1* was one of only two airships using this unique construction method (the second being the uncompleted McMechan anti-Zeppelin airship of 1915). Succeeding Schütte-Lanz airships were built using the Zeppelin form of main transverse frames and longitudinal girder framework, although still employing wooden framework.

As a private undertaking in airship construction, the *SL1* was a technological tour de force and a brilliant achievement; in sharp contrast to the incompetent official British government's involvement in airship construction, as represented by the failure of the contemporaneous *Mayfly* project.

In Austria the two officers took a flight in another Parseval non-rigid whilst in France they went aloft in an Astra-Torres tri-lobe, both of which impressed them with their abilities and performance. Subsequently, upon their return, they recommended the Royal Navy purchase airships of this type.

In their detailed report to the committee, Sueter and O'Gorman were anxious to communicate their knowledge of the alarmingly advanced state of aeronautical preparedness they had discovered in Germany, including Germany's plans for the protection of their North Sea coastline during war, by stating:

> That the German airships have by repeated voyages proved their ability to reconnoitre the whole of the German coastline of the North Sea, and that in any future war except

Schutte-Lanz Military Airship S.L.1 - 1911
Luftschiffbau Schutte-Lanz Gmbh, Mannheim-Rheinau

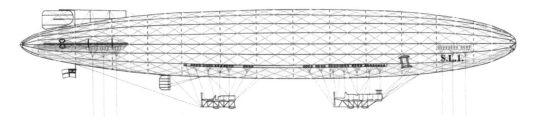

100 Feet

730,000 Cubic feet capacity
420' x 59' x 73'
20.47 tons gross lift
4.8 tons useful lift
2 x 240 h.p. 8 Cylinder Austro-Daimler
Speed 38 m.p.h.
Range 700 miles
Crew 12

in stormy or foggy weather no British warship or torpedo craft could approach its coast without their presence being discovered and reported.

These prophetic words were soon to be proved correct. On many occasions during the coming war during their forays into the German Bight the Royal Navy's activities were frequently detected early and reported by scouting Zeppelin. After further consideration the committee, rescinding their previous edict, instructed the Admiralty of the desirability to resume construction of the rigid type of airship for fleet purposes and the reconstitution of the naval airship service.

Developments abroad at this time continued at an accelerated pace, with larger Zeppelins of improved performance delivered to DELAG such as the *Sachsen* of 1913 of 742,000cu ft capacity. This craft, with a total lift of 21 tons, demonstrated by its regular passenger flights the practicality of the rigid and of Germany's mastery of this type of aerial craft.

The German high command was by 1908 sufficiently confident in the future of the airship for war purposes, seeing it as an ideal weapon in a support and tactical role, and for use in bombing and reducing frontier fortifications and fortresses. By the outbreak of war in 1914 they had eight Zeppelins and other large airships in service, the German navy were slower in adopting this type of craft, insisting that the performance of contemporary Zeppelins did not meet the vigorous demands of the navy's requirements. Although in Britain at this time it was assumed that Germany were already using airships with their fleet.

Despite their misgivings, and with the encouragement of Admiral Tirpitz, in June 1912 the German navy took delivery of its first Zeppelin, designed to operate as a scout for the High Seas Fleet.

The *L1* of 776,700cu ft was 518ft in length with a beam of 48ft, powered by three Maybach CX engines of 270hp and capable of carrying a crew of twenty for thirty hours at a speed of 50mph with a full war load. The *L1* displaced 25 tons and was possessed of a useful lift of 9 tons, a figure in terms of useful load that was not to be exceeded by a British airship until the arrival of the *R27* in 1917.

After entering service with the High Seas Fleet in June 1912, the *L1* performed much useful work in a training role and accumulating experience of working with the fleet. On 3 September 1913 on her sixty-eighth flight, a scouting mission in the German Bight, the *L1* ran into severe weather conditions to the east of Heligoland. At 6.30 in the evening she encountered a fierce storm front with strong, vertical gusts that caused the ship to rise out of control to over 5,200ft, with a resultant loss of large quantities of gas through her automatic valves.

The *L1*, now heavy, began to fall towards the sea at increasing speed, where despite the engines being run at flank speed to increase dynamic lift all efforts to hold the craft in the air failed and she crashed into the stormy sea. Only six survivors were picked up from the crew of twenty, which included the loss of the chief of the Naval Airship Division, Korvettenkapitän Metzing.

Despite the good work previously performed by *L1*, she was now considered by the German navy department to be too small for effective ocean-going work involving long over-water patrols. The navy department instead entered into discussions with the Zeppelin Company to consider the construction of an alternative design, which while still able to fit within the existing sheds was of greater capacity than the existing class of Zeppelin.

This new type, incorporating a series of novel and untried constructional techniques, would provide greater lift, range, and improved performance. The navy department prepared a design of their own under the direction of the brilliant naval architect Felix Peizker, who upon evaluating the performance of the current Zeppelin airships came to the conclusion that in their present form they were unsuitable for the conditions they would encounter over the North Sea, and must be of greater capacity in order to carry out their duties efficiently.

German Naval Airship L2 (LZ 18) - 1912
Zeppelin "i" Type
Luftschiffbau Zeppelin GmbH, Friedrichshafen

950,000 Cubic feet capacity
515'x 53'x 61'
27.7 tons gross lift
10.2 tons useful lift
4x 160 h.p. C-X six cylinder in line
Max.speed 47 m.p.h.
Range at cruising speed 1400 miles
Ceiling 9000 feet
Crew 18

100 Feet

Peizker initially submitted a proposal for a six-engine craft of 1,223,00cu ft capacity to be built at a cost of £150,000, however this design was considered to be too revolutionary for the Zeppelin Company and was turned down by them as impractical. Peitzker then submitted a less ambitious plan, although still of an innovative nature, and by using his considerable influence with the Admiralty he was able to persuade the Zeppelin Company to build the alternative design against their better judgement as they had many reservations concerning the airship they were constructing.

Here we can see an echo of official governmental interference, as with the *Mayfly*, in the complicated equation of aeronautical expertise, which was best understood at the time by those possessed of the limited knowledge then available. The builders of these highly experimental craft had, not for the last time, to submit to the demands of government departments who felt competent to dictate design policy in an area where they had no experience, and so sowed the seeds that were to lead to a spectacular disaster.

The resulting airship, the *L2*, was of 990,000cu ft capacity with a total lift of 30 tons. It contained many new features including an internal keel corridor, a separate control car attached directly to the underside of the hull, with four powerful Maybach motors mounted in gondolas attached close to the hull. So close, in fact, that the gap was filled at the front and sides by a windscreen; this last aspect of the design was to prove a fatal mistake.

After the initial trial flights the giant airship was readied for its tenth flight, which was to be an altitude test on 17 October 1913, with a full naval crew including the chief of the Admiralty's aviation department, Korvettenkapitän Behnisch, and the airship's designer Marineschiffbaumeister Pietzker from the air station at Johannisthal near Berlin.

The morning was bright and warm, and the airship lay out in the field for a considerable time as some trouble was experienced starting one of the motors. While in the warm sunshine, the gas in the cells began to expand due to superheating and began to valve off gas through the automatic valves. Eventually, the faulty motor burst into life and the *L2* lifted off, rising almost vertically for what would be its last flight.

On reaching around 1,000ft the dirigible began to gather way, as it did so the airflow past the windscreens drew down the valved hydrogen gas that had accumulated in the keel before take-off. This gas was drawn into the partly enclosed engine compartment of the fore car that was seen to be emitting a stream of sparks and smoke from the engine exhaust. Here, the lethal oxy-hydrogen mixture was immediately set off.

Within seconds the entire ship was ablaze, exploding into a fireball which sent it crashing vertically into the ground with the loss of all on board.

In Germany this double tragedy was stoically borne by the naval airship service under its newly appointed commander, Korvettenkapitän Peter Strasser. He was a man of great organisational ability who over the years was to inspire his men by example and was devoted to developing the Zeppelin as an instrument of war to serve the Fatherland.

At Friedrichshafen work continued on the *L3*, the design of which reverted to the tried design of *L1* but with more powerful engines. Three naval airship crews were meanwhile under training using the hired DELAG Zeppelin *Sachsen*.

Since the visit of Sueter and O'Gorman to Germany and France in mid-1912 the results of their findings had been published. This, together with additional intelligence gathered from other continental sources, led to the feeling in official circles that Britain was again being left behind in the vital field of aeronautical development, and that prompt and decisive action was required to redress the balance.

This came in September 1912 when, after long deliberation, the Committee of Imperial Defence sanctioned the order of a large rigid comparable in size with *L1*, which was to be built by Vickers Ltd to be known as naval airship *No. 9*, or *R9* as she was later designated.

In addition to this order in July 1913, following further discussions in Cabinet and at the request of the First Sea Lord Winston Churchill, an additional £1 million was sought from the Treasury to be included in the naval estimate. This was to provide for the construction or purchase of six large non-rigid airships of around 300,000cu ft capacity from foreign builders at an estimated cost of £240,000, and the building of two 750,000cu ft rigid airships with a gross lift of 22 tons and priced at £92,000 each – these being additional to the rigid *R9* ordered previously.

German Naval Airship "L3" (LZ24) - 1914
Zeppelin "m" Type
Luftschiffbau Zeppelin GmbH, Friedrichshafen

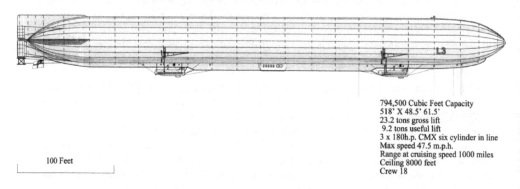

100 Feet

794,500 Cubic Feet Capacity
518' X 48.5' 61.5'
23.2 tons gross lift
9.2 tons useful lift
3 x 180h.p. CMX six cylinder in line
Max speed 47.5 m.p.h.
Range at cruising speed 1000 miles
Ceiling 8000 feet
Crew 18

Also included in this ambitious programme was the establishment of a large airship constructional facility at Hoo on the north side of the Medway estuary. This included the building of a large double shed to house the two new vessels together with the workshops and all necessary stores and equipment.

The station at Hoo was later renamed RNAS Kingsnorth and was destined to become the main constructional and operational base for Sea Scout and Coastal class airships, which were produced in quantity after 1915.

The original intention had been to acquire the six non-rigids within a year, with the two additional rigids following into service as quickly as possible. Some preliminary design work for the two rigids was put in hand, these ships no doubt would have been of a similar arrangement to the projected *R9* then being prepared in draft format by Vickers. The contracts for these two ships were to have been awarded to Vickers and Armstrong Whitworth and were to be designated as *No. 14* and *No. 15*. The initial enthusiasm for the project soon waned, however. Presumably, the grandiose announcement of the proposed programme having served its political purpose of allaying public anxiety at the time, it could now be conveniently forgotten.

With both ships both being smaller than the *R9* and possessing less lift, the contracts were soon cancelled after the preliminary draft. On the purchase of the large non-rigids a more determined effort was made to acquire them for the navy, with three types considered to be suitable for naval purposes: the Astra-Torres from France, the Parseval from Germany and the Forlanini from Italy.

As mentioned earlier, senior officers from the Royal Navy had previously made trips to the Continent to inspect and evaluate the products of these companies over the past two years. Following negotiations between the Parseval Company and the Admiralty in late 1912 the purchase of the latest model, the *PL18*, was concluded in May 1913. Captain Sueter, together with Commander Boothby, found himself in Germany as the official Admiralty representative for the trials of the new airship at Bitterfeld, which were conducted successfully.

The *PL18* represented an advanced design for the period, being very clean aerodynamically. She was of 300,000cu ft capacity, being 279ft in length with a beam of 48ft and a gross lift

Parseval PL18, Naval Airship No 4 - 1913
Parseval Luft-Farzeug Gmbh - Bitterfeld, Nieder Sachsen

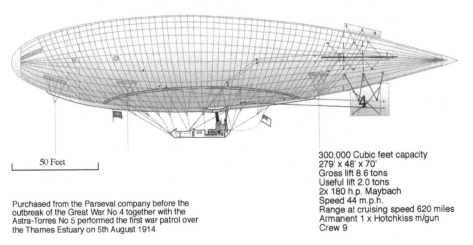

50 Feet

Purchased from the Parseval company before the
outbreak of the Great War No 4 together with the
Astra-Torres No 5 performed the first war patrol over
the Thames Estuary on 5th August 1914

300,000 Cubic feet capacity
279' x 48' x 70'
Gross lift 8.6 tons
Useful lift 2.0 tons
2x 180 h.p. Maybach
Speed 44 m.p.h.
Range at cruising speed 620 miles
Armanent 1 x Hotchkiss m/gun
Crew 9

14. Passenger Zeppelin *LZ-10 Schwaben* at Friedrichshafen, 1913.

of 8.6 tons, capable of achieving a speed of 44mph under the power of her two 180hp eight-cylinder Maybach engines.

An advanced feature of this airship was the use of the controllable pitch, four-bladed wooden propellers, which could be adjusted to suit the most economical engine speed or, if required, to give reverse thrust.

Horizontal and vertical steering planes were augmented by the provision of two air ballonets, one of which was situated near the bow, the other placed behind the centre of lift. These were used in conjunction with the elevators to change the trim of the ship to raise or lower the bow.

On 30 June, under the command of the company's chief pilot Kapitän von Stelling and with Sueter and Boothby on board, the *PL18* was flown to Farnborough via London where the ship circled the Houses of Parliament in a flight of eighteen hours.

The Parseval airship was commissioned as naval airship *No. 4* and was employed for training purposes until the outbreak of the war. It appeared together with the Astra-Torres *No. 3*, *Gamma* and *Delta* before the king at the naval review in July 1914 at Spithead.

The new Parseval exceeded expectations in terms of her ease of handling and all-round performance, and an order for a further three similar airships were given to the German company. However, at the outbreak of war these ships, then under construction, were appropriated by the German government. Vickers, who were already in possession of detailed constructional plans for these ships, agreed to build replacements, but for various reasons their delivery was delayed until late 1916, by which time they were outdated.

In company with the Astra-Torres *No. 3*, the Parseval *No. 4* performed the first aerial patrol in defence of the capital over the Thames Estuary on the night of 4/5 August 1914, this being the first operational air patrol by either side in the war. During the transport of the BEF to France in August and September the Parseval, now armed with a Hotchkiss machine gun and again in company with the Astra-Torres *No. 3*, patrolled the convoy routes across the Channel operating from Kingsnorth while the crews endeavoured to spot for mines and submarines.

The Parseval *No. 4* incurred many adventures during her service, including an occasion during August 1914 when she had the misfortune to shed a propeller blade in flight, fortunately replaced in the air by two crew members from spares carried in the car. In the meantime, however, the airship had drifted as a free balloon over Belgium and the front line at night. With repairs effected, *No. 4* made her way safely back to base.

Later, *No. 4* was relegated to a purely training role as more advanced types of airships were supplied for naval service. In November 1915 the airship suffered serious damage when it collided with its shed in fog, although the damage was repaired and it continued its training duties until finally deleted in June 1917 after a busy life of four years.

The Astra-Torres *No. 3* was of a unique design, being of the tri-lobe cross-section design of the Spanish engineer Señor Torres Quevedo, and so incorporating an ingenious method of car suspension. Viewed in cross-section, two circular envelopes were positioned side by side with a third superimposed above to produce a trefoil section, joined at the arc section to form an inverted triangle.

This type of construction had great rigidity when inflated and had the additional advantage that, by fixing suspension wires internally to the upper corners of the triangular cloth curtains, loads could be more effectively spread. This allowed the car to be attached closer to the envelope than otherwise, and reduced the number of suspension wires etc, which in turn reduced the drag and head resistance

Another novel feature was the means of achieving ascent and descent, which relied on moving the car backwards or forwards by a series of pulleys relative to the centre of gravity, thereby pitching the airship up or down.

The Astra-Torres *No. 3* was of 280,000cu ft capacity with a gross lift of 8.2 tons, driven through the air by two powerful 200hp Chenu engines and having the distinction of being the fastest airship of that period with a top speed of 51mph. *No. 3* possessed a large, fully enclosed car, which greatly improved crew comfort on long patrols and although smaller than the Parseval it had a greater range.

On her first flight in Britain in June 1913, manned by her French crew, she lost envelope pressure due to an engine failure and had to make a forced landing. On being shedded for repairs to the engines, the opportunity was taken to modify and simplify her rudder and elevators. In addition, an experimental armament consisting of a Vickers 1.5lb semi-automatic shell gun was mounted in the car for a short period, but tests showed this type of weapon to be unsuitable and subsequently a Hotchkiss machine gun was substituted.

These modifications were completed by September and her trials were continued. In October 1913, still in the hands of the Astra-Torres company pilot, she completed a six-hour flight around the Isle of Wight. Following this flight the ship was accepted by the navy, wireless apparatus was installed and she was used for mooring mast experiments at Farnborough in early 1914, undertaking further training flights with the Parseval *No. 4*.

In late August 1914, like the other naval airships, *No. 3* was dispatched to France to the RNAS station at Dunkirk from where she operated from a short portable mast. During this period she carried out a series of hazardous reconnaissance missions into occupied Belgium to determine the limit of the German advance.

Due to the danger posed by anti-aircraft and rifle fire (often from their own side!) *No. 3* was soon withdrawn to England, as were the other airships once their vulnerability had

Astra -Torres No XIV, Naval Airship HMA No 3 - 1913
Astra Societe de Constructions Aeronautiques, Billancourt, Seine

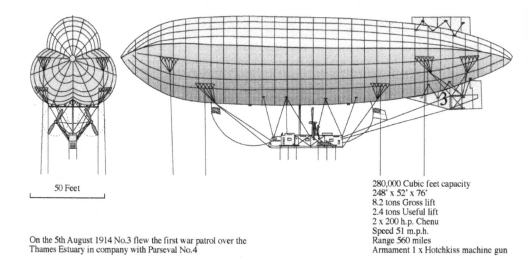

50 Feet

280,000 Cubic feet capacity
248' x 52' x 76'
8.2 tons Gross lift
2.4 tons Useful lift
2 x 200 h.p. Chenu
Speed 51 m.p.h.
Range 560 miles
Armament 1 x Hotchkiss machine gun

On the 5th August 1914 No.3 flew the first war patrol over the
Thames Estuary in company with Parseval No.4

been established, and was instead employed in night flights over the capital providing the searchlight crews with some much-needed practice.

No. 3 was taken out of service in mid-1915 as more Sea Scout airships became available to take over her duties, allowing her to be overhauled. She made her last flight in August 1915 and was finally deleted in May 1916.

The Astra-Torres had proved to be an effective and efficient ship, providing both crews and the ground-handling teams with valuable experience. Three other Astra-Torres airships (*No. 8*, *No. 10* and *No. 16*) were ordered and delivered from the French company and also gave sterling service, but, more importantly, the Astra-Torres trefoil design was developed by the Admiralty as the basis for the highly successful Coastal and North Seas classes of long-range patrol airships that followed as the war progressed.

During this period there also was much interest shown in airships in Italy, with the Ministry of War setting up a special unit of the Corps of Engineers for this purpose, under the direction of Captain Crocco and Captain Ricaldoni. The semi-rigid form of construction was developed, from the outset, to a high state of efficiency until government support was finally withdrawn following the loss of the polar airship *Italia* in 1928.

The first design of Crocco and Ricaldoni was in 1908, being of 90,000cu ft capacity and capable, it was claimed, of the respectable speed of 35mph (the British *Nulli Secundus* of the same date could only manage 22mph).

This effort was followed by the development of two types, the P or Piccolo class (small), of 180,000cu ft, and the M or Medium class of 430,000cu ft. Both types were semi-rigid with a substantial keel structure attached directly to the envelope, and a cone of bamboo stiffeners at the bow and the multiple division of the gas compartments to prevent gas surging in the event of the envelope being ripped.

Both types showed great promise and conformed to the latest thinking in the provision of good streamlined form, while other features included automatic gas valves and variable-pitch propellers. These potentially useful P and M class airships were employed by the Italian forces throughout the war period. Due to Italy's inability to produce airships in sufficient quantity, however, later in the war Britain had to supply airships to her Allies while purchasing in return a single example of an M class ship, the *SR1*, late in 1918.

As a colonial power Italy had been engaged in a war since 1911 in Tripolitania, then part of the crumbling Ottoman Empire, and had established themselves in Libya. Here, in the course of continuing operations against the Turks and their Arab allies, the Italian army employed two airships of the P class during 1912 to bomb Turkish bases and fortifications.

These bombing raids were the first use of this form of aerial warfare, which involved attacks on both military and civilian targets, together with reconnaissance flights and what can be considered as tactical strikes on camel supply caravans bringing ammunition up to the front.

Italian Forlanini Military Airship F3 -1913
British Naval Airship No 11*
Forlanini Construczioni di Aeronautiche - Roma

(La Societa Leonardo da Vinci)

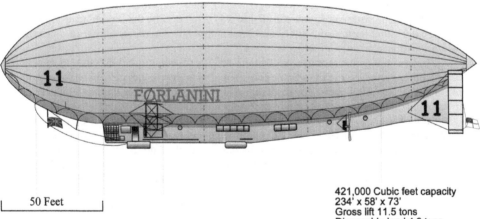

50 Feet

* This airship was being built to a British Admiralty order in 1913, but was appropriated by the Italian government at the outbreak of war in 1914.
Two further airships of this design were to be built under licence by Armstrong Whitworth, but were not proceeded with.

421,000 Cubic feet capacity
234' x 58' x 73'
Gross lift 11.5 tons
Disposable load 4.2 tons
2 x 180 h.p Isotta Fraschini V5
8 cylinder in line
Speed 40 m.p.h.
Range 600 miles
Crew 10

The Italians' experience of successfully using airships in warfare bestowed a great deal of kudos upon them in the eyes of the British War Office, and were considered worthy of consideration as suppliers of suitable and proven airships for British military and naval use.

Accordingly, Captain Murray Sueter was dispatched to Rome in 1913 to evaluate another large semi-rigid of revolutionary design, a Forlanini airship.

The first of Signor Forlanini's airships, the *Leonardo da Vinci*, was launched in 1909 with a capacity of 132,000cu ft and powered by a steam engine. The second airship was a more sophisticated design of large size that suited the semi-rigid method of construction, with a capacity of 421,000cu ft the *Citta de Milano* was equipped with two 80hp Italia petrol engines, giving a still air speed of 40mph (a very credible performance on a relatively low horse power).

This airship possessed an excellent streamlined form, though somewhat spoilt by the multiplicity of steering planes and rudders at the stern which were typical of Italian airship design of the period. The design was innovative and showed great promise, however, the ship's career with the Italian army was cut short when she suffered a disastrous crash due to elevator control failure. It caused her to nose into the ground during manoeuvres in 1913, resulting in the airship being damaged beyond repair.

A conning tower was situated at the bow connected to a deep, V-shaped keel, which allowed access to the eight separate gas bags and the engine cars in an arrangement similar to that employed in the Zeppelins. The gas bags were in turn enclosed within an annular air ballonet positioned between the gas bags and the outer cove. This 2in air space served the double purpose of acting as a ballonet in the usual manner and as an insulating air cushion to minimise the effects of external temperature changes on the hydrogen gas.

Following Sueter's visit and his recommendation as to the suitability of the type, an order was placed with the company for a similar semi-rigid of 500,000cu ft capacity with two more to be built under licence by Armstrong Whitworth once they were in possession of the appropriate drawings.

Construction of the vessel got well under way on the airship to be known as naval airship *No. 11*, but at the outbreak of war the airship was taken over by the Italian army for its own use. The two other Forlaninis, *No. 12* and *No. 13*, also failed to materialise from Armstrong Whitworth due to pressure of other war work.

The Italian airships operating in the Adriatic did little to distinguish themselves in what little action they saw, and later in the war they were supplied with British Sea Scout airships in order to make up the shortfall caused by the Italians' inability to quickly build home-produced craft in sufficient quantity.

The non-delivery to the Royal Navy of the Forlanini early in the war was a lost opportunity, as the early delivery of these ships would have given the navy useful experience in the handling of large airships prior to the acquisition of the big rigids in late 1916. There is also the possibility that the Italian ships would have been employed in a more determined manner by British crews at an early date in the fleet's scouting role, the desired objective of this type of craft.

In England design work for HMA *R9* began in April 1913 at Vickers, Barrow. However the work had not proceeded far when an incident of great importance for the constructors occurred. The *ZIV*, the brand-new Zeppelin on a delivery flight to the army at Metz, made an accidental landing in foggy weather on a French military parade ground at Lunéville in French territory.

During the period of the airship's internment, which lasted some hours, French army officers entertained the German crew while frantic diplomatic efforts were being made by the Germans to secure its release. At the same time, French army engineers were able to make detailed drawings and measurements together with photographs of the dirigible's framework

French Military Airship "Spiess - Zodiac X111" - 1913
Societie Zodiac, St Cyr, Paris

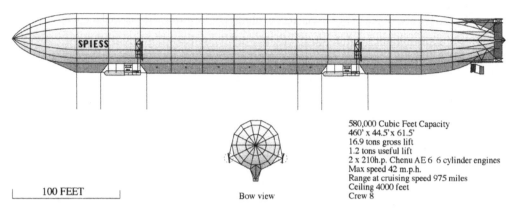

580,000 Cubic Feet Capacity
460' x 44.5' x 61.5'
16.9 tons gross lift
1.2 tons useful lift
2 x 210h.p. Chenu AE 6 6 cylinder engines
Max speed 42 m.p.h.
Range at cruising speed 975 miles
Ceiling 4000 feet
Crew 8

100 FEET

Bow view

and equipment, concentrating particularly on the girder work, which held the key to the combination of lightness and strength essential to building a successful airship.

After some hours the embarrassed Germans were allowed to leave and the accumulated data was passed on by the French military authorities to the Zodiac Company, where the information was utilised by them in the Spiess wooden-framed rigid they were then building. The Spiess-Zodiac *No. XIII* was already far advanced in its construction at this time but was now modified, to an extent, in light of the new information. It was, for instance, lengthened; however, being of wooden construction it could have benefited very little from the technological windfall obtained from the aluminium-framed *ZIV*.

The Spiess dirigible was completed within six months of the incident, flying in late 1913 and immediately entering army service. The data obtained by the French from the *ZIV* was, however, speedily communicated to the British government in the spirit of the entente cordiale. Here the results were even more usefully incorporated in the construction of the *R9*, to the extent that the finished product bore more than a passing similarity to the *ZIV*. Nevertheless, as the *R9* was for various reasons not completed until late 1916, any advantage which may have been obtained in acquiring the information was by then superfluous.

Russia is a country that is not often mentioned in the history of lighter-than-air development, but in the immediate pre-war period the Russians produced a remarkably advanced semi-rigid design that incorporated many novel and original features. Indeed it placed them in the forefront of aerostatic technical development at that time.

The *Gigant* was built by engineers A. Kovanko and A. Shabskogo at the Russo-Baltic Wagon Works, south-east of St Petersburg, between 1912 and 1915. It was of 724,060cu ft capacity, measuring 374ft in length with a diameter of 56ft, giving a gross lift of 21 tons and a most respectable useful lift of 8.8 tons. As originally designed, the power arrangements consisted of four 215hp Austro-Daimler eight-cylinder petrol motors, later reduced to two such units mounted in streamlined nacelles positioned amidships above the keel.

The airship made its first flight in February 1915, where after less than an hour in the air the keel collapsed amidships, perhaps because of the loads imposed on this part of the structure by the rearranged disposition of the engine cars. The *Gigant* fell comparatively slowly into a forest, allowing the crew to escape uninjured whilst the airship was subsequently recovered and repaired.

Due to the non-availability of sufficient quantities of hydrogen the *Gigant* did not fly again and this promising airship was scrapped in late 1916.

6

The Second Navy Rigid: *R9*

D espite the loss of both *L1* and *L2* within a month of each other in the autumn of 1913, the German Admiralty were now more than ever convinced of the value of the scouting airship. This confidence was based on the experience gained with *L1* in more than sixty-five flights over the Baltic and the German Bight during the year the airship was in service.

The German navy was, and would remain throughout the coming war, weaker than the Royal Navy in terms of numbers of ships and, in particular, of Dreadnoughts. In consequence the protection of these valuable vessels by airships operating in conjunction and ahead of the fleet, reporting the number, course and speed of an enemy fleet, offered a great advantage to a fleet conceived to fight a defensive and retiring battle plan.

This strategy was in-keeping with the 'Fleet in Being' theory propounded by the American naval theoretician Rear Admiral Alfred Thayer Mahan (1840–1914) in his book *Influences of Sea Power upon the French Revolution and Empire* published in the 1890s. This book and other works of his had a profound influence in countries which in the late nineteenth century were expanding their naval strength. The German kaiser in particular was a keen adherent to his theories, and he insisted that Mahan's works were recommended reading for young naval offices in training.

In his works, Mahan contended that a weaker navy with fewer overseas imperial commitments could influence the course of naval warfare merely by the possession of a powerful but smaller fleet that could be used to pose a threat to both the enemy fleet and its merchant vessels.

This effect could be accomplished either by the fleet remaining in harbour – requiring the opposing navy to tie up warships and resources on blockade duties – or by putting to sea at a favourable opportunity and, by eluding contact with the enemy's main force, endeavouring to intercept and overwhelm a detached and inferior portion of the enemy fleet.

Additionally, small squadrons of fast warships could be dispatched at regular intervals to attack enemy harbours or coastal installations and harry and disrupt the enemy's trade routes.

In at least the first two of these situations, reconnaissance by Zeppelins had a key role to play that would, in theory, give the German fleet a great advantage. Accordingly, and at the personal intervention of the kaiser, a five-year plan of airship development was initiated. It called for the establishment of two squadrons of five ships apiece to fulfil the requirements of the High Seas Fleet, with the necessary funds and facilities being made available at once.

In England this news was received with disquiet and fostered the need for urgency in dealing with matters of aerial defence. On the positive side, since the visit of Sueter and O'Gorman to Europe in mid-1912, action had been taken to secure the air defence of Great Britain in the event of war, first with the formation of the Royal Flying Corps in April 1912, and then with the government undertaking to acquire suitably large, fast airships for the navy from abroad.

15. Vickers' rigid naval airship *No. 9* landing at Walney Island, Barrow-in-Furness, 1916.

In April 1913, the remaining army airships came under the control of the navy, though not without a struggle with the War Office who were reluctant to relinquish this aspect of aviation in which they had taken a leading role.

In the long term these changes of responsibility proved beneficial, for although the army lost a certain number of skilled officers and technicians to the naval service, it allowed the army to concentrate on the development of the aeroplane for army service.

As a part of the new construction and purchase plan announced in the naval estimates for 1913, design work began on the larger of the three promised rigids, *No. 9* (the two proposed 750,000cu ft rigids were never to get beyond the initial planning stages).

No. 9 was to incorporate the lessons and experience gained on *Mayfly*, particularly in the area of working with Duralumin alloy, which had again been chosen for the structural components. The task of assembling a design group was a difficult problem as so many of the experienced artisans had moved on to other jobs, but eventually, under the supervision of H.B. Pratt as chief designer, a team was formed at Barrow, amongst whom was also a young designer named Barnes Wallis.

At the same time work was progressing on the new shed, to be built wholly on land this time as it was realised that the attempt to copy the early Zeppelin practice involving floating sheds was in error. The new shed on Walney Island was completed at a cost of £100,000, the structure being 540ft long by 150ft wide and 98ft high. At the time Vickers considered that they had acquired a useful asset that would leave them well placed to receive further government orders. However such was the rapidity of aeronautical development in this field that within a very short time the shed would prove to be too small for building the later and larger class of airship.

Vickers discovered, to their detriment, that by acquiring a shed of limited dimensions at such an early date that they later lost out to their competitors for building the larger R33 class and her successors.

The order for *No. 9*, or *R9*, was finally placed on 10 June 1913, with the Admiralty specification requiring that the airship should be capable of 42mph for eighteen hours at full speed with a range of 800 miles, or 1,200 miles at an operating speed of 32mph.

A disposable load of 5 tons was specified, together with a ceiling of 2,000ft. In addition, the original requirement called for the *R9* to carry a 1-pounder Pom-Pom in each car, together with two Maxim machine guns mounted on a gun platform on top of the hull. At this period no provision had been made for any bombs to be carried, although consideration was given to adapting naval shells for this purpose.

It was envisaged that in service *R9* would operate as an adjunct to the battle fleet, occupying a particular position in the van or on the wings of the battle squadrons. It was also expected to operate at a low altitude of under 1,000ft, and was required to keep in visual contact with the fleet thereby being under the effective control of the admiral in the same way as the surface ships.

These requirements would of course nullify the airship's primary function and demonstrated a complete lack of understanding of the unique possibilities effective deployment of this new weapon in an independent distant scouting role offered.

Even as late as 1917 the Coastal and North Seas classes of airship were expected, when detailed to accompany units of the Grand Fleet on periodic North Sea sweeps, to keep in close proximity of the surface ships and keep station with them.

It is interesting to note, as will be described later, that the projected performance figures expected of the *R9* and its actual load figures were to be achieved by both the Astra-Torres and the Parseval non-rigids, with capacities of only one third of the rigid. Both the non-rigids, unlike the rigids, could be built more quickly; an advantage in a time of rapidly advancing aeronautical progress, particularly during the pre-war period where it was almost impossible to plan long-term projects that had to accommodate the results of new research and development accumulating with such rapidity.

As mentioned previously, the design of the *R9* was profoundly affected by the information derived from the accidental landing at Lunéville, France in April 1913 by the German army

Naval Airship H.M.A. No9 (R9) -1916
Messrs Vickers Ltd - Barrow-in-Furness

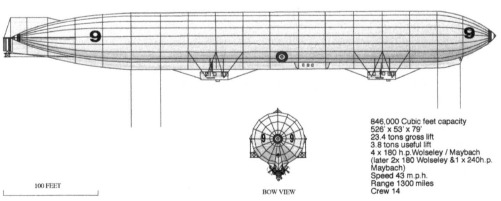

100 FEET

BOW VIEW

846,000 Cubic feet capacity
526' x 53' x 79'
23.4 tons gross lift
3.8 tons useful lift
4 x 180 h.p.Wolseley / Maybach
(later 2x 180 Wolseley &1 x 240h.p. Maybach)
Speed 43 m.p.h.
Range 1300 miles
Crew 14

Zeppelin *ZIV*. Indeed the subsequent drawings of her made by the French army engineers were of immense importance to the Vickers design staff.

The final design of *R9* emerged as a ship of 866,000cu ft capacity giving a total lift of 26 tons with a disposable of 5 tons. The dimensions were length 526ft, beam 53ft and overall height 72ft, while the power was supplied from four six-cylinder Wolseley-Maybach engines of 180hp. These engines were mounted in tandem in two aluminium watertight cars slung close under the keel, driving four-bladed swivelling propellers mounted on either side of the car on outriggers.

In the light of the experience with *Mayfly* proving to be too heavy to fly, particular attention had been paid to lightening the structure in order to increase the useful lift. The most notable feature was the reduction in the number of main transverse frames from the forty-ring girders employed in *Mayfly*'s construction, to eighteen frames in *R9*.

The hull structure generally followed Zeppelin practice, consisting of eighteen main transverse frames constructed from 8in by 12in triangular section girder work forming a series of seventeen sided, wire-braced polygons set at 30ft intervals. Two light tubular alloy intermediate frames were in each bay. The mainframes were connected at each corner to seventeen 12in-deep triangular latticed longitudinal girders that each ran the full length of the ship.

Again the British design showed an advance in aerodynamic form over the Zeppelin original, with a finer entry and a longer streamline to the rear section of the hull. Also in this design, the radius of curvature for bow and stern was determined by the aerodynamic calculations propounded by the American Professor Zahm.

As originally conceived, the *R9* was to have bracket propellers mounted high on the hull as used on *L1*, but at an early stage swivelling propellers were incorporated in the design to be mounted close above the cars, where their effect acting nearer to the longitudinal axis of the ship would aid stability and power of manoeuvre.

Once the contract was signed in March 1914 construction began almost at once. Trouble was again experienced with the rolling of the Duralumin parts required to fabricate the girders, however successful methods of manufacture were soon devised by the Vickers engineers and girder-assembly began.

The work continued at a steady rate up to the outbreak of war in August 1914, an event which prompted a rash of modifications to the original design. At the same time there was increased pressure on Vickers with a rush of additional war work, such as ship repair, which created the need for skilled tradesmen to be transferred to other areas and causing a slowdown on the construction of *R9*.

By the end of 1914 a considerable portion of *R9*'s structure had been fabricated, and erection of the frame commenced. This work continued at a leisurely rate until, following talks at the Admiralty, a decision was taken by the First Sea Lord Winston Churchill to abandon work on the *R9*, with this order coming into effect in March 1915.

The reason put forward for this course of action was that as the expected mass-attack by Zeppelins on the British Isles had not yet occurred, it indicated that airships were not the threat they had been seen to be. Secondly, it was believed in Whitehall that the war would be brought to a conclusion by the end of 1915, and that any further work on airships was an unnecessary expenditure.

Vickers had put a great deal of money into airship work in the hope of obtaining future Admiralty contracts, and now it would seem that the government was not fully committed to the idea of their development. This governmental policy of vacillation on the subject was the cause of grave concern to the company and the future of their investment, and in turn caused the company to wonder if their own commitment to building rigid airships was misplaced.

However, the Royal Navy's encounters with Zeppelins at sea, and the public outcry following the bombardment of east coast towns by German battlecruisers of the High Seas Fleet in December 1914, vividly illustrated the dangers of allowing an enemy to approach the British coastline undetected. This again directed attention to the vital need for scouting airships to guard against this form of incursion.

Accordingly, additional defensive measures were taken, which included the strengthening of the Channel fleet and the Harwich Light Cruiser force to protect the coastal towns in order to counter this threat. Additionally, in May 1915, with Arthur Balfour (now the new First Sea Lord following the resignation of Winston Churchill in the wake of the Dardanelles debacle) called for the reinstatement of an expanded rigid airship programme with work to recommence in August of that year.

Once again the scattered *R9* design team were reassembled, with Wallis now serving with the Artist Rifles in France and Pratt with the Royal Navy, while other members of the group were serving king and country at the front or in other service or civilian work.

The assembly of the girder work, completed before cancellation for the mainframes, got under way on the shed floor at Walney Island immediately, while the fabricating shops began producing more latticed girders and continued work on fitting out the engine cars. As each frame was completed it was raised and positioned on the wooden cradlework at 30ft intervals, to be connected to the seventeen lattice section longitudinals at the corners to form a bay, and so on for succeeding frames as they were added.

Wire bracing was then applied in the plane of the transverse frame, and the entire surface of the bay was criss-crossed with further bracing wires and netting to contain the gas cells. In the adjacent fabric workshops the seventeen gas cells, or gas bags as the British referred to them, were being made from finely woven Egyptian cotton fabric rendered gas-tight, as previously described, by lining the interior surface with goldbeaters' skins made from the large intestine of cattle.

After being cured and scraped, the skins were glued to the cotton backing of the cell, a process which in order to complete required the skins of some 20,000 cattle per cell. The outer cover was made from stout doped cotton, forming an impermeable and watertight covering.

This was in contrast to the German practice where the outer cover was of a porous material. The reason for this difference being that in the German craft the automatic gas valves were positioned at the base of the cell, with the resultant discharge of gas being allowed to disperse into the keel area and through the outer. This practice of course presented a potential danger due to the proximity of the engines and electrical systems in the lower hull.

The British, recognising this danger, installed in both *R9* and the earlier *Mayfly* and Parseval types automatic valves positioned on the top of the hull. It is surprising that the Germans persisted for so long with the practice particularly after the *L2* disaster.

In later ships gas discharge shafts were fitted from the base of the cells to a vent at the top of the ship; an innovation first used by the Schütte-Lanz concern in 1911 and employed on contemporary British ships, but not used by the Zeppelin Company until the launching of the *L32* in mid-1916.

By the autumn of 1915 the greater part of the hull structure of *R9* was in a completed state, with engine cars attached and gas cells, petrol, ballast and electrical systems installed. Problems were being experienced with the gas cell retaining wiring, and at a late date alterations to the keel structure were called for together with numerous minor design and system changes that militated against an early completion date.

Eventually, in late November 1916, *R9* was completed. It made its first flight on the 27 of that month, thus becoming the first British rigid to fly. It soon became apparent, however, that *R9* was well over her design weight, and in tests was unable to lift the required contract disposable load of 3.1 tons. This figure had already been reduced early in construction from the original proposed figure of 5 tons disposable. In consequence it was now refused acceptance by the Admiralty inspectors and deemed unfit for service.

Additionally, the initial trial flight disclosed control problems in both the lateral and vertical planes, whilst the set of auxiliary rudders originally fitted abaft the rear engine car were ineffective and interfered with the effective airflow over the main rudders. At Wallis' suggestion they were removed, which at once improved lateral control.

The ship still proved difficult to control in the vertical plane, resulting in a somewhat erratic flight path which was to remain a feature of her flight characteristics for as long as the ship was in service. Vickers immediately put measures in hand to remedy the problem of the low disposable load, which included the removal of the gas cells to be replaced by lighter ones and the relatively heavy twin-engine rear power car, including the swivelling propeller gear, to be taken out.

This arrangement was replaced with a single 240hp Maybach HSLu engine, complete within the original wing car taken from wreckage of the Zeppelin *L33* brought down at Little Wigborough, Essex in September 1916. The Maybach engine was considered to possess a far higher compression ratio to be suitable for low-altitude work, and the engine was rebuilt with shortened pistons to remedy this, although the modification was unnecessary as the Maybach high-altitude motors did not come into service until late 1917.

With these changes and a disposable lift now of just over 4 tons the *R9* was eventually accepted by the Admiralty on 10 December 1916. Thereafter being sent to join the rigid airship trial flight at Howden, Yorkshire, which was under the command of Wing Commander Masterman in April 1917.

The *R9* had been specially strengthened to allow for the handling of inexperienced flight and ground crews, particularly for gaining practice in the tricky art of manoeuvring her on the ground, withdrawing her from her shed in varying weather conditions and re-housing her in stiff crosswinds when occasion demanded.

The results of her speed and turning trials conducted at various speeds served to indicate her erratic flight characteristics: with turns to port at a particular airspeed and rudder setting being performed at a tighter radius than a similar turn to starboard; where a turn to port at 24 knots described a circle of 890yds and taking five minutes to complete, while a turn to starboard would require 1,080yds taking eight minutes.

In either case, such a craft would be at a distinct disadvantage in attempting to attack a nimble and elusive U-boat while describing a circle of some 1.5 miles in circumference. From a tactical point of view, as an offensive weapon for use against the U-boat, it left much to be desired (the later *R29* being the only British rigid ever to make a successful attack on a U-boat in 1918).

Later, in 1917, *R9* was for a short time stationed at Pulham and again involved in training flights. She was employed in a series of war flights over the southern North Sea – armed with bombs while on anti-submarine duties – with one of these patrols being of thirty-three hours' duration.

During October 1917 her bow and lower keel were badly damaged in an accident while being re-housed in her shed with the assistance of a tank towing her in. Repairs were effected and she subsequently carried out mooring tests in February 1918 in the Wash, where she was moored out at sea using anchors and buoys, engaging in exchanging crews from surface vessels. She was also used in further ground-mooring experiments at Pulham using the three-wire system.

This was a comparatively simple solution to the ground mooring problem, which involved the use of three wires set at 120° to each other with the ship attached to the apex of the wires at the lower keel near the bow by a swivelling tackle. This method allowed the ship to vane in the wind while allowing her to be anchored with a degree of safety some 30ft above the ground.

During April 1918 she again sustained further damage to her bow girders against the shed entrance while being re-housed. In this damaged condition she was partially deflated and slung from tackles in the roof where she remained until May 1918, when, after some deliberation to refit her for further experimental work, the decision was taken to scrap her. *R9*'s contribution to the British rigid airship programme was in the area of training of flight and ground crews unfamiliar with this type of craft, but by the time she entered service in 1917 she was hopelessly outmoded, being built to a design that was by then five years old.

By 1918 she could in no way compare in performance with the contemporary German 2 million cu ft Super Zeppelins with a disposable load of 37 tons, or nine times that of the British craft. These Super Zeppelins could now be built in three months, compared to the three years required for building *R9*.

Despite her shortcomings she served her purpose in her short service career and had flown for a total of 198 hours up to her scrapping in July 1919.

7

War at Sea: Sea Scouts vs the U-boat

A s early as March 1915 it had become apparent that a serious threat was posed to British merchant shipping by the depredations of German undersea craft operating along the North Sea coasts and the English Channel.

This threat was made all the more dangerous with the seizure by German forces of the harbours of Ostend and Zeebrugge in October 1914, which presented the enemy with submarine bases in dangerous proximity to Britain's southern shores, and from where they could threaten maritime trade.

The Admiralty, in its planning, had seriously underestimated the potential danger posed by this novel form of warfare, together with its effect on British shipping. The navy had until then regarded the submarine as merely a form of coastal defence craft of limited range, useful perhaps to protect anchorages and harbours but not to be considered as a serious offensive weapon at sea.

Within a few short months of the start of the war, however, the U-boat was shown to be an efficient silent predator that could launch its attack unseen and was almost invulnerable to counter-attack, being possessed of the ability to slip away into the depths of the ocean undetected. It soon became apparent that this unforeseen aggressive deployment of the U-boat imperilled the passage of British troops and munitions to the front, and most importantly the supply of foodstuffs and raw materials so essential to the British Isles in time of war.

The German Admiralty, contrary to the British, had for some years encouraged the development of the U-boat, recognising it to be a potential key weapon in the struggle against the numerically superior surface fleet of the Royal Navy and as the ideal means to effect disruption of merchant trade.

A German Admiralty order dated 18 February 1915 announced that the waters around the British Isles should henceforth be declared a war zone, wherein all British and Allied ships were liable to be sunk without warning, and neutral shipping sailed at their peril. Any neutral merchant vessels sailing within the controlled zone would, the order decreed, be intercepted and subject to boarding and confiscation of contraband cargo as determined by the Imperial German Navy.

The fears of the unseen dangers posed by the mine and the torpedo were graphically illustrated with the loss in October 1914 of the modern Dreadnought *Audacious* off the Irish coast to mines laid by a German auxiliary minelayer. This loss occurred in an area to which the battle fleet had been dispatched for safety at the start of the war, whilst the anchorage at Scapa Flow was being fortified to render it immune to submarine attack.

A further blow was struck in the early hours of 1 January 1915, when the pre-Dreadnought *Formidable* was lost to torpedo attack from a submerged U-boat off Start Point in the English Channel. The battleship sunk some twelve hours after being struck, with the loss of 35 officers and 512 men. This tragic loss of life, together with the increasing losses of merchant ships over the past five months, further underlined the dangers to be faced in the war zone by vessels in the English Channel, or as they plied the broader waters of the North Sea or the Western Approaches.

The Royal Navy suffered heavily during the war from attack by mine and torpedo, losing twelve capital ships and numerous smaller craft; a continuing loss which encouraged the Admiralty to seek a remedy in the development of the airship for mine spotting and anti-submarine duties.

The locomotive torpedo invented by Whitehead had been in service with the worlds' navies since the 1870s, where progressive developments had increased its range, speed and accuracy to produce a formidable weapon when used in conjunction with submersible craft. A further violent demonstration of just how destructive this combination could be was shown in September 1914, with the sinking of three large armoured cruisers, the *Hogue*, *Aboukir* and *Cressy*, in the Broad Fourteens off the Dutch coast as they steamed in line ahead at 10 knots.

Their nemeses were the torpedoes of Kapitän Otto Weddigen's *U9*, a comparatively old boat, sinking first the *Hogue* then the other two ships as they stopped in turn to pick up survivors from the previous attacks, eventually sending all three ships to the bottom with heavy loss of life.

The British public were at the time outraged by the 'unsporting' tactics of the U-boat commander in attacking ships that were picking up survivors. This, together with the apparent

16. Sea Scout prototype flying at Capel, Kent, April 1915.

Willows Airship No IV - 1912
Naval Airship HMA No 2
Messrs E.T. Willows, Birmingham

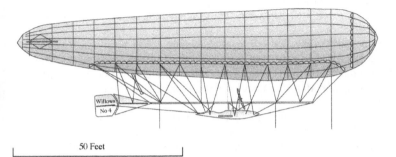

Original configuration June 1912
(With oiled silk envelope)

Capacity 24,000 cubic feet
110' x 18'6'' x 32'
Gross lift 0.70 tons
Disposable lift 0.1 tons
1 x 35 hp Anzani
Speed 30 mph
Crew 2

50 Feet

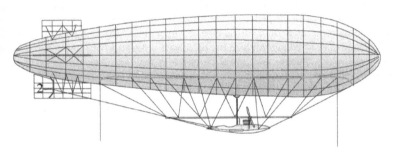

As modified by Admiralty August 1912
(With new rubberized fabric envelope)

Capacity 38,000 cubic feet
110' x 23' x 35'
Gross lift 1.1 tons
Disposable lift 0.2 tons
1 x 40 hp Renault
Speed 35 mph
Crew 3

ease with which such an attack could be accomplished, only served to heighten the fear and awe in which these weapons were held.

As a consequence of these alarming developments an urgent conference was called at the Admiralty in February 1915, presided over by Lord Fisher, to address the problem and find an immediate solution to the submarine menace.

Following their lordships' deliberations, proposals were made for the provision of more fast, coastal motor boats and other patrol craft, together with a strengthening of the light cruiser and destroyer flotillas based in the Channel and on the east coast, and aided by patrols of seaplanes in coastal areas. Lord Fisher additionally summoned Wing Commander Masterman and Commander Neville Usborne to the Admiralty to evaluate the possibilities of using small pressure airships as submarine hunters and as spotters to co-operate with surface warships.

After receiving an assessment of the practicality of such craft from the two officers, Fisher issued instructions for the immediate production of a number of small, fairly fast airships that could be handled after a minimum of flying instruction by a midshipman and two ratings. These were to be used for the purpose of hunting down and destroying enemy submarines within coastal waters. It says much for Fisher's determination and organisational ability that, within the space of three weeks of receiving this order, the first three SS or Sea Scout airships were ready for their trial flights.

The first ship, *SS1*, utilised the existing *Willows No. 4* envelope, combined with a BE2c aeroplane fuselage. This makeshift prototype proved to be an adequate craft for the task in hand, and on test proved capable of patrolling for six hours carrying an offensive load of 160lb. Encouraged by these results manufacture of the SS class began, with everything possible being done to ensure their rapid construction utilising BE2c or Farman fuselages, and later Armstrong Whitworth cars.

Later airship cars were specially designed and were constructed by firms of furniture and cabinet makers, while the waterproof envelopes were provided by obtaining the rubber-proofed material from mackintosh manufacturers, who supplied the thousands of square yards of material necessary.

By the end of 1915 some thirty of the new Sea Scout airships fitted with larger envelopes of between 60–70,000cu ft capacity had been built at Kingsnorth and Wormwood Scrubs. The first batch of SS ships were quickly delivered to the newly established RNAS stations at Capel, Polegate, Pembroke, Anglesey and Marquise near Dunkirk where they immediately took up convoy protection work.

Their presence brought about an immediate improvement in the situation, with the number of attacks on shipping falling in areas where they were on patrol. Their crews quickly settled to their arduous tasks and proved they were able to offer this form of protection in all but the very worst of weather conditions.

In the original form the Sea Scout was fitted with either a BE2c fuselage with an 80hp Renault, an RAF engine driving a tractor propeller, or a Maurice Farman aeroplane body with a 75hp Renault pusher. The approximate cost of the SS type airship was £2,800 each.

With the 18 gallon fuselage tank fitted as standard the BE2c gave a duration of up to eight hours in its role as an airship car, while an additional 32-gallon petrol tank was also fitted to the underside of the fuselage, which allowed flights of longer duration to be undertaken.

"SEA SCOUT" Class Naval Airship SS16 -1915
R.N.A.S. Constructional Facility, Wormwood Scrubs

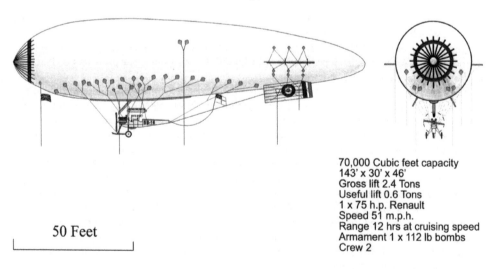

70,000 Cubic feet capacity
143' x 30' x 46'
Gross lift 2.4 Tons
Useful lift 0.6 Tons
1 x 75 h.p. Renault
Speed 51 m.p.h.
Range 12 hrs at cruising speed
Armament 1 x 112 lb bombs
Crew 2

50 Feet

17. Coastal airship *C26* landing, 1915.

A record flight of forty-four hours fifty minutes was made by the *SS16* in September 1916, which must have been a tremendous feat of endurance for the crew concerned considering the rudimentary and exposed accommodation in the cramped car.

The early SS ships were fitted with two horizontal steering planes and two combined rudder planes on the lower side of envelope, set at an angle of 40° to each other in order to improve lateral stability. Later, the majority of Sea Scouts were modified so that both fins were positioned one behind the other on the lower side of centre line, which reduced drag without affecting stability. The rudder pedals were retained for steering and an elevator wheel fitted for pitch control; a crew of two was carried with the pilot in the rear seat with the ballast and gas controls and the wireless operator/observer in front.

By the time production of the type ceased in late 1916, some forty-nine Sea Scouts had been built, together with ten examples being supplied to Britain's Italian allies and two each to France and Russia. The total number of airships built during the war is somewhat confused by the fact that each airship was supplied with a complete set of spares, which included an extra envelope.

In some instances a spare or an old envelope was rigged to a surplus car to create an additional airship that sometimes would not be included in the official naval list. This was the case of a SS Zero airship assembled by US crewmen at Howden using an old envelope and a spare car that had been discarded to create an unnumbered airship, which the Americans intended to use for instruction.

In August 1918 this airship was in the shed with two other SS class airships, the Parseval *P6* and the rigid *R27*. Whilst testing the wireless set the American airship caught fire in the shed, destroying the *R27* and the two other SS ships in the process (with *P6* luckily escaping destruction). This unnumbered ship would have gone unnoticed in Admiralty lists had it not been for the dramatic nature of her denouement.

At other times a deleted ship was completely replaced, for instance after being lost in the Channel the *SS10* was replaced by a new *SS10a*, which was assembled from spares (the number of a naval airship always remained with the car).

These products of typical British improvisation were to prove to be very versatile craft, being easy to handle both in the air and on the ground and within their limitations of their lift and range were well suited for patrol work in coastal waters. After the initial batch of SS airships had been built at Kingsnorth the majority of the remainder were constructed at Wormwood Scrubs, in the *Clément-Bayard* shed paid for in 1910 by the *Daily Mail* public subscription.

Such was the pressure on the facilities at Scrubbs Lane to rapidly build the new airships, contracts were also placed with Vickers at Barrow, and at Cranwell RNAS air station.

In January 1916, following eight months of fruitless sacrifice on the beaches of Anzac Cove and Suvla Bay, and after losing upward of 140,000 men together with several capital ships in an effort to force the Dardanelles, the British and ANZAC troops were successfully evacuated by night to safety. The British had established military and naval bases on the nearby Aegean islands of Imbros and Lemnos, where prior to the evacuation the Admiralty set up the Airship Expeditionary Force at the airfield of Murdos on Imbros.

Initially one airship, the *SS7*, was sent out to sea in September 1915 where it was employed in mine-spotting duties in the Dardanelles and was also used to spot the fall of shot for the pre-Dreadnought HMS *Venerable* as she bombarded the Turkish forts with other warships; an activity that was not without danger to the airships crew.

Over the next two years six further airships were sent out to the Aegean in spite of the extremely unfavourable conditions for airship operations, their performance severely reduced by the prevailing high temperatures, which adversely affected the already small useful load of the SS type. Despite these drawbacks, the SS airships performed their main duty of mine spotting successfully, and, on occasion, were employed spotting for the naval guns with great credit.

An additional hazard was that, as the islands were within range of Turkish and German aeroplanes from the mainland, the base was subjected to almost daily bombing attacks on their airship shed at the base at Murdos, which received so much damage it became known as the 'Pepper Pot'.

During the course of operations in the Aegean no airships were lost to enemy action and only one was lost to mechanical failure when the *SS17* came down in the sea off Imbros due to engine failure.

Of the forty-nine SS airships built, the majority had long service lives, working between two or even three years whilst operating under the most arduous conditions – proof of the rugged construction of the type. No SS airships were lost to enemy action during the course of the war but around thirteen SS airships, or 25 per cent of the total, were lost as a result of accident, largely caused by loss of gas pressure, collisions with sheds or engine failure.

Naturally, any loss of envelope pressure over the sea from engine failure invariably resulted with the loss of the ship, and in some cases the loss of crew members. Engine failure was a frequent occurrence, often requiring the crew to make repairs whilst free ballooning, causing the observer or engineer to climb out on the skid to swing the propeller in mid-air to restart the engine – certainly not a job for the fainthearted.

The *SS17* in late 1915, for instance, had the misfortune to have both engine failure and her rudder control cables part simultaneously while operating out of Luce Bay in western

Scotland, from where she drifted out of control with the wind across the Irish Sea as a free balloon. The crew were unable to effect repairs in the air but made a successful landing in Ireland. Here, after being repaired, she was returned to service, eventually being sent out to Murdos in the Aegean, finally being deleted in 1918.

The *SS34* was damaged when the shed she was housed in blew down in a gale, but she too was subsequently repaired, re-rigged and used for mooring experiments in Cavendish Dock, Barrow.

Perhaps the most hazardous task was that assigned to *SS40*, which was equipped with a larger black-painted envelope of 85,000cu ft and a specially silenced engine to be employed on night reconnaissance missions over enemy lines in the Somme area during August and September 1916. There is some evidence to indicate that *SS40* may have been employed on clandestine operations to drop agents behind the lines, before being withdrawn from this dangerous work to serve in the Aegean, presumably with her black envelope replaced by a standard silver one. *SS40* remained in service in the eastern Mediterranean until deleted at Murdos in October 1918.

Much of the work performed by these little airships far out to sea was by its very nature dangerous, monotonous and unrewarding; the crews often flying for many hours in conditions of extreme cold and discomfort as they scanned the grey waters below for the signs of a U-boat or floating mines, ever watchful of the weather and the possibility of engine failure.

An example of the hazards they faced is that of *SS42*. In September 1916, whilst returning from patrol near Lundy Island in bad weather, *SS42* was damaged by a gust when landing which tore away the port side suspension ropes, throwing out the engineer and wireless operator. Relieved of this weight and the broken trail rope the *SS42* shot skyward, carrying the pilot Flight Lieutenant E. Monk clinging precariously to the upturned car.

On reaching 7,000ft most of the forward suspension ropes also gave way, causing the car to hang vertically below the envelope. Here Monk, with great presence of mind, managed to scramble on to the undercarriage structure, to which he clung to for some three hours before the airship began to fall, eventually landing some 100 miles away at Ivybridge, Devon where he escaped with only minor injuries.

Mention should also be made of the two SS envelopes that were modified as Airship Planes or anti-Zeppelin airships, designed to carry a complete BE2c to a suitable patrol height to be released when a Zeppelin was spotted. (See Chapter 9 for details.)

The SS class airships were very successful, within the limitations of their performance, and the majority of them were retained in service for over two years, flying on average 1,000 hours and covering thousands of miles on vital convoy, patrol and mine-spotting work.

Excellent though the SS class were, the need for a larger airship of greater range and all-round performance was evident if the needs of the fleet for true patrol airships were to be met. Accordingly, in early 1915 the design team at Kingsnorth opted for utilising the unused envelope of the 150,000cu ft Astra-Torres *No. 10*, combined with two Avro 519 seaplane fuselages joined together end to end to form a four-seater car, with two 150hp Sunbeam, one tractor and one pusher mounted at either end.

Large rectangular stabilising fins were fitted to the rear of the envelope, which proved to be most effective in operation, allowing a greater degree of lateral and vertical control than had been the case in earlier British airships.

In order to resist air pressure at high speed pressing in the bow, stiffening bamboo battens were fitted into cloth pockets in the nose, this being the standard method of addressing this problem. The choice of the tri-lobe form of the envelope, with its internal suspension system, allowed a greater stiffness of the envelope and for the car to be attached much closer to the envelope. This, together with fewer external suspension wires, led to a corresponding reduction in drag.

The trial flight of Coastal *C1* took place at Kingsnorth on 26 May 1915. There followed a long period of development and modification to the original design, with *C1* being used for various experiments, including being towed by HMS *Craysfoot* in May 1916 and refuelling experiments from surface craft.

The second Coastal *C2* was a larger craft of 170,000cu ft with a length of 196ft (this being the standard capacity for production Coastals). First flying in September 1915, this ship was powered by one 150hp Sunbeam and a more powerful 220hp Renault pusher. The armament consisted of two machine guns, four 112lb bombs and a crew of four.

After trials and development work *C2* was dispatched to her war station at Mullion in Cornwall, and from here she maintained continuous patrols in the Channel and Western Approaches for the next two and a half years, totalling over 2,500 hours' flying time before being deleted in October 1919.

The production of Coastals began in earnest with twenty-seven airships (plus two replacements) being constructed at Kingsnorth during 1916. The last of these, *C27*, completing trials in December of that year.

The efficient manner in which these airships were assembled was a commendable effort, requiring considerable organisational ability on the part of the constructional department controlling the steady flow of envelope fabric, materials, engines and other equipment to ensure a production rate of two airships per month.

A further four Coastals were also built to a Russian government order, being sent by sea to Sabastopol in October 1916 where they were assembled with the assistance of a Royal Navy team who also acted as instructors for the Russian crews.

The Russian airships were named *Albatros*, *Chaika*, *Delfin* and *Chernomor*. However, *Albatros* was quickly lost at sea and *Chaika* was destroyed in a hangar fire the following March, whilst the other two ships do not seem to have been used to any great degree in the Black Sea theatre.

One other Coastal, *Coastal E*, was also supplied to the French navy where it was renumbered *AT1* and was equipped to carry twin 20mm cannon in the bow.

Each Coastal cost approximately £7,000 to build and, although initially not liked by their crews, were found with experience to be most reliable and possessed of good flying characteristics, a fair turn of speed and, importantly for long patrols, greater crew comfort than the earlier SS ships.

Once in service various modifications were made in the light of operational experience, including in some cases the fitting of a fully enclosed and glazed control position, as in the case of *C25*. In later Coastals the air scoop to the ballonets was repositioned from behind the front propeller to abaft the rear propeller. This repositioning was necessary as it was found that, in its former position, the scoop could damage the fore propeller in the event of a heavy landing and also interfered with the forward vision of the coxswain and the pilot.

The disposition of the crew in the car was, from the front, coxswain, pilot, wireless operator and engineer. Later a fifth cockpit was installed for an observer/gunner in the space previously occupied by the water ballast bag, the deletion of which was compensated for by utilising petrol or bombs as ballast in an emergency.

As more Coastals were built the power of the Sunbeam engines was increased to 160hp, but they were plagued with mechanical breakdown. In an effort to solve this problem the engine layout often comprised of the more reliable 220hp Renault together with either a 100hp Green or Berliet.

Operating under normal war conditions a Coastal would carry two to three machine guns and up to 1,000lb of bombs or depth charges, and was capable of flying at a maximum speed of 45mph. While at cruising speed these ships had an endurance of over twenty hours. Two machine guns were carried in the car, while a Lewis gun position was mounted on top of the envelope accessible through a canvas tube and ladder running through the envelope.

18. A Coastal airship prepares to depart on patrol.

19. Interior of shed at RNAS station Kingsnorth, showing a Sea Scout on the left.

Although little used at first, after the loss of *C17* in April 1917 standing orders required that the position should be manned at all times if the airship was operating more than 3 miles out to sea.

With the Coastals coming rapidly into service in 1916 they were distributed to the main war stations, namely East Fortune to operate directly with the Battle Cruiser Fleet, Howden in Yorkshire to protect the northern North Sea convoy routes and Pulham in Suffolk to cover the southern North Sea.

The importance of the port of Liverpool as a main artery for the supply of food, material and munitions from the United States dictated that other Coastals should be based at Pembroke and Luce Bay to serve the Irish Sea, and at Mullion in Cornwall to cover convoys in the English Channel and Western Approaches.

The arrival of the C class airships in these areas considerably improved the range of protection that could be offered to incoming and outward-bound convoys, and throughout the course of the war the Coastals were the work horses of long-distance and convoy-escort protection work.

The Coastals on convoy duty often worked in conjunction with surface craft, summoning destroyers when U-boats were spotted, and it has been claimed that no convoy escorted by these craft was ever attacked.

"COASTAL" Class Airship C23 - 1916
R.N.A.S Constructional Facility Kingsnorth, Kent

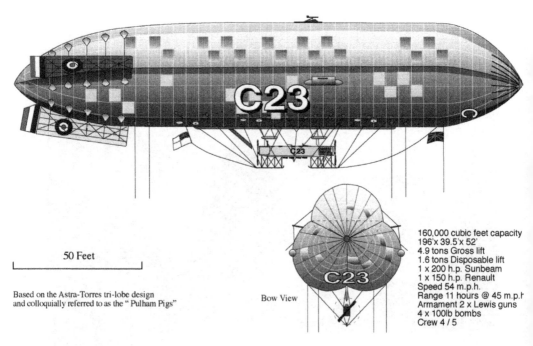

50 Feet

Based on the Astra-Torres tri-lobe design
and colloquially referred to as the " Pulham Pigs"

Bow View

160,000 cubic feet capacity
196'x 39.5'x 52'
4.9 tons Gross lift
1.6 tons Disposable lift
1 x 200 h.p. Sunbeam
1 x 150 h.p. Renault
Speed 54 m.p.h.
Range 11 hours @ 45 m.p.h
Armament 2 x Lewis guns
4 x 100lb bombs
Crew 4 / 5

The *C9*, based at Mullion, had a long and adventurous career, flying over 70,000 miles between her commissioning in July 1916 and the end of the war. Typical of the arduous nature of her duties was an incident on 3 October 1917, when *C9* was forced by a rising gale to leave the convoy it was escorting and return to base.

As Flight Commander Struthers fought against the rising wind he observed an Italian ship some 6 miles astern on fire. He immediately turned *C9* and raced downwind at 90mph, reaching the ship in a matter of minutes. Struthers sighted the U-boat responsible and attacked it as it submerged, dropping his full bomb load on the spot and at the same time calling up destroyers who attacked with depth charges.

The *C9* then again turned for her base, into the face of a full gale and with her engines running at flank speed. It took six hours to cover the 45 miles to her base, where sometimes en route she was held stationary by the wind.

Out of the twenty-seven Coastals built, ten survived the war after active lives and were deleted in October 1918. Of the remainder, seven were lost at sea due largely to engine failure, whilst four were destroyed in ground accidents and the rest were deleted as unserviceable after long service lives.

C11a, commissioned in July 1916 as *C11* and subsequently wrecked near Scarborough in April 1917 without fatalities, had been rebuilt at Howden as *C11a*. From here, whilst on patrol over the Humber estuary, *C11a* burst into flames, probably as a result of a battery fire, and was lost with all hands.

War patrols by German seaplanes from bases in the Friesian Islands against the east coast were an ever-present danger to the airships, and on 21 April 1917 *C17*, her commander Sub-Lieutenant E. Jackson and a crew of four, were attacked 60 miles east of Great Yarmouth by a flight of four Brandenburg sea-monoplanes from Norderney. After a brief engagement *C17* fell into the sea in flames; a photograph taken from one of the attacking machines exists in German naval archives dramatically showing the doomed airship's last moments.

A second Coastal, the *C27* commanded by Flight Lieutenant Dixon, was also lost whilst on patrol from Pulham when, on 11 December 1917, she was observed by a patrol boat to fall in flames to the north-east of the Outer Gabbard light vessel with the loss of all her crew. It was later confirmed that *C27* had also fallen victim to German seaplanes.

C25 left East Fortune on 31 July 1918 on patrol looking for a reported damaged U-boat, the airship last being heard of at 6.30 p.m. some 60 miles east of Aberdeen. The following day searching aircraft found an airship propeller, and it was assumed that *C25* had indeed found the damaged U-boat on the surface and successfully attacked it with bombs, whilst the U-boat in turn had used its deck gun scoring a hit on the airship. Captain Hopperton and the *C25* crew were presumed lost. The U-boat's loss was confirmed by German sources as it failed to return from patrol.

As a result of this loss a brother officer at Pulham, Flight Lieutenant Kilburn aboard *C26*, obtained permission from the station commander the next day to search for *C27* and her crew. After a long, fruitless search, which had carried Kilburn deep into Dutch waters, the airship reluctantly turned to return to base with fuel running low. As *C26* fought against a rising headwind she finally exhausted her fuel, causing the airship to drift as a free balloon over the Dutch coast where she finally made a crash landing near Dordrecht. The uninjured crew were interned for the duration.

From July 1917 to August 1918 the Coastals operated with the Grand Fleet or the Battle Cruiser Fleet on their sweeps into the North Sea, and to a limited extent at last provided the early warning scouting function so desired by naval planners over the past four years. When favourable weather conditions allowed the airships to operate, they greatly extended the scouting line of the fleets' cruisers, although by late 1917 the navy were increasingly coming to rely on the use of the first generation of aircraft carriers or aircraft carried by cruisers for this form of reconnaissance.

During the latter stages of the war (July 1917–October 1918) RNAS airships flew in excess of 1.5 million miles; escorted 2,200 convoys; flew 10,000 patrols; sighted 50 U-boats and attacked 27 (the majority in concert with surface craft); and sighted 200 mines, destroying 75 from the air.

Casualties to the RNAS airship service during the war included 21 officers and 38 men killed in action or accident, plus another 5 officers and 170 men injured in ground accidents.

8

War at Sea: Airships with the Grand Fleet

It was said at the time of Admiral Sir John Jellicoe, the commander-in-chief of the Grand Fleet, 'that it was within his power to win or lose the war in a single day'.

It was widely believed on both sides that when the long-awaited clash between the Royal Navy and the High Seas Fleet came it would have a dramatic and decisive effect on the outcome of the war. In the event, when the two fleets did meet off the Jutland coast in the late misty afternoon of 31 May 1916, the resultant battle fought on a heroic scale and involving some 300 warships of both nations was inconclusive.

With the onset of night, full of confused action, both fleets withdrew; the Germans exultant at their narrow escape from destruction and the British dejected at their failure to more fully punish the High Seas Fleet. In the end, despite the heavy loss of life and ships on both sides, the aftermath of the great sea battle did little to alter the immediate course of the war.

For the British people the loss of three of their most modern battlecruisers, three armoured cruisers and numerous other smaller craft was a severe shock to a nation that believed implicitly in the invincibility of the Royal Navy in battle. In higher government circles this feeling of failure was compounded by the knowledge that poor communications and tactical ineptitude on the part of the British commander-in-chief had allowed Admiral Scheer to extricate the High Seas Fleet from the initial trap set by Jellicoe.

Admiral Scheer's three 180° reversals, when the German battle line was exposed to the full might of the Grand Fleet's guns, allowed the German ships to escape almost certain destruction and eventually to be able to slip away into the safety of the night.

Despite the losses in ships and men to the British, together with the knowledge that the Germans had won a tactical victory, the command of the sea still lay with the Royal Navy. Immediately following the battle the battered ships of the Grand Fleet were able to report 'Ready for sea within 24 hours', as opposed to the many months required for repairs to be carried out to the ships of the High Seas Fleet.

The Germans, whilst losing fewer ships, had suffered enormous material damage from the 12in and 13.5in guns of the Grand Fleet, and it was only due to the integrity of their warships' watertight sub-division that many of their ships survived at all.

While the British had, in the eyes of the neutral nations, suffered a considerable loss of prestige, morale was still high amongst the British crews. In contrast, a deep psychological blow had been dealt to the High Seas Fleet, demoralising the German crews and in turn their naval staff. This served to make their naval policy evermore cautious, to the extent that never again did the High Seas Fleet venture into the North Sea in any great force to contest the control of the seas and engage the Royal Navy in a comparable major surface action.

20. An SS Zero non-rigid signals to the convoy she is escorting, 1916.

There can be no doubt that the severe mauling received from the heavy guns of British battle squadrons had a profound effect on the morale of the German crews. Dissension and discontent festered in the barracks, where political agitators found fertile ground to spread their defeatist message as the great warships swung uselessly at their anchors in harbour.

This discontent grew over the next two years and was ultimately to contribute to the seamen's revolt of November 1918, when the seamen of the High Seas Fleet refused to put to sea to engage the Grand Fleet in support of the beleaguered and retreating army on the Western Front. The German high command had hoped that a successful foray against the Royal Navy would put the German government in a stronger negotiating position at the Armistice talks only a few short days away.

The British failure to achieve a decisive outcome at Jutland in 1916 further enforced the continuing necessity to maintain the bulk of the Grand Fleet at Scapa Flow, to control the North Sea and guard against the possibility, remote though it seemed, of the German fleet once again emerging to do battle.

In order to protect the Grand Fleet during this blockade, many ships, especially the light cruiser squadrons and destroyer flotillas, were tied up at Scapa Flow and Rosyth; ships that possibly could have been better employed in countering the U-boat menace.

The availability of scouting airships to the opposing fleets during the Battle of Jutland was a one-sided affair, as the Coastals were at that time only just getting into production: only *C1* and *C2* were completed, while *C3* was not commissioned until 16 June.

Consequently, no suitable long-range airships were on hand to serve the needs of the Royal Navy on the fateful 31 May and, in addition, only one British aircraft was launched during the course of the battle. The High Seas Fleet, on the other hand, deployed eleven Zeppelins during the course of the operation for reconnaissance duties. Yet despite the best efforts of their crews their contribution on the day was negligible.

Due to the poor weather conditions, with decreasing visibility prevailing at the time and battle not being joined until the late afternoon, such reports made by the Zeppelins were largely of an inaccurate nature with regard to ship unit identification, position and course. The confused signals that were passed on to Admiral Scheer failed to give him the vital information that would enable him to plot the disposition of the enemy fleet during the course of the battle.

Admiral Beatty, however, was convinced that the Zeppelins had been the key element in allowing Scheer and Hipper to evade their nemesis that, on two occasions, faced them during the battle under the combined guns of Jellicoe's battle squadrons and his own Battle Cruiser Fleet.

The High Seas Fleet in fact owed its survival to Jellicoe's caution, the poor communication of the developing situation from scouting cruisers to their commander-in-chief, and the tactical genius of Scheer.

During the course of this great sea battle the airship, and for that matter the aeroplane, had failed to provide the vitally important reconnaissance function for which it had been designed.

The Germans, with their greater technical ability and expertise, were able by 1916 to provide their fleet with large and effective rigid airships, capable of fulfilling not only the strategic bombing role but also the majority of long-range reconnaissance and patrol duties in the North Sea and the Baltic.

The British, on the other hand, despite their best efforts were unable to produce rigid-framed craft that in any way could be comparable in performance to contemporary Zeppelins, or for that matter in terms of useful lift to pre-war commercial Zeppelins.

However, British expertise in the field of pressure airship, an area where they were pre-eminent, began to pay dividends with the development of a series of efficient classes of blimps. These were to be built in large numbers, proving to be rugged and effective weapons systems, with each succeeding class incorporating advances which were to contribute to countering the U-boat menace throughout the remainder of the war.

The SSP Class

By the end of 1916 twenty-six Coastals had been delivered and were being usefully employed in the increasingly intense war against the U-boat. But the need for yet still more airships focused attention on an attempt to improve the performance and range of the basic Sea Scout by the design team at Kingsnorth.

The result was the SSP or Sea Scout Pusher, which employed the standard 70,000cu ft envelope to which was attached a purpose-built, ash-framed, plywood and canvas-covered car of roomier proportions than the Maurice Farman car on which it was based.

A 75hp Rolls-Royce Hawk was initially fitted at the rear of the car, although at this stage of its development the Hawk was somewhat unreliable and was later replaced by the proven 100hp Green on most ships of this class. A crew of three was carried in comparative comfort as with the pusher layout the crew were not subjected to the propeller wash of the earlier tractor design.

The introduction of the longer car with its increased side area required the mounting of an additional fin ahead of the standard lower fin to counteract the effect.

The performance of the SSP class represented a definite improvement over the SS, having a speed of 53mph and a range increased by 30 per cent. Originally it was proposed to build some thirty SSPs, in the event however only six SSP ships were actually built, this being due

"SSP" Class Naval Airship SSP.2 - 1917
R.N.A.S. Constructional facility, Kingsnorth Kent

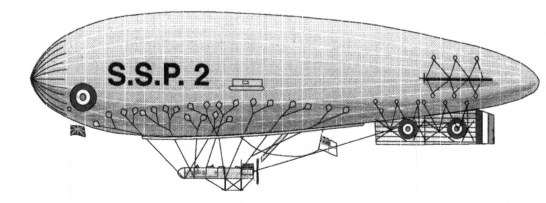

70,000 Cubic feet capacity
149' x 34' x 56'
Gross lift
Useful lift
1 x 75 h.p. Rolls Royce Hawk
Speed 53 m.p.h.
Range 15 hrs at cruising speed
Armament 2 x 110 lb bombs + 1 Lewis gun
Crew 3

50 Feet

to the promise being shown by the new SS Zero design then being developed at Capel air station near Folkstone.

The SSP airships had mixed fortunes, with *SSP3*, *SSP4* and *SSP6* being lost at sea, while *SSP2* and *SSP5* were fitted with engine silencers and black envelopes for clandestine operations over enemy lines on the Western Front during 1917. This was of course a highly dangerous undertaking, but they successfully accomplished it, dropping agents and collecting secret intelligence from spies behind the front line. The wonder is that both airships came through these hair-raising adventures unscathed.

The SS Zero Class

The next class of airship was the highly successful SS Zero non-rigid, of which at least seventy-seven examples were built. This was the result of an independent and unofficial initiative by a group of naval officers at Capel air station, led by Commander A. Cunningham who designed a streamlined, watertight aluminium car with a V-shaped hull that was suitable for landing on either land or water.

This car provided for a crew of three: wireless operator in the front cockpit, pilot and engineer aft – where he tended the now fully reliable 75hp Rolls-Royce Hawk. This imparted a speed of 53mph and a rate of climb of 1,200ft per minute with a duration of over twenty hours. Along with a machine gun mounted in the fore cockpit, the Zero also carried two 110lb anti-submarine bombs on racks below the engine.

The prototype was submitted for Admiralty trials, which it successfully completed in September 1916, with the Admiralty recognising the superior performance characteristics inherent. The type was ordered into quantity production, although Cdr Cunningham was taken to task for 'making unauthorised modifications to Admiralty equipment'.

A total of seventy-seven SS Zeros were brought into service between late 1916 and January 1919, when the last of this class was delivered to Pulham.

Due to the relatively high numbers produced the SS Zeros were well utilised, operating on convoy protection work mainly over the Irish Sea and the Western Approaches. Here they performed sterling work and proved to be extremely handy ships, with an excellent reputation for reliability, largely thanks to the excellent 75hp Rolls-Royce Hawk engine.

These airships were often moored out in clearings cut into woods, sometimes for weeks at a time. This afforded a measure of protection from the weather and the airships were able to operate from these limited facilities most effectively, which demonstrates their sound design and rugged construction.

A story that illustrates the excellent handling and flying characteristics of the Zero is represented by the final exploit of the *SSZ73* whilst stationed at Anglesey under the command of Major T. Elmhirst (who was later to rise to air marshal in the RAF). Following an Armistice night party in November 1918, Elmhirst flew his airship under the central span of the Menai Bridge, having first reconnoitred the bridge, and determined the height of his approach by lowering a sandbag on a rope to trail in the water. He then opened up to full speed to ensure full elevator control and successfully accomplished the daring manoeuvre.

SS Zero – Mullion Development of the Rubber Eel

At Mullion air station in Cornwall during 1917, experiments were conducted using acoustic methods for the detection of submerged U-boats. During these trials a group of hydrophones were contained within a rubber sleeve, known colloquially as the 'Rubber Eel', which could be lowered from a Zero non-rigid airship. From the airship the operator could then make an assessment of the range and track of a submerged U-boat from the received signal, and plot the results on to a chart. Experience demonstrated that with two or more airships so equipped working in concert, a more precise 'cut' and position could be made.

These experiments were very successful and showed great promise, but for reasons that are not clear this promising device was not proceeded with. If it had been fitted to the efficient and manoeuvrable Zeros, which were probably the handiest of all Allied airships, they could have made a major contribution to anti-submarine warfare.

It is interesting to speculate that, had this effective detection system been employed under wartime conditions, the employment and development of British naval airships would almost certainly have continued throughout the interwar period, albeit on a much reduced scale due to financial stringency. Had further development taken place during the interwar period it would have provided an effective airborne detection system, capable of giving much needed protection to merchant shipping in the Channel and Western Approaches in the early years of the Second World War.

As it was, it was not until the 1950s, with the advent of airborne sonar systems deployed from helicopters, that the ability to track underwater craft from the air became possible and was once more available to the navy.

21. North Seas class airship *NS9* at Longside RNAS station, 1918.

The North Seas Class

In January 1916 a new Admiralty specification was issued for a larger non-rigid of twice the capacity of the Coastal and possessed of sufficient power and range to accompany and scout for the fleet on its North Sea sweeps, until larger rigids became available.

The first of the North Seas class, *NS1*, was delivered from Kingsnorth on 7 February 1917. It made a flight of forty-nine hours in April, to be followed by *NS2* and *NS3* in June and July respectively.

The North Seas were again tri-lobes, being 262ft in length and of 360,000cu ft capacity, giving a gross lift of almost 11 tons, which was a major advance over the proceeding class. Six ballonets were incorporated in the envelope and, after initially arranging the fuel tanks externally on the envelope, they were later mounted internally supported on the rigging.

In the original Admiralty design the long car was divided into a spacious control car with separate navigation and wireless compartments, behind which were the crew quarters. The two engine cars were mounted on outriggers, reached by a wire walkway from the car and included a small streamlined car on the centre line for the engineers to service both engines in flight. An offensive load of 1,380lb of bombs or depth charges could be carried, plus a defensive armament of up to five machine guns.

These ships carried a crew of ten divided into two watches, and crew comfort was greatly increased by an enclosed car, the provision of bunks for off-duty personnel and cooking facilities. Initially these airships were powered by 250hp Rolls-Royce Eagle engines, which had proved to be most reliable power plants in many installations in airships.

However, problems were encountered with the long shaft drives of the propellers, to the extent that the earlier ships of this class were plagued with transmission problems and breakdowns that rendered them almost useless for service

In order to remedy this situation the *NS 3* was taken in hand at East Fortune. Here, under the direction of Flight Commander J.S. Wheelwright and Engineer Officer Lieutenant Commander A. Abel, the control and engine cars were rebuilt into a more practical layout whilst the engine installations were replaced by direct drive 240hp Fiats. These changes had an immediate improvement on the type's serviceability and performance.

Due to the success of the reconstruction carried out at East Fortune, the Admiralty approved the new layout and sanctioned a further modification to the car layout for several of the other airships still building to a layout devised by the Kingsnorth design team. In the Wheelwright modification the control and engine cars were combined into a single car, with the direct drive Fiats mounted outboard, whereas in the Kingsnorth Admiralty modification the control and engine cars remained separate, but also included the direct drive Fiat engines.

"SS ZERO" Class Naval Airship SSZ-27 - 1917
R.N.A.S. Constructional Facility, Wormwood Scrubs, London

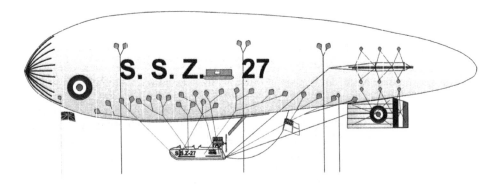

50 Feet

70,000 Cubic feet capacity
149' x 34' x 47'
Gross lift 2.4 Tons
Useful lift 0.7 Tons
1 x 75 h.p. Rolls Royce Hawk
Speed 53 m.p.h.
Range 15 hrs at cruising speed
Armament 2 x 110 lb bombs + 1 Lewis gun
Crew 3

"NORTH SEAS" Class Airship N.S.11 - 1918
R.N.A.S. Constructional facility, Kingsnorth Kent

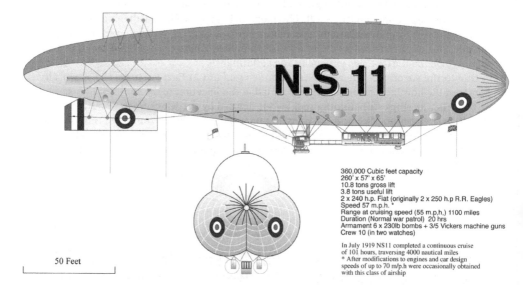

N.S.11

360,000 Cubic feet capacity
260' x 57' x 65'
10.8 tons gross lift
3.8 tons useful lift
2 x 240 h.p. Fiat (originally 2 x 250 h.p R.R. Eagles)
Speed 57 m.p.h. *
Range at cruising speed (55 m.p.h,) 1100 miles
Duration (Normal war patrol) 20 hrs
Armament 6 x 230lb bombs + 3/5 Vickers machine guns
Crew 10 (in two watches)

In July 1919 NS11 completed a continuous cruise
of 101 hours, traversing 4000 nautical miles
* After modifications to engines and car design
speeds of up to 70 m/p.h were occasionally obtained
with this class of airship

50 Feet

Throughout the remainder of their service the North Seas performed well, operating with the fleet on their North Sea sweeps and conducting long patrols, many being of several days duration.

The North Seas class in its modified form had a top speed of 62mph with a rate of climb, under normal conditions, of 1,000ft per minute and a ceiling of 10,000ft. Additionally the NS class were handy craft having a turning circle of forty seconds, which under favourable weather conditions made the airship capable of dealing with submarines effectively.

Of the fifteen or so North Seas airships that reached naval service, four succumbed to weather. The *NS2* was wrecked near Stowmarket in June 1917 only a week after her first flight, while *NS3*, after attacking a submarine in the North Sea, was also wrecked in a gale on her return near Dunbar next morning. The airships *NS7* and *NS8* did, however, have the honour of escorting the surrendered German High Seas Fleet into Rosyth following the Armistice in November 1918.

A long-standing endurance record was set by the *NS11* when, in February 1919 during a routine mine patrol and clearance flight over the North Sea, she covered over 4,000 miles in a flight of 101 hours. This airship, accompanied by *NS12*, also made a long flight to Norway of over twenty-four hours, searching for minefields whose positions were reported by wireless to minesweepers. Airship *NS11* was later lost off Blakeney on the Norfolk coast when the ship was struck by lightning whilst valving gas.

With the war's end the last three airships were completed but not delivered to the navy, the *NS13* (later renumbered *NS14*) was sold and shipped to the United States; and six of this class are referred to as completing at Vickers to an order for the US, but whether this order was completed is in doubt. The final airship, *NS16* delivered in January 1919, undertook an experimental programme involving sweeping for mines from the air with a towed sweep.

The last NS in service was the *NS7*, which was seconded to the US navy for crew training for the American detachment destined to fly the Atlantic aboard British-built *R38*. She made her last flight in October 1921.

In September 1917 an Admiralty conference proposed a larger successor to the North Seas class, apportioning the designation K class. It can be assumed that the proposed class would have been of 500,000cu ft capacity with a gross lift of 15.5 tons, and would be powered by three or possibly four Rolls-Royce Eagles.

Such a large and powerful non-rigid, benefiting from the experience learned from operating the preceding North Seas class, would possibly have been the ideal airship to operate with the fleet in the North Sea. In terms of handiness, speed, endurance and load carrying capacity it would have exceeded the performance of the earlier rigids.

The Coastal Star Class

The introduction of the Coastal Star in January 1918 was in response to the engine and transmission problems that were besetting the new large North Seas class tri-lobes, which was delaying their entry into naval service.

In order to speedily find an interim replacement for the urgently required North Seas ships, the Coastal *C12* was withdrawn from service at Polegate in August 1917 and sent to Kingsnorth where it was rebuilt with a new envelope to become Coastal Star *C*1*.

The Envelope of the C* was aerodynamically superior to the preceding C class, being of better streamlined form with the envelope enlarged to 217ft and a capacity of 210,000cu ft. Six ballonets were installed which aided longitudinal stability.

22. Coastal Star class non-rigid airship *C*1*, 1918.

"COASTAL STAR" Class Airship   C★10 - 1918

R.N.A.S. Constructional Facility Kingsnorth, Kent

TRI-LOBE

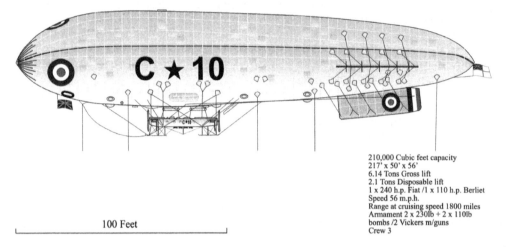

100 Feet

210,000 Cubic feet capacity
217' x 50' x 56'
6.14 Tons Gross lift
2.1 Tons Disposable lift
1 x 240 h.p. Fiat /1 x 110 h.p. Berliet
Speed 56 m.p.h.
Range at cruising speed 1800 miles
Armament 2 x 230lb + 2 x 110lb
bombs /2 Vickers m/guns
Crew 3

Again, crew comfort was improved with a plywood covering to the open cockpit, replacing the canvas of the earlier class. It contained a crew of five, parachutes being fitted to the side of the car. Four 85-gallon petrol tanks were suspended within the envelope, allowing for an endurance of over twenty hours at cruising speed – this often being exceeded, as with a flight of thirty-four and half hours by Captain Cleary in May 1918 on mine patrol.

Motive came from a 220hp Renault aft and a 110hp Berliet forward. In later ships the Renault was replaced by a more effective 240hp Fiat water-cooled, six-cylinder engine. The increased power of this, together with their improved aerodynamic form, allowed them to reach speeds of up to 56mph on occasion.

An offensive load of two 230lb and two 100lb anti-submarine bombs were carried, as well as two Lewis guns mounted in the car. In this class the upper gun position was suppressed.

It must be said of British anti-submarine bombs of the period that they were far from effective; the pressure fuses had a failure rate of some 40 per cent and the impact fuses were often not sensitive enough to explode on contact.

Originally twenty Coastal Stars were ordered, but this was reduced to ten in July 1918 as the later SS Twin was deemed to be a superior and more effective design. The C*s were all built at Kingsnorth between January and July 1918, effectively filling the gap until the North Seas class difficulties were resolved successfully by mid-1918. Subsequently, they went on to enjoy long service lives with just one being damaged to accident or enemy action. This was the *C*9*, which was damaged by an accidentally dropped bomb whilst landing at Howden. Nevertheless, she was repairable and continued flying until deleted with the majority of naval airships in 1919.

These airships were delivered to East Fortune, Mullion and Howden where they provided reliable support to the fleet and on convoy protection work, totalling more than 5,000 hours' flying time. The *C*1*, the original prototype, was based at East Fortune from February 1918 and flew a total of 868 hours before being deleted in October 1919. The rest of the class all individually exceeded 500 hours in service.

A proposal was made in June 1918 to equip Coastal Stars with two 18in circling torpedoes, primed to a shallow depth setting these would be used to attack U-boats operating at periscope depth or on the surface. Although some experiments were conducted, however, the war's end precluded any deployment of this promising weapon.

A further suggestion was to employ the Coastal Stars as airborne ambulances, shuttling severely wounded soldiers from the Western Front to hospitals on the south coast. Although considering the enormous amount of causalities involved it is difficult to see what criteria would have been used in selecting deserving patients.

The SS Experimental and SS Twin Classes

The final two classes of non-rigids built during the war were the SS Experimental (SSE), of which only three were built, and their immediate successor the SS Twin (SST), which were the final expression of British development in this field.

With the reliability and effectiveness of the SS Zero well established, the Admiralty decided to improve the design during 1917 to produce an enlarged version with a higher all-round

"SS TWIN" Class Naval Airship SST-12 1918
R.N.A.S. Constructional Facility, Wormwood Scrubs

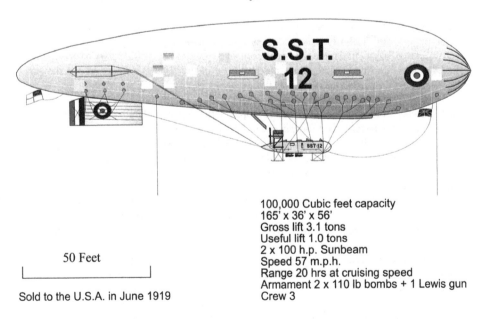

50 Feet

Sold to the U.S.A. in June 1919

100,000 Cubic feet capacity
165' x 36' x 56'
Gross lift 3.1 tons
Useful lift 1.0 tons
2 x 100 h.p. Sunbeam
Speed 57 m.p.h.
Range 20 hrs at cruising speed
Armament 2 x 110 lb bombs + 1 Lewis gun
Crew 3

"SS.E"(Experimental) Class Naval Airship SSE-3 - 1918
R.N.A.S. Constructional Facility, Wormwood Scrubs

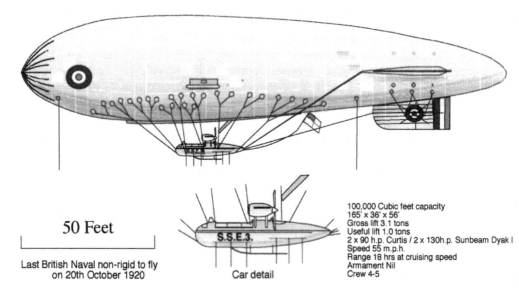

50 Feet

Last British Naval non-rigid to fly
on 20th October 1920

S.S.E.3.

Car detail

100,000 Cubic feet capacity
165' x 36' x 56'
Gross lift 3.1 tons
Useful lift 1.0 tons
2 x 90 h.p. Curtis / 2 x 130h.p. Sunbeam Dyak I
Speed 55 m.p.h.
Range 18 hrs at cruising speed
Armament Nil
Crew 4-5

performance. *SSE1* and *SSE3* prototypes were produced at Wormwood Scrubs. They were noteworthy for having highly streamlined semi-enclosed cars, envelopes of 100,000cu ft capacity and two 90hp Curtiss engines (which ensured greater range and safety of operation).

The *SSE3* was later used to train the American crew of the ill-fated *R38* (or *ZR2* as she was called by the US navy) and had the distinction of being the last British naval airship to fly, on 20 October 1920, on a final training flight.

The *SSE2* was built at Mullion and was equipped with a less complicated and roomier slab-sided car that was to be used as standard on the succeeding class.

The SS Twins were slightly smaller – the envelope being of 90,000cu ft capacity with four ballonets – giving a gross lift of 3 tons and a disposable of 1 ton. The envelope was 165ft in length and of 49ft beam, power was provided by either two 75hp Rolls-Royce Hawks, or one 100hp Sunbeam.

The performance of the Mullion-built *SST2* so impressed the Admiralty inspectors that they proposed an order for 120 SS Twins be placed immediately. Plans for quantity production were well in hand when the order was cancelled at the time of the Armistice, and in the end only thirteen airships of this class were built, coming rapidly into service during early 1918.

The Twins carried a crew of four or five. The airships' superior range and endurance was demonstrated in May 1918 when the *SST14*, under the command of Captain S.E. Taylor DSC, made an extended flight of fifty-two hours whilst spotting for mines in the North Sea.

In 1919 the *SST9*, *SST11*, and *SST12* were sold and shipped to the United States. Of the remaining airships only one of the class was lost, when the *SST6* had the misfortune to catch fire in the air, with the death of the crew of four when the car fell from 400ft.

All the other airships of this class were deleted with the closing of the airship service for financial reasons in October 1919.

Italian Semi-Rigid Airship *SR1*

The Admiralty purchased the M class semi-rigid *SR1*, which appeared to be suitable for long-range operations in the North Sea, from Italy in October 1918.

The *SR1* was of 441,000cu ft capacity, 269ft in length and 59ft at its greatest diameter, giving a gross lift of 12.8 tons with a useful load of 4 tons. Power was initially provided by two 200hp Italia petrol motors and a 200hp SPA-6A mounted on top of the control car. This arrangement gave a top speed of 52mph and a range at cruising speed of 1,600 miles.

The airship left Rome on the delivery flight to Pulham on 28 October 1918. Under the command of Captain George Meager with a crew of ten it took three and a half days to complete the difficult voyage against strong headwinds. *SR1* took stops at Aubagne, Lyon and St Cyr to refuel and repair the SPA motor, which at one stage caught fire and, but for the swift action of the crew in shutting it down and throwing the red-hot exhaust pipe overboard, would have spelt disaster.

On arrival in England the airship was taken in hand for modification, but the Armistice interfered with these plans and the *SR1* was little used, undertaking only a few North Sea patrols including one of twenty-five hours and a Victory loan flight over London.

Naval Airship HMA SR1-1918
Italian "M" Class
Corps of Engineers - Ciampino, Rome

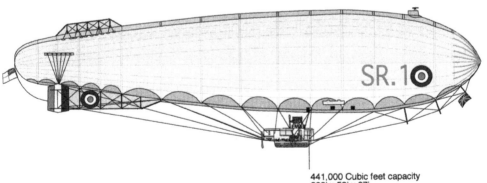

441,000 Cubic feet capacity
269' x 59' x 87'
12.8 tons gross lift
4.2 tons useful lift
2 x 200 h.p. Italia + 1 x 220 Maybach
Speed 52 m.p.h.
Range at full power 960 miles (18 hrs)
Armament 6 x 230lb bombs 2 x Lewis guns
Crew 10

50 Feet

23. SS Twin non-rigid at RNAS air station Pulham, Suffolk.

Her most important task took place on 20 November when, together with the *R26*, she oversaw the surrender of German U-boats, meeting them at sea and escorting them into Harwich.

During her active life with the Royal Navy the *SR1* flew for a total of only ninety-one hours (including the delivery flight) and was finally decommissioned and deflated in October 1919.

9

Air Defence of Great Britain:
Opening Moves

Plans for the protection of naval dockyards, forts and magazines against the menace posed by aerial craft had long been considered by the Committee of Imperial Defence prior to the outbreak of war. As early as 1910 this committee had recommended the manufacture and provision of a large number of high-angle, quick-firing guns in order to guard such installations.

The recognition of the possibility of aerial attack on the British Isles prompted the promulgation of the Aerial Navigation Act, put before Parliament in 1911, which established the right of sovereignty to the Crown of the air space over Britain. This act additionally prohibited the over-flying by foreign and civilian pilots of certain restricted areas, including defended forts, armament factories and wireless stations.

Under the existing defence act the overall responsibility for the protection of Great Britain devolved on to the army, although they themselves, given the unusual nature of the threat, had not at that time evolved a coherent strategy for dealing with any form of aerial attack.

The long-established inter-service rivalry between the War Office and the Admiralty only added to this sense of confusion, with the latter insisting on the right to protect all Admiralty property and installations whilst the former were adamant that they alone should bear the responsibility for the protection of the whole of the British Isles.

The main weapon of defence was seen, initially, to be in the use of high-angle artillery, with searchlight and balloon barrage defences in passive support. In the event, and despite the forward thinking of the government of the day, only a few inadequate 1lb Pom Poms dating from the Boer War, supplemented by a number of obsolete 3in naval guns, were available to protect the Royal Dockyards and the Thames forts.

In theory the Pom Poms were capable of throwing a shell to a height of 10,000ft, which should have been more than adequate to deal with the early low-flying Zeppelins. However, in practice deficiencies in the ranging techniques employed nullified their relatively high rate of fire in the hands of inexperienced gun crews.

The deployment of an aeroplane corps to supplement this form of defence was only in the earliest stages of consideration, the problem being compounded by the lack of what was then considered to be the most suitable type of aircraft to employ for the purpose, together with the absence of any idea as to how to best arm and utilise them in this new and novel defensive role. The Director General of Military Aeronautics at the War Office, General Hamilton, had estimated that a minimum of 170 aircraft would be required to protect the east coast and the inland cities, including London, from aerial attack.

Needless to say, between that time and the outbreak of war in 1914 no such defensive measures were in place, despite the protestations in Parliament of the First Sea Lord Winston Churchill that 'any flying machine attacking this country, would be met by a swarm of hornets'.

Once the war had begun, the existing army air squadrons of the Royal Flying Corps were quickly accounted for in supporting the BEF in northern France and Belgium, with the task of supplying aircraft for the home defence squadrons at that time clearly beyond the capability of the War Office. Accordingly, the responsibility for the aerial defence of the British mainland passed by default to the Royal Navy, with the army still maintaining the control of the fixed gun defences.

It would be many months before the army were in a position to augment the naval aircraft with army night flying machines and to undertake the construction of a ring of landing grounds around the capital.

In anticipation of mass airship raids on the capital, and hastened by the recent news of a dramatic night attack on Antwerp by army Zeppelins, the first ordinances were issued in September to govern restrictions on the lighting of domestic and commercial properties in the Metropolitan area. These measures were quickly followed quickly by further restrictions covering trains, trams, buses and street lighting alongside the institution of official blackout times and fines for non-compliance.

The first moves to organise an effective home defence network involved the setting up of a Defensive Protective Zone around Whitehall, Charing Cross and Buckingham Palace; areas

24. Wreckage of Zeppelin *L31*, shot down at Potters Bar, Hertfordshire by Lieutenant Wulfstan Tempest, October 1916.

deemed to be of special importance. Within this designated zone, guns and searchlights were installed, manned by a small number of regulars assisted by civilian volunteers and special constables.

Of more immediate concern to the authorities towards the end of 1914 was in the receipt of intelligence indicating that the German army had now established new airship bases in occupied Belgium, thus bringing the threat of attack closer to the capital.

The first enemy aircraft to fly over the British mainland was on Christmas Day 1914, when an Albatross seaplane flew up the Thames Estuary, passing over Gravesend, Dartford and as far as Tilbury to reconnoitre the Thames forts. In its passage the lone biplane dropped two high-explosive bombs at Cliffe, these being the first bombs to fall on this country, before making out to sea. Several RNAS aircraft took off from Eastchurch air station in an attempt to intercept the raider, but failed to come within range.

In response to the failure of defending machines to engage enemy aircraft over Britain, in this and subsequent early raids, the Admiralty conceived a bold plan to strike at the Zeppelin bases in occupied Belgium and on the North Sea coasts, with the intention of destroying the airships before they could be used.

On the same day the German seaplane flew over Tilbury Fort, Commodore Tyrwhitt's light cruiser and destroyer flotillas of the Harwich Force made a raid deep into the Heligoland Bight to launch an attack from seaplane carriers against the Zeppelin sheds at Cuxhaven. The ships themselves came under attack from both Zeppelins and seaplanes, while German light forces put to sea to confront the attackers.

Although all the aircraft were successfully launched and reached the German coast, they failed to locate the sheds due to deteriorating weather conditions, with all of the aircraft returning to their carriers. However, one of the seaplanes with Lieutenant Erskine Childers, the author of *The Riddle of the Sands*, as observer successfully reconnoitred the Schillig Roads, identifying seven battleships, three battlecruisers, various other cruisers and destroyers, and observing heavy concentrations of enemy shipping in the Wesser Estuary.

This successful example of aerial reconnaissance resulted in the larger portion of the High Seas Fleet being moved up the Kiel Canal into the Baltic, to avoid further attention from the RNAS.

A more successful attack had previously been made by the RNAS squadron based at Dunkirk, when, on a second attempt on 9 October 1914, Commander Spencer Grey and Flight Lieutenant Marix, flying Sopwith Tabloid biplanes, set out to bomb the Zeppelin sheds at Cologne and Düsseldorf.

After a flight of 100 miles Marix located the Düsseldorf shed and in the face of heavy machine-gun fire dropped his two 25lb bombs, scoring a direct hit and destroying the army Zeppelin inside. This was the first success of its kind by the RNAS.

Meanwhile, Spencer Grey, frustrated by thick fog over Cologne and being unable to locate the sheds, had to settle for dropping his bombs on the main railway station instead.

A further setback to German plans was dealt by a third and even more daring raid carried out from Belfort near the Alsace–French border that was aimed at the very heart of the Zeppelin empire at Friedrichshafen. On the morning of 21 November 1914 three Avro 504s, each carrying four 25lb bombs, flew 125 miles through the mountains on a route designed to carefully avoid over-flying Swiss territory to emerge on the south side of lake before climbing to attack the Zeppelin works.

A British airman claimed a direct hit on one of the main sheds, however German sources to this day continue to claim that this raid caused no damage. The testimony of Swiss workers at the factory and other neutral observers dispute this claim and maintain that a partly built airship was in fact destroyed.

At this time the fear of a mass attack by Zeppelins filled the publics' imagination; in reality the situation was very different across the North Sea. The German navy, in point of fact, had only one operational airship in August 1914, the *L3*, although five more were to enter service before the end of the year. On the other hand, the German army initially possessed ten airships, which were quickly augmented by the appropriation of the DELAG airships *Hansa*, *Sachsen* and *Viktoria Luise*.

Of this force the army, however, soon lost the Z6 in an attempt to bomb the frontier fortress of Lutetia in Belgium, while the Z7 and Z8 were lost on similar low-level missions against heavily defended positions to rifle and artillery fire.

In January 1915 the German Naval Chief of Staff Admiral von Pohl proposed a joint attack by army and naval airships on London, claiming that due to its surrounding forts it was a legitimate target. 'A defended place', as described in the Hague Convention of 1899 relating to the rules of warfare.

The German chancellor, Bethmann-Hollweg, objected to the scheme on the grounds that an understanding could still be reached at that stage of the war with Great Britain, leading to an early peace on favourable terms for Germany. After considerable wrangling the kaiser, on 9 January, signed the document authorising raids on London but stating that such action should be 'confined to the dock area east of the Tower'.

The first Zeppelin raids on Britain took place on the night of 19/20 January 1915, when the *L4* (Kapitän-Leutnant Count Magnus von Platen-Hallermund) accompanied by the *L3* (Kapitän Johann Fritz) and *L6* (with Oberleutnant von Buttler-Brandenfels in command) left Fuhlsbüttel naval base in the early morning to attack England. In freezing weather *L3* and *L4* crossed the coast together near Ingham, Norfolk at 8 p.m. Von Buttler-Brandenfels' *L6* carried the leader of naval airships, Korvettenkapitän Strasser on board, but was forced to turn back with engine trouble some 60 miles out to sea.

The commander of *L3*, Fritz, made his landfall at Ingham and, turning south, headed for Great Yarmouth where he dropped eight high-explosive bombs causing widespread damage to buildings, killing two persons and injuring several others before heading safely out to sea.

Meanwhile, *L4* passed over Bacton, Cromer, Sherringham and Brancaster, flying on to Kings Lynn where the majority of her 1,000lb bomb-load fell, causing most of £7,000 worth of damage to property in the raid and killing or injuring seventeen persons including three children. The *L4* then crossed Norfolk flying over Norwich and passed out to sea at Great Yarmouth. Both airships returned safely to their base after being in the air for twenty-two hours.

This first raid caused great excitement and resentment from the populace, who were justly outraged that German airships could fly at will over the darkened countryside, bringing death and destruction to those below without any effective defensive countermeasures by the army or navy. The civilian deaths, particularly those of the women and children, were seen to be barbaric by the British public. This, together with the stories of the atrocities that had been committed by the German army on civilians in Belgium, resulted in the German airmen being branded as 'The Hun Baby Killers'.

If the British were lacking in the material wherewithal in regards to an effective counter to this airborne threat, one area in which they excelled was that of intelligence and the dissemination of the data so gained. At the outbreak of war Sir Alfred Ewing was appointed controller of intelligence at the Admiralty, with Commander Alastair Denniston acting as his ADC. The centre of this intelligence-gathering network was Room 40, Old Building, at the Admiralty in Whitehall; the undertaking being overseen by the redoubtable Admiral 'Blinker' Hall, an officer of great energy and the ability to inspire his staff to a remarkable extent.

Here, throughout the war, a team of specialised cryptographers, drawn from a wide variety of civilian and service occupations, worked tirelessly on wireless code: intercepts of German military and naval traffic; telephone; information garnered from neutral sources; commercial and business information; plus just plain spying activities. By itself this information, randomly gathered, meant very little, but in the hands of the Admiralty experts each piece was made to fit the complicated jigsaw that helped to understand the military intentions of the Central powers.

In contrast, the Germans and their allies paid scant heed to the requirements of an effective military or naval intelligence service to aid their fighting services throughout the war. Their own efforts were fragmentary, lacking a single unified central department for sifting the huge amounts of the varied information they received. In addition, their intelligence services were riddled with inter-service rivalries that were of an order of magnitude greater than that found in the British services.

The German high command failed to appreciate the value of the information available to the British naval and military command, and little realised to what extent their own efforts were being compromised by the highly efficient British secret service and the untiring work of the skilled analysts in Room 40.

The Admiralty also possessed a further great advantage over the Marine Luftschiffe Abteilung (German Naval Airship Service) when, in August 1914, the German light cruiser *Magdeburg* was sunk by Russian cruisers in an engagement in the Gulf of Finland. Russian divers were subsequently able to dive on the wreck and recover a haul of secret German naval codebooks, which included the important SKM code that was the main cipher then in use.

In an uncharacteristic spirit of co-operation on the part of the Russians, these codebooks were immediately dispatched to the Admiralty in London, arriving on the desk of Admiral Sir Reginald Hall, chief of naval intelligence, in Room 40.

These codebooks and others, such as the GN and VB codes similarly retrieved from the German destroyer *T119* sunk at the Dogger Bank action, gave a valuable insight into not only the current codes but also the system employed in the construction of new codes. This was a piece of good fortune that was to allow the Admiralty to read German wireless traffic throughout the whole course of the war.

A further piece of good fortune occurred in the autumn of 1914, when Winston Churchill was approached by two English wireless amateurs, Barrister Russell Clarke and Colonel Richard Hippisley. Using their own private civilian equipment they informed him that they had for some time been picking up German military and naval traffic, and that these intercepts included wireless bearings issued to Zeppelins in the air and German warships at sea.

Churchill, recognising the significance of this revelation, immediately installed the two men at a remote coastguard station near Hunstanton on the Wash to listen in on the German frequencies. A further refinement was added in February 1915 when, following successful wireless direction-finding experiments on the Western Front by the Marconi Company, Admiral Hall ordered the establishment of fourteen further listening stations sited from the south coast to the Wash, each staffed by handpicked GPO personnel.

These stations were equipped with the latest equipment, capable of taking cross-bearings on Zeppelins and pinpointing their positions accurately as they approached the British coasts. The data provided by these reports allowed their tracks to be plotted at Room 40 in Whitehall from the moment they left their bases on the other side of the North Sea, and for the duration of their flight.

Due to a further oversight on the part of the Germans, an early warning of a raid on the British Isles was provided by the departing airship sending the invariable wireless message 'Only HVB on board'. HVB – or *Handelsschiffverkehrsbuch*, the standard German Merchant Service codebook long compromised by British Intelligence – indicated to the Admiralty that, because the more secret codebooks had been left behind, the airship was on a raid on the British mainland rather than a scouting mission.

A very efficient and sophisticated air-raid warning and tracking system was rapidly developed to deal with this unprecedented threat from the skies, involving the services of police, local authorities and many civilian volunteers working in concert to produce an air-defence-warning network that was second to none.

This well-planned and effective organisation was to become the forerunner of the Radar-controlled early warning and defence system that was fundamental to victory over the Luftwaffe in September 1940.

Lightships at sea provided the first link in the visual spotting network, reporting Zeppelins up to 60 miles out to sea through telegraph cables linked directly to the Admiralty. Onshore, a chain of observer posts manned by army and civilian personnel often sited near local police stations were set up along the coast and deep inland to track the course of hostile aircraft across the country.

These observers would plot this information on grid-reference maps, indicating the position, speed, direction of flight and approximate height of enemy aircraft. The results of this intelligence was then transmitted by land-line to the Admiralty, where within three minutes of receiving the first sightings the airships' positions were plotted on the master map. This map was then used to follow the course of the raid as it developed across the numbered squares overlaying a map of the country.

By reference to this map the controllers at the Admiralty could then inform the defensive guns on the path of the raiders' track as they entered their zone of fire, while similar instructions were also sent by telephone to the night-flying airfields around London. The aircraft at these airfields had already been put on standby and warmed up in anticipation of a raid following the first wireless contacts.

The Room 40 analysts could provide course and speed information, together with approximate height to the station commanders, from where they could vector their aircraft to the last recorded position of the raiders.

As well as the use of wireless, acoustic methods of detection were also employed during the war. Parabolic concrete reflectors were set up along the coast that could detect the sound of approaching aircraft up to a range of 15 miles under ideal conditions. This method of detection was masterminded by Lieutenant Colonel A. Rawlinson, who shared with Sir Percy Scott the responsibility for the gun defences of the capital.

Sound detectors suffered interference from extraneous noise and background clutter, particularly from breaking waves, which rendered them of limited use. Nevertheless they continued to be employed throughout the interwar period in various forms until the advent of radar rendered them obsolete.

The earliest versions of the sound location apparatus consisted of a pair of gramophone horns set at the extremities of a long horizontal pole, in order to provide a long-ranging base similar to that employed in an optical rangefinder. This array was then mounted on a tall vertical pole in an attempt to remove it from residual ground clutter. Tubes leading from the gramophone horns were then connected to a standard doctor's stethoscope worn by the operator, who would read off the bearing on an attached compass card by directing the horns toward the direction from which the sound was emanating, together with an estimate of the approximate range.

One of the many problems associated with this primitive 'Audio Radar' was the wide variation of range and bearing obtained by different operators tracking the same target. Rawlinson hit on the idea of employing only blind operatives for this purpose, as he reasoned that their more highly developed sensitivity of hearing would ensure a more accurate plot. And so it was to prove, as immediately a greater degree of accuracy was achieved by the blind operatives than their sighted counterparts, a unique and touching experiment where those without sight were able to contribute to the safety of the nation through their special gifts.

On the afternoon of 17 February, the *L3* and *L4* were dispatched to locate British warships reported off the coast of Denmark. After a fruitless search that carried them almost to the Norwegian coast, the two ships found themselves battling against fierce headwinds on the return trip that had not been predicted by the German weather service. *L3* and *L4*, unable to make progress due to partially disabled engines and lack of fuel, were eventually both lost by stranding on the Jutland coast, where their crews were interned by the Danish authorities.

In March the new naval airship *L8*, whilst on patrol in the southern North Sea, was driven by contrary winds over the Belgian front line where it was promptly shot down by the Flemish gunners. These fortunate accidents of weather offered a temporary respite to the home defences, but the next raid aided by better weather was a more determined attempt.

Following an abortive attempt in April 1915, Hauptmann Erich Linnarz, aboard the army airship *LZ-38* and accompanied by *LZ-39*, left the base at Gontrode on 10 May bound for England. The two airships reached as far as Southend, where incendiary bombs were dropped, before turning back. Several RNAS aeroplanes from Eastchurch attempted to intercept the raiders but failed to come within range. During the raid a searchlight crew did, however, manage to illuminate one of the Zeppelins in its beam; this being the first occasion a raider had been caught by a searchlight.

In the early evening of 31 May 1915 Linnarz, again aboard the *LZ-38*, left the Belgian coast close to Steenbrugge and oriented his ship with two searchlights shining vertically into the night sky to establish a bearing that would bring him to a point at the mouth of the Thames Estuary. By 10 p.m. the *LZ-38* was standing off North Foreland preparing to launch its attack on the largest city in the world.

The *LZ-38* was the latest product of the Zeppelin factory. Her yellow, doped envelope containing 1 million cu ft of hydrogen was 536ft long and driven through the air at 53mph by three engines totalling 960hp. Her total lift was 36 tons, allowing her to carry a war load of 2,500lb for a duration of over twenty-four hours.

The *LZ-38* was by the standards of the day a potent weapon of war, able to out-climb with ease any pursuing aircraft by dropping water ballast, which would enable it to ascend at a rate of climb that left the defensive planes standing. Even an object as large as an airship was difficult to locate in the cold winter sky above a darkened London, so the chances of interception were very slim indeed.

Passing over Shoeburyness, Linnarz continued his course north over the darkened Essex countryside, where an apprehensive population listened fearfully to the unfamiliar drone of the airships engines and the beating of her airscrews. Changing his course to the west, *LZ-38* then passed over Stoke Newington, Shoreditch and Whitechapel following the Great Eastern Railway line to the capital's centre and dropping his cargo of bombs at intervals as he went, before turning eastwards towards Waltham Abbey and finally out over the coast.

In her wake the *LZ-38* had left considerable damage to property, estimated at £18,600, together with seven dead and thirty-five injured. Again, all efforts to come to grips with the raider had come to nothing and even more surprisingly, although it was a clear night, it transpired that no one had even seen the airship overhead.

In his report Linnarz claimed that he had bombed only docks and military installations. Although the material damage was slight, the sense of shock and outrage was profound to a nation that had for so long lived behind the protecting bulwarks of the Royal Navy, where such a visitation was seen as a portent of still more terrible things to come. It is difficult at this distance in time to fully appreciate the powerful effect Linnarz's daring raid of the heart of the empire had on all sections of society, and the deep impression it made on the uncommitted neutral nations.

At home the War Office was under tremendous pressure from both the Cabinet and the people to strike a blow against these seemingly unassailable monstrous and frightening weapons of modern warfare. Retribution was, however, swift in coming to the army Zeppelins.

On the night of 6 June *LZ-37*, together with *LZ-38* and *LZ-39*, again set out for England. All was going well when several hours into their flight they received wireless instructions to cancel the raid due to unfavourable weather conditions over southern England. They were directed instead to bomb a railway communications centre to the south of Calais.

This wireless message was picked up by the listening station at Hunstanton and passed on to Room 40 at the Admiralty. Here it was quickly decoded and the information was immediately communicated to the commander of the RNAS air station at Dunkirk, Wing Commander A. Longmore, instructing him that his machines should attempt to intercept the returning Zeppelins at dawn.

At 12.30 a.m. a further phone call from the Admiralty confirmed that three Zeppelins were nearing the Belgian frontier. Accordingly, at 12.40 a.m. Longmore ordered two of his pilots, Flight Lieutenant R.J.A. Warneford and Flight Lieutenant J. Rose, to take off in their Morane parasol monoplane scouts, patrol the area around Ghent and seek out the returning airships.

The Morane scout was armed with a single Vickers machine gun firing directly through the airscrew arc, which was fitted with steel deflector plates and carried six 20lb anti-Zeppelin bombs. Powered by an 80hp LeRhone nine-cylinder rotary engine the little aircraft had a top speed of a little over 90mph. The rate of climb was only 350ft per minute (only a third of that at which a Zeppelin could rise), it was, therefore, essential that the pilots got above the airship in order to deliver an attack.

Longmore also ordered Flight Lieutenant J. Wilson and Flight Sub-Lieutenant J. Mills into the air in their heavier Henri Farman bombers, with orders to attack the Zeppelin sheds at Evere near Brussels. In the pre-dawn darkness Linnarz had landed the *LZ-38* and it had just been walked into the shed when Wilson and Mills appeared over the base. Despite fierce anti-aircraft fire and searchlights they laid their bombs with accuracy on the shed.

Their attack completely destroyed the Zeppelin *LZ-38* inside, and Linnarz and his crew were fortunate to escape from the inferno with their lives. Meanwhile, Rose's Morane had to make a forced landing due to engine trouble, leaving Warneford to continue the search alone.

At about 1.50 a.m. he at last sighted a Zeppelin some miles to the west over Ostend. Putting his machine into a shallow dive he gave chase, closing on her. He was turned away in the face of heavy Maxim fire, but not before he had read her number: *LZ-37*. Climbing hard to 11,000ft, Warneford stalked the airship as it made off north-east heading for its base, when suddenly the airship nosed downwards. Warneford seized his chance and attacked from above and astern, flying along the length of the hull he released six bombs on to her back.

The Zeppelin immediately exploded with great violence, throwing Warneford's machine skyward on to its back. Warneford struggled to regain control of his little plane as the doomed blazing airship crashed to earth, falling on the convent of St Elizabeth in a suburb of Ghent. In the ensuing conflagration two nuns and two children were burnt to death and several others injured; of the Zeppelin crew, all perished apart from the coxswain who had what can only be described as a miraculous escape.

After falling 5,000ft in the control car of blazing airship, the coxswain was thrown out when it struck the convent roof. He then crashed through the tiles and ceiling to end up on a nun's bed, from where he was later to make a full recovery in hospital.

Warneford's aircraft was flung on its back by the force of the explosion and he was forced to land behind enemy lines when his engine cut out, but was able to effect emergency repairs. After fixing a faulty petrol pipe and swinging his own propeller, Warneford leapt on board and made his escape just before a German cavalry patrol could capture him.

For his bravery Warneford was awarded the Victoria Cross and the French Legion of Honour, but was killed two weeks later together with his passenger when the Farman he was flying broke up in the air.

With two Zeppelins destroyed by the RNAS pilots from Dunkirk on the same morning of 7 June, the general staff of the German army were forced to realise that their recently acquired Belgian bases, for which they had such high hopes, were now too vulnerable to be used.

Reluctantly they withdrew the remaining airships across the border to safer havens on German soil, the Belgian bases now only being used in emergency. The army airships now had to fly much further in order to reach London.

10

Air Defence of Great Britain: Nemesis

T he destruction of two Zeppelins in one day by the RNAS was a major success for the Allies, although the effect on the British public was somewhat muted as the action had taken place unseen and far from Britain. The announcement of the airships' destruction was treated with a degree of scepticism by a population who had become accustomed to believe that the German airships were almost invulnerable to attack, such was the awe in which these weapons were held by the general public.

The effect on the German army, however, was to immediately remove their airships from the vulnerable Belgian bases to the security of the Fatherland, using the westerly sheds in future only in emergency and lessening, by this action, the threat posed to the British Isles. The German army was soon to recognise the unpalatable fact that their airships were now very vulnerable to attack by both aircraft and increasingly better organised ground defences.

From mid-1915 the German army began to operate with greater circumspection, moving most of their remaining airships to the Eastern Front, the Balkans or the Baltic, where they were able to undertake reconnaissance and bombing missions with a reasonable degree of safety due to the poorly organised opposition they faced in those theatres. In contrast to the army, Strasser continued to undertake reconnaissance in the North Sea and to lead squadron raids against the east coast during the early part of 1915. Here, apart from increasing the anxiety of the public, such raids were ineffectual, both from the point of view of the raiders and the defences alike.

The marginal performance of the existing M type Zeppelin of 794,000cu ft capacity and 630hp limited their ability to carry out long-distance missions and dictated the need for a more powerful class of airship. Throughout the summer, when the long hours of daylight made raids too hazardous, Strasser was to take delivery of the greatly improved P type of the L10 class, with a capacity increased to 1,126,000cu ft and an engine power of 840hp, which boosted the top speed to a useful 60mph. At last Strasser had airships that could effectively carry out offensive operations against the British Isles, with every chance of success.

By September, with the return of longer nights and the start of the raiding season, Strasser now had seven of these new airships at his disposal, whilst the more cautious army had five more added to their own strength. The L10 class were, in hindsight, the most effective type of airship employed by the German navy during the course of the war. Their robust construction and ease of handling both in the air and on the ground, combined with an adequate turn of speed, a reasonable reserve of power, a heavy military load and a large radius of action, made them ideal for the purpose in hand.

The success of this model is demonstrated by the fact that in the course of the war the L10–L20 class were responsible for almost 80 per cent of all the attacks made by airships on the British Isles. In great part, however, this seeming success during 1915 and early 1916 was as a result of the ineffectual defences then in place.

25. Kapitänleutnant Heinrich Mathy, commander of the *L31*, 1916.

Following a series of successful raids on the capital and other cities that caused great alarm, more strenuous efforts were put in place to defend Britain's night skies against the aerial menace. In early 1915, for instance, Winston Churchill had established a line stretching from the Humber to the Thames Estuary that would be patrolled by RNAS aircraft, which would hopefully detect and attack any hostile aircraft.

The dangers of night flying at this early stage of aviation were only too obvious; the aircraft were fitted with the most rudimentary of instruments, lacking even an artificial horizon, hardly sufficient for flying in daylight let alone in dark or foggy weather. Again various ingenious schemes were suggested, which included flare paths for take-off (which proved successful) and lowering a wire below the aircraft, which set off a buzzer in the cockpit when it made contact with the ground (which was less successful). Despite the dangers of flying at night, the pilots of the RNAS, and later the RFC, unflinchingly continued to go aloft whenever the warning came through.

Linnarz's raid on London in the army Zeppelin *LZ-38* in May had galvanised defences, and a further ring of landing grounds were hurriedly established around the capital. Apart from high-angle artillery and defensive balloon aprons, offensive measures were centred on the yet unproven fighting aeroplane to carry the war against the Zeppelin raiders.

With the RFC now more firmly established in France they could afford to provide a small number of aircraft at Hornchurch and Hainault Farm Ilford to augment the RNAS airfields at Chingford, Hendon, Isle of Grain and Maidstone. The machines, mostly superannuated BE2a or RE5s, were equipped with a variety of weapons which included Hales bombs, explosive grapnels towed behind aircraft and assorted rifle armament – all of doubtful offensive capacity. At one stage it was even suggested that if all else should fail, and the opportunity presented itself, pilots should ram the enemy airships.

At the same time the gun defences were reorganised under the energetic leadership of both Lieutenant Colonel Rawlinson and the equally effective Sir Percy Scott, a naval gunnery expert who had done much to improve long-range gunnery in the fleet. Between them they got rid of the ineffectual 1lb Pom Poms and installed 100 new guns, plus fifty additional searchlights that could reach up to 15,000ft.

Despite these measures, throughout the spring and early summer the German airships were seen off the east coast and made a series of probing attacks on the mainland, roaming over the eastern counties and approaching the environs of London.

GERMAN NAVAL AIRSHIP L13 (LZ-45) ZEPPELIN "P" TYPE - 1915

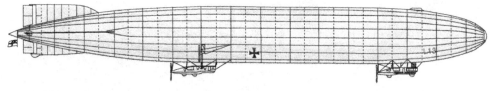

1,126,400 Cubic feet capacity
536' x 61' x 79'
36.5 tons gross lift
15.2 tons useful lift
4 x 210 h.p. Mabach C-X six cylinder in line
Speed 60 m.p.h.
Range at cruising speed 1800 miles
Crew 18
Ceiling 10,000 feet

100 FEET

The most effective of these early raids was the audacious attack made on London on the night of 8/9 September 1915 by a squadron of naval airships, led by the resolute Kapitän Mathy. In this raid only Mathy's airship, the *L31*, reached London. Approaching from the north-west at high speed he was able to sail unopposed over the empire's capital, dropping high-explosive bombs at will.

This raid created a great sense of outrage both in the public's mind and that of government. After a meeting with the prime minister, Lord Kitchener, the Secretary of State for War, demanded that 'something must be done to halt these air raids', and that 'effective action must be taken immediately to curb these attacks'.

Under instructions from Kitchener, Director General of the Army Sir David Henderson summoned Lieutenant-Colonel A. Rawlinson, an army gunnery expert, giving him carte blanch to procure by any means a large number of mobile, high-angle rapid-firing guns of the type that were then being employed in Paris. Rawlinson, acting with all due speed and thanks to his contacts in the French Ministry of War, had an example of the latest '75mm Automobile Cannon' on Horseguards for inspection by Prime Minister Balfour and Kitchener by 19 September, with the promise of other guns to follow.

Less than a month later, on 13 October just after 6.55 p.m., the Admiralty were informed from their listening stations that a fleet of six Zeppelins had crossed the coast in the vicinity of Yarmouth, and were headed for London. Rawlinson set the new gun up near to the Bank of England and opened fire on the *L15* when she appeared over Hyde Park, already held in the beams of several searchlights.

26. Rotating double airship shed at Nordholz near Cuxhaven, circular track is still in evidence.

McMechan Anti- Zeppelin Airship - 1915
Marshall-Fox-MacMechan A/S Syndicate - Barking, Essex

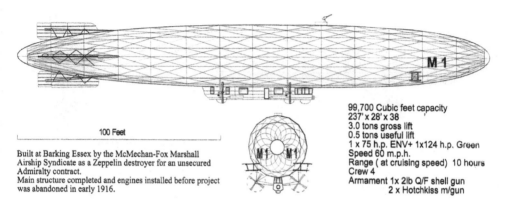

99,700 Cubic feet capacity
237' x 28' x 38
3.0 tons gross lift
0.5 tons useful lift
1 x 75 h.p. ENV+ 1x124 h.p. Green
Speed 60 m.p.h.
Range (at cruising speed) 10 hours
Crew 4
Armament 1x 2lb Q/F shell gun
2 x Hotchkiss m/gun

100 Feet

Built at Barking Essex by the McMechan-Fox Marshall Airship Syndicate as a Zeppelin destroyer for an unsecured Admiralty contract.
Main structure completed and engines installed before project was abandoned in early 1916.

The *L15*'s commander, Kapitänleutnant Joachim Breithaupt, was taken by surprise by the shells exploding close abroad and immediately dropped most of his water ballast, opening up the engines to climb away from the danger. However, unknown to Breithaupt a further danger faced him in the form of Second Lieutenant John Slessor (later to become Marshal of the Royal Air Force), who had taken off from the mist-shrouded Sutton's farm airfield near Hornchurch in his BE2c.

Climbing hard in an attempt to intercept the raider, Slessor sighted the Zeppelin gleaming in the searchlights' beams and, surrounded by bursting shells over Dartford, the young pilot opened the throttle and turned in pursuit. As Slessor struggled to gain altitude and was closing on his quarry he suddenly saw the *L15*'s water ballast cascade out of her undersides and immediately the airship's nose tipped up and she rose rapidly into a cloud bank to be lost from his sight, much to his chagrin.

A shaken Breithaupt's lucky escape from both aeroplane and anti-aircraft fire must have signalled to Strasser the improved effectiveness of the defences and the increasing vulnerability of his airships. Despite this, Strasser continued to send raiding squadrons of Zeppelins against England with increasing frequency during the spring and early summer months of 1916. With these raids that ranged widely over the British Isles came a string of casualties to the raiders. Casualties that were not always attributable to the defences but often to weather, engine failure or from the sheer exhaustion of crew and ship in flights that could last over twenty-four hours.

The first such loss occurred in August 1915, when of one of the newer airships, the *L12*, after being damaged by gunfire over Dover, came down in the sea off Zeebrugge. Here, despite attacks by RNAS aeroplanes from Dunkirk, it was successfully towed into harbour by a torpedo boat. Whilst the airship was being salvaged, however, it unfortunately exploded when being lifted from the water and became a total loss.

"AP-1" Anti-Zeppelin "Airship-plane" -1915
R.N.A.S. Constructional Facility, Kingsnorth

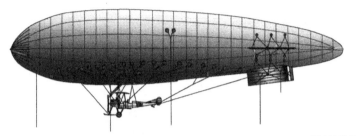

50 Feet

The "Airship-plane was designed to hover on patrol at 10,000 awaiting the arrival of raiding Zeppelins when it would be released from the envelope to make it's attack.
Following a successful unmanned trial in August 1915, during a manned trial in April 1916 the aircraft became entangled in the release gear and both occupants were killed, ending any further trials

70,000 Cubic feet capacity
143' x 30' x 46'
Gross lift 2.1 Tons
Useful lift 0.6 Tons
1 x 75 h.p. Renault
Speed (Combination)40 m.p.h - (.A/C) 90 m.p.h
Range 6 hrs at cruising speed
Armament 10 x 20 lb A/Z bombs
1 x Vickers machine gun
Crew 2

A further loss occurred in September, when *L10*, returning from a patrol, exploded in the air over the island of Neuwerk as she was coming into land at Nordholz, with the loss of her entire crew of nineteen.

A particularly cruel end befell the *L19* when, on 1 February 1916 after crossing the coast in the vicinity of Sheringham, Norfolk and bombing Burton-on-Trent, the airship experienced severe engine trouble. After then drifting over the Dutch island of Ameland she was fired upon by the Dutch gunners.

The unfortunate airship was now without power and was blown westward out to sea again, until eventually coming down in the water some 95 miles north of Spurn Point. Here, a British fishing trawler the *King Stephen* came upon the derelict airship, with the crew huddled on the upper hull. The skipper refused to rescue the Germans, however, for fear of them taking over his boat. The *King Stephen* then stood off the wreck and sailed off, leaving the crew to their watery fate.

On 31 March Breithaupt's luck finally ran out when on a raid in company with *L9* and *L13*. Whilst approaching the eastern outskirts of the capital he was attacked again by both aeroplane and anti-aircraft fire over Dartford. One 3in shell exploded inside cell 11, emptying it of hydrogen, whilst adjacent cells were also leaking gas rapidly. After dumping all unnecessary equipment, including tools, bombs and machine guns, and after sending out a last call for assistance, the *L15* set a course for the Belgian coast.

It soon became obvious that they would not make it, and just after midnight the *L15* broke her back and fell into the sea 1 mile from the Kentish Knock lightship in the Thames Estuary. Armed trawlers were on the scene and promptly opened fire on the wreck; the firing quickly ceasing when it was realised that no resistance was being offered. The Zeppelin crew, apart from one man who had drowned, were taken off by destroyer to be landed at Chatham Naval Base as prisoners of war. The destroyer *Vulture* attempted to take the wreck in tow, but the *L15* sank off Westgate next morning, although considerable portions of her structure were later recovered for analysis by Admiralty experts.

During May the *L20*, under Kapitänleutnant Stabbert, was also lost whilst attempting to raid the naval anchorage on the Firth of Forth. After being blown well to the north of his intended landfall by a strong southerly wind, the *L20* left the Scottish coast in the region of Peterhead, but Stabbert found he was unable to gain his home base of Tondern due to shortage of fuel. In desperation, and to save being blown out to sea, the *L20* eventually came down on the Norwegian coast in the waters of the Hafrsfjord near Stavanger.

An unusual fate befell the older *L7* during the raid on the airship shed at Tondern in May 1916 by RNAS aircraft launched from aircraft carrier HMS *Furious*. As she observed the British ships she was shot down by the 6in guns of the accompanying light cruisers *Galetea* and *Pheaton*, who surprisingly managed to rescue seven survivors from the wreckage.

This rate of attrition deterred, to an extent, further raids during the late summer of 1916, as Strasser awaited delivery of the even more advanced L30 class of airship from the Zeppelin factory in the autumn.

In his book *Zeppelins and Super Zeppelins* published in early 1916, the British author R.P. Hearne, a jingoistic patriot, claimed that the only way to defeat the Zeppelin was to 'build a fleet of British Super Zeppelins, and Super Aeroplanes, manned by our own incomparable airmen'. The author outlined a formidable battery of weapons to arm these craft, including shell guns, aerial torpedoes and asphyxiating gases. Hearn visualised the British airships being clearly superior to the German Zeppelin in terms of speed (suggesting that speeds of 100mph were possible), range and fighting ability, concluding that a fleet of only five of these high-speed British Super Zeppelins would ensure immunity to future attacks on this country.

One of the earliest attempts to meet the Zeppelin in her own element was a daring plan conceived by two naval officers for an 'Anti-Zeppelin Airship Plane', comprising of a fully operational BE2c aeroplane slung beneath a Sea Scout envelope. The Airship Plane was designed to hover on patrol at 10,000ft awaiting the arrival of the attacking Zeppelin, whereupon it would be released from the envelope to make its attack. Yet during a manned flight in April 1916, following a successful unmanned trial in August 1915, the aircraft became entangled in the release gear and both occupants were killed, ending further trials of this promising idea.

At Barking in Essex during 1915 an 'Anti-Zeppelin Airship' was being built by the McMechan-Fox Marshal Airship Syndicate as a Zeppelin destroyer for an unsecured Admiralty contract. The airship was built on the lines of the Schütte-Lanz *SL1*, having a wooden geodetic structure of excellent streamlined form. Power was provided by one 75hp ENV and one 100hp Green, which it was optimistically claimed would give a speed of 60mph and an endurance of ten hours, while carrying a crew of four and an armament that included a 2lb quick-firing shell gun.

The airship was inspected by the Admiralty design department who came to the conclusion that although well engineered, due to its small size of only 99,000cu ft capacity and a limited gross lift of a mere 3 tons, it was an impractical design. The main structure was largely completed and the engines were installed before the project was finally abandoned in early 1916.

A further example of the Zeppelin destroyer principle was the extraordinary Pemberton-Billing PB31c *Nighthawk*. Built in 1916 this was a huge, twin-engine aeroplane equipped with four sets of wings, a fully enclosed cockpit and even a sleeping berth for an off-duty crew member. The armament consisted of two Lewis machine guns and a 1.5lb Davies recoil-less quick-firing cannon as the main anti-Zeppelin weapon, aided by a large searchlight mounted in the nose. Yet due to the low power of the engines employed in this revolutionary super quadruplane the concept was abandoned in 1917, leaving it to more conventional aircraft to finally defeat the Zeppelin menace.

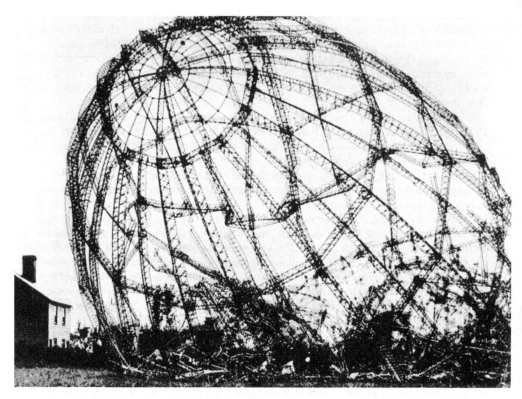

27. Wreckage of the *L33* at Little Wigborough on the Essex coast, 24 September 1916.

At the request of the German Admiralty, in early 1915 the Zeppelin Company and Schütte-Lanz pooled their accumulated knowledge to produce optimum designs for a 2 million cu ft airship that would fit the largest North Seas sheds, and that would adequately fulfil all fleet requirements.

The Schütte-Lanz design, for various reasons, was not completed until almost a year after the Zeppelin Company design, and so played only a minor part in this stage of airship development.

The first of Strasser's improved Super Zeppelins, the 2 million cu ft Zeppelin R type L30 class, was delivered to Nordholz in May 1916, with the ageing Count Zeppelin himself onboard for the delivery flight. The R type marked a great leap forward in airship design, with a useful lift of 25 tons, a ceiling of 17,000ft and a much extended range. Strasser was now convinced that, at last, he had an airship that could operate with impunity over the enemy defences; one that was immune from anti-aircraft fire and could out-climb any attacking aircraft.

The L30 class were impressive craft, being 649ft in length, with a diameter of 78ft and powered by six 240hp HSLu motors. They were capable of driving through the air at 62mph and on raids could carry almost 5 tons of bombs. The hull structure of the L30 class was of a better streamlined form with a shorter parallel hull section than the preceding class, employing a new girder design larger in cross-section, whilst the mainframes were also of deeper section and incorporating king post bracing, increasing the rigidity of the structure. In addition, the mainframe spacing was increased from 8m to 10m, with a lighter intermediate frame set between.

Several of the Schütte-Lanz advanced features were employed in the L30 class, including the first use of wing-mounted engine cars; gas shafts; the introduction of an impermeable outer cover; and the axial cable running from bow to stern through the gas cells connected to the cells radial wiring, serving to reduce the loading on the structure.

Strasser began working up his new airship with a series of endurance and altitude trials that gave him great encouragement as to its qualities, causing him to write to Admiral Scheer on 10 August from Nordholz that:

> The performance of the big airships has reinforced my conviction that England can be overcome by means of airships, inasmuch as the country will be deprived of the means of existence through increasingly extensive destruction of cities, factory complexes, dockyards, harbour works with war and merchant ships lying therein, railroads etc.

The Imperial German navy's establishment for its North Sea scouting requirement was nominally eighteen front-line airships, a number which was considered sufficient for this purpose.

Strasser's proposal to extend the strategic bombing capability of the service would require, initially, four additional airships, followed by a further expansion of raiding squadrons to prosecute the intensive bombing programme he envisaged. Such an expansion was viewed as being outside the Imperial Navy's main remit, and would require a great investment in new ships, sheds and crew training that militated against the service's main purpose: scouting for the battle fleet.

In England whilst these developments were taking place, evermore serious thought was being given to the problem of destroying Zeppelins in the air. Despite the British having an extensive spy network operating in Germany, and indirectly in the Zeppelin works itself, much of the intelligence gathered was misleading or even downright inaccurate.

One of the myths associated with the Zeppelins was the belief that the explosive hydrogen was protected by a double-walled cell into which nitrogen or the engine exhaust gasses were introduced to form a barrier against the lethal oxy-hydrogen mixture forming. So strong was this view (no doubt encouraged by the difficulties attacking aircraft had up to then encountered in attempting to set Zeppelins on fire by machine-gun fire) that early promising efforts to produce an effective incendiary bullet were temporarily discontinued.

However, wiser counsels prevailed and by combining the work of the Pomeroy explosive bullet of 1915 with the Buckingham phosphorous and the explosive bullet of Commander F. Brock, a mixed armament was produced for the .303 Vickers machine gun. Early tests of the mixed ammunition proved highly effective against captive balloons and all three went rapidly into production, so that by the early summer of 1916 over 1 million rounds were available to the defence forces.

Following the Battle of Jutland on the last day of April 1916, where the Imperial German Navy had received such a fearful battering from the Grand Fleet, Admiral Scheer resolved to restore morale by leading a portion of his repaired fleet on an operation against the English coast in August 1916. In his operational orders Scheer noted, 'For the undertaking, extensive airship reconnaissance is a prerequisite'; with the airships performing the important task of covering the fleet and warning of the approach of enemy warships.

On the evening of 18 August squadrons of the High Seas Fleet left harbour and set course for Sunderland, with the intention of bombarding the harbour works and military installations, together with causing disruption and destruction of maritime trade. However, Room 40 at the Admiralty had once again intercepted and decoded the High Seas Fleet wireless traffic, which allowed the alerted Admiral Jellicoe to lead the Grand Fleet to sea on the afternoon of the 18th; seven hours before Scheer had left Wilhelmshaven at 9 p.m. the same day.

The Battle Cruiser Fleet under Beatty departed Rosyth in the late evening, followed by the Harwich force of light cruisers and destroyers commanded by Admiral Sturdee at 11.30 p.m. the same evening.

Strasser ordered out twelve Zeppelins to support the raid, including three of his latest L30 class airships. Unfortunately for the German navy these airships did not reach their designated patrol lines until late on the morning of the 19th, by which time the Grand Fleet and the Battle Cruiser Fleet had met 100 miles north-north-east of the Firth of Forth, to form a combined fleet of thirty-three Dreadnoughts, which were steaming southward to engage the German fleet.

Meanwhile, the Harwich force was pushing northward to cut off the German retreat to the south. The scene was set for a perfect trap.

The first contact was made by *L13* when she encountered the Harwich force, which at that time were temporarily steaming in a southerly direction, and reported their presence of light cruisers to the commander-in-chief aboard the fleet flagship SMS *Friedrich der Grosse*. Scheer was of the opinion that these ships were bound for the Belgian coast and would not interfere with the operation, continuing instead on course for Sunderland.

The *U52*, shadowing battlecruisers in the north, attacked and sank the light cruiser *Nottingham*, causing the Grand Fleet to turn north. This manoeuvre was observed by the *U53*, who reported 'heavy units off the Farne Islands steering north', again this warning was ignored by the German fleet commander.

At 12.30 p.m. the *L31*, with Kapitänleutnant Mathy in command, flying in squally cloudy weather brought his ship down to 650ft and identified the British fleet. But on coming under fire climbed to 6,000ft and shadowed them for some time, sending a series of wireless reports to Admiral Scheer. At 3.35 a.m., following further reports of a conflicting nature from airships and submarines, Scheer decide to abandon the raid and head for home.

On her return trip the *L31*, cruising at 2,000ft, was surprised by a trawler that opened an accurate fire on the airship, which managed to escape with minor damage. Whilst on the British side the light cruiser *Falmouth* was torpedoed by the *U66*.

Despite having the element of surprise the ships of the Grand Fleet failed to bring the Germans under their guns, whilst the carefully planned scouting and reconnaissance mission of Strasser's Zeppelins failed to provide the German commander-in-chief with the detailed and accurate information he needed to safely carry out the intended raid on Sunderland.

This was the only time naval airships attempted to support the High Seas Fleet in a major operation, and although they largely failed in this task, British and American analysts incorrectly interpreted the undertaking as a success for the reconnaissance capabilities of the airship.

After a series of raids in August against the British mainland that were attended with mixed fortune, but no losses to the raiders, in early September Strasser resolved to make a 'maximum effort' against London in pursuance of his policy of delivering a hammer blow to the capital.

Twelve Zeppelins were available to him in the North Sea bases, to which he added a further two wooden-framed Schütte-Lanz airships by temporarily recalling the *SL8* and the *SL9* from their Baltic bases. The first raid of the augmented fleet took place on 24 August. Again the

British were alerted to the raid by wireless traffic and warships were standing by to fire on the incoming airships, once more without success.

Flying across the Ruhr and Belgium, *L32* managed to attack Dover before turning for home. Mathy in *L31*, following a similar route, pressed on up the Thames to bomb the south-eastern portion of the city, eluding the defences to return via Belgium. On return to Ahlhorn she damaged her rear engine car in a heavy landing, causing her to be out of service for almost a month and so to miss the next raid that marked the turning point for airships in the war.

On 2 September, encouraged by a favourable weather forecast, Strasser ordered out twelve Zeppelins against 'England south main target London', the raid to be undertaken in conjunction with four army Zeppelins, *LZ-90, LZ-97, LZ-98* and the new *SL11*; sixteen airships in all, the largest airship raid of the war. In the early afternoon the naval and army airships left their North Sea bases heading westward along the Friesian Islands or over Belgium on course for London. Despite the optimistic forecast the weather did not favour the attackers, with a strong south-westerly wind and icing conditions at attack altitude.

By the early evening both the army airship *LZ-97* and the navy *L17* had already turned for home after experiencing adverse conditions and severe icing that militated against a successful attack. Oberleutnant zur See Ernst Lehmann, aboard the army airship *LZ-98*, pressed on, however; crossing Belgium at low altitude before climbing to his attack altitude of 8,600ft as he approached Dungeness around midnight. From here he headed north, crossing the Thames at Gravesend, where he was heavily fired on by the guns of Tilbury fort, and Dartford, causing him to unload his main cargo of bombs in the belief that he was over the London docks.

Relieved of her bomb load the *LZ-98* rose rapidly to 14,000ft, and at 60mph made off to the north-east. As she turned away she was briefly pursued by Second Lieutenant William Leefe-Robinson who had tried to close on the Zeppelin but, by sacrificing speed in his attempt to gain altitude, lost the airship in cloud before he could overhaul her.

Leefe-Robinson, the disappointed pilot, had been born in India in 1895 and was commissioned into the Worcester Regiment at the end of 1914, where he was wounded at the Battle of Loos. Whilst convalescing in England he transferred to the RFC where he gained his wings as a pilot and was subsequently posted to the Home Defence Squadrons protecting the capital. On the night of 3 September 1916, Second Lieutenant Leefe-Robinson of 39 Squadron was on duty at Sutton's farm near Hornchurch on the north-east outskirts of London.

At 11 p.m., after receiving orders from the War Office to 'take air raid action', Robinson's BE2c was trundled out of its canvas hangar on to the mist-shrouded field, where after several attempts the engine spluttered into life. Soon the frail aircraft was bouncing across the airfield to climb with painful slowness into the starry night, taking almost an hour to reach 10,000ft.

The BE2c he flew was possibly the most suitable type of aircraft for the hazardous business of night flying then available. It was an inherently stable machine, powered by the reliable Royal Aircraft Factory 90hp engine and armed with a single Vickers .303 machine gun mounted on the upper plane, which was loaded with the new ZPT mixed tracer and explosive ammunition.

At 1.10 a.m. Leefe-Robinson, now flying at 11,000ft, sighted an airship held in the beams of a searchlight south-east of Woolwich with shells bursting around it, but in an effort to gain extra height Robinson lost his intended victim in cloud. Resuming his patrol he climbed slowly to 12,800ft, where at 1.50 a.m. he turned toward a fiery glow to the north, sighting as he did an airship illuminated by a single searchlight at his own level.

The airship was Schramm's *SL11*, which was just turning away northward after dropping bombs on the Hertfordshire villages of North Mimms, Littleheath, Northaw and Hadley Wood. Determined not to repeat his earlier error, Robinson dove on his prey at full speed, raking the hull from bow to stern with a full drum of incendiary ammunition, but without effect. Robinson hauled his machine round, turning after the fleeing airship where he distributed a second drum of ZPT along the port side, again without any sign of the airship taking fire.

Again Robinson closed in on the airship, moving in from below to within 500ft of the hull he concentrated his fire on one spot at the stern. Suddenly the area aimed at began to glow a dull red, and in an instant the whole of the stern burst into flames. Pitching down by the nose with the whole structure engulfed in flames, the *SL11* took almost two minutes to fall to earth; the brilliance of the flames illuminating the streets and buildings below. It was a spectacle that was witnessed by almost the whole population of London.

Leefe-Robinson fired off red and blue signal flares in celebration and made his way back to Sutton's farm to a hero's reception.

The effect on the German air fleet was immediate and unanimous. Peterson's *L32* with Strasser on board was at Tring heading towards the docks from the north, but on seeing the fireball close to port Peterson hauled off sharply to the east at full speed voiding his bomb load as he went, without receiving any censure from the silent and shocked Strasser. In his war diary Schutze, in *L11* and 50 miles to the north-east, later wrote: '2.15 a.m. an enormous flame over London, slowly sinking below cloud horizon, gradually diminishing. A burning airship.'

Lehmann aboard *LZ-98*, already heading north-east, was to write:

Looking back the way we had come, I saw a huge ball of fire high up at our own level, it must have been forty miles distant and over the edge of the City.

The flaming mass hung in the sky for more than a minute, and we could see parts breaking loose and falling faster than the main body.

Poor fellows they had no chance at all when the ship caught fire.

Similarly, Gayer in *L16* coming in over St Albans, Frankenburg aboard *L21* at Hitchen and Boeker in *L14* over Thaxted were horrified by the fiery vision they had witnessed in the night sky. They immediately reversed courses and ran for home.

Kapitänleutnant Guido Wolff aboard *SL8*, sister ship to the doomed *SL11* flying almost 50 miles to the north-east, immediately recognised the significance of the distant fireball, portending an unseen and terrible death; he also turned away. Further to the east, von Buttlar in *L30* over Lowestoft, together with the remaining naval airships *L24*, *L11*, *L23* and the army airship *LZ-90* all standing off the east coast, were similarly affected by seeing the *SL11*'s destruction, and at full speed made for home.

Only far to the north did Kapitänleutnant Martin Dietrich commanding the *L22*, and Kapitänleutnant Prolss in *L13*, make determined attacks on the docks at Hull and Retford respectively before making out to sea from the Humber.

Of the remaining airships all reached their bases without further loss, but the terrifying sight of *SL11* being blasted out of the night sky had a devastating effect on the morale of the remaining crews. The largest raid on London had ended in disaster.

Throughout London almost the whole population had witnessed the destruction of the feared raider which had illuminated the entire city with a fierce light that could be seen well over 100 miles away. Cheering, almost hysterical, crowds poured out into the street singing, dancing and shouting with sheer relief and celebration, and shrieking factory hooters and ship sirens on the river rent the night air; nothing like this had been seen in London since Mafeking night fifteen years before.

My own mother, who was 10 at the time and living in Poplar, recounted her experience of the event:

I ran outside with my sister, and immediately we could see the blazing airship above the houses at the end of the street slowly sinking down the sky like a Chinese lantern.

Everyone was shouting and cheering, I just stared open mouthed as I watched the doomed airship sinking down the sky. I was asked by one of the neighbours, 'Why are you crying?' To which I replied, 'All those poor men, they are all being burnt,' and was answered with, 'They're Germans you silly cow!'

The *SL11* had fallen into a field in the Hertfordshire village of Cuffley behind the church, the remains burning for several hours fed by the remaining fuel.

Before dawn people began arriving at the crash sight and by first light thousands had descended on the quiet village, blocking the roads and hindering the RFC working parties that were sent to secure and recover the wreckage.

At first the authorities were surprised by the large amount of wood employed in the structure, and initially issued a statement that this was due to a shortage of aluminium in Germany because of our naval blockade. This statement was retracted later that day after experts had advised Lord French, commanding home forces, that in fact the doomed raider was not a Zeppelin but a wooden-framed Schütte-Lanz airship.

Leefe-Robinson visited the crash site twice the next day and by his second visit had been identified as the victor of the air combat. He was so beset by cheering well-wishers that he had to take refuge at the nearby Plough Inn. On 5 September the *London Gazette* announced that His Majesty the King had bestowed the Victoria Cross on Second Lieutenant William Leefe-Robinson for 'conspicuous bravery ... having attacked an enemy airship under circumstances of great difficulty and danger'.

Following the destruction of the *SL11*, Strasser ordered that in future, particularly over London, the Zeppelins would now need to fly at between 15,000 and 16,000ft to gain any sort of immunity from attack by aircraft. The navy's earlier types of Zeppelin would have difficulty in reaching these altitudes, and would have to be detailed to raid the less well-protected targets in the Midlands. As if to add to Strasser's woes, on 16 September two School airships, the *L6* and *L9*, were lost in a fire in their shed at Fuhlsbüttel whilst being inflated.

The next large-scale attack was set for 23 September, with the *L30*, *L31*, *L32* and *L33* detailed to attack London from the south-east, whilst seven older Zeppelins shaped course for industrial targets in the English Midlands. The four L30 class Zeppelins came in over Belgium, with Kapitänleutnant Horst von Buttlar claiming to have bombed Gravesend and the eastern dock area of London. This claim is not substantiated by the official British record, which in fact shows the *L30* cruising along the north Norlfolk coast dropping bombs in a haphazard fashion as she went.

Bocker, commanding *L33* after crossing the Belgian coast, came in over North Foreland, across the mouth of the Thames to make landfall at the River Crouch at 10.15 p.m. having climbed to 13,000ft. At 11.15 p.m. the *L33* was over Wanstead Flats, adjusting her course to the north-east to West Ham and Bow, and dropping bombs at intervals as he went.

Here, Bocker was held by the searchlights and came under intense and sustained artillery fire, receiving splinter damage to the gas cells from shells exploding around her and causing a severe loss of gas. Bocker, realising his ship had been badly damaged, turned away due east heading for the coast. Unfortunately his troubles were far from over, when at 12.55 a.m. the *L33* came under attack from Second Lieutenant de Bath Brandon near Chelmsford, who fired three drums of ZPT into the airship at close range which failed to take effect.

De Brandon's attack had, however, riddled the gas cells still further, with precious hydrogen pouring from her cells, reducing her lifting capacity to dangerous level and severely reducing the chance of the airship getting home again. With his ship sinking lower, Bocker endeavoured to keep her in the air by throwing all unnecessary weight overboard: machine guns, fuel tanks, tools and the wireless station all went over the side.

German Naval Airship L33 (LZ-62) - 1916
Zeppelin "R" Type
Luftschiffbau Zeppelin Gmbh, Friedrichshafen

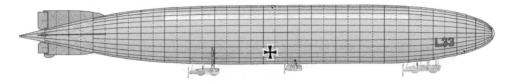

1,949,000 Cubic feet capacity
649'x 79' x 92'
61tons gross lift
24 tons useful lift
6 x 240 h.p. Mabach HSLu six cylinder in line
Speed 62 m.p.h.
Range at cruising speed 5,000 miles
Crew 19
Ceiling 13,000 feet

100 FEET

Bocker put out to sea over Mersea Island but quickly realised that to go on would mean the death of his crew, and reluctantly headed back to the safety of land. At 1.20 a.m. the *L33* slowly fell to earth close by a group of cottages at Little Wigborough, with all twenty-two crew members escaping uninjured from the wreckage. After warning the householders, Bocker set fire to the little remaining hydrogen in the cells, then, assembling his crew, he attempted to march to the coast with the intention of seizing a boat. They were soon apprehended, however, in an Essex lane by Special Constable Edgar Nichols on his bicycle, to whom they surrendered without a struggle.

The *L31* and *L32* had come in together over the coast at Rye, before separating at Tunbridge with Mathy's *L31* heading north dropping his bombs over Mitcham, Stretham and Brixton. Mathy came under fierce anti-aircraft fire as he followed the River Lee north heading to the coast. After leaving Mathy, Peterson in *L32* turned north-east near Tunbridge Wells, at 12.15 a.m. crossing the Thames at Dartford dispersing a regular pattern of bombs as he went.

North of the river where the atmosphere was clearer the *L32*, held in the beams of several searchlights and surrounded by exploding shells, was sighted by Second Lieutenant Fredrick Sowery of 32 Squadron. Sowery, aged 23, had been wounded on the Western Front whilst serving with the Grenadier Guards and had, like Leefe-Robinson, subsequently transferred to the RFC and gained his wings at the same time.

Bringing his BE2c up under the stern of the Zeppelin, Sowery fired two drums of ZPT into the envelope, with no result. Hurriedly attaching a third drum he again swooped on the airship. As the drum ran out the *L32* suddenly rose momentarily upward, it burst into flames then the hull broke into two halves and fell with increasing speed towards the ground. Sowery had to corkscrew his aircraft out of the way to avoid the blazing wreckage.

The *L32* fell at Snail's Hall Farm near Billericay, where amongst the items recovered next day was a badly burnt but usable copy of the latest German naval codebook.

Far to the north-east was Kapitänleutnant Ganzel, who later noted in his war diary: '1.15 a.m. Bright glow of fire towards Thames mouth, airship falling in flames.' At this time the *L32* was over Lincoln, almost 150 miles distant.

Mathy's *L31* with Strasser on board was by now well to the north on the way homeward as they too witnessed the destruction of Peterson's Zeppelin; it must have been a terrible blow for Strasser to realise that his much vaunted new airships were still vulnerable to aeroplane attack. Meanwhile the effect on the crews of the double loss of the most modern ships caused an air of gloom to descend on the airship bases, but the crews continued to bravely man their ships with an air of fatalistic resignation.

Strasser, however, was determined to keep up the pressure and ordered the next raid for 25 September, with the *L31* and *L30* detailed for London, whilst the *L14*, *L16*, *L21*, *L22* and *L23* were to attack the English Midlands. In his orders Strasser stipulated that 'extreme caution should be observed over England, and that in the case of clear weather over London, the attack should be abandoned'.

The *L30* under von Buttlar came in over North Foreland where despite, or because of, the excellent visibility he turned away from the capital and claimed to have bombed Ramsgate and Margate, although no bombs could be traced overland.

Meanwhile, Mathy in *L31*, crossing the coast at Dungeness under a clear starry night, also abandoned any attempt on London, deciding instead to raid Portsmouth. At 11.50 p.m., after

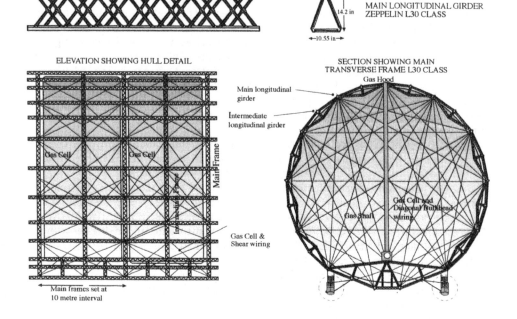

FRAMEWORK AND STRUCTURAL DETAIL
L30 CLASS ZEPPELIN AIRSHIP - 1916

following the English Channel coast, he came inland over the great naval base where he claimed to have dropped his heavy cargo of 8,000lb of bombs on what he believed to be the naval dockyard, although this was not substantiated by British authorities in the dockyard.

Of the older airships, Martin Dietrich's *L22* reached Sheffield where he dropped his bombs, whilst another three of the older airships, after experiencing adverse conditions further north, turned back before the coast. Of these, Kapitänleutnant Ganzel in *L23*, who was suffering from severe stress neurosis, abandoned his attack after making three attempts to make a safe landfall, and veered away from the English coast. With his nerve gone Ganzel turned for home, whereupon he was judged to be unfit and relieved of command, to be transferred to a unit of the surface fleet.

The remaining airships pressed home their attacks on targets of opportunity, but they were largely unaware of the locations they were bombing. Only Dietrich achieved anything of note by bombing John Brown's engine works at Sheffield, causing severe damage to the workshops with twenty-eight killed and fourteen wounded in the attack. All the airships returned without loss from this raid.

At Ahlhorn on 1 October, Strasser ordered out eleven Zeppelins including von Buttlar's *L30* and Mathy's *L31* for a squadron raid on England. Once again only *L30* and *L31* were to attempt an attack on London if conditions were favourable, whilst the other airships were ordered against the Midlands.

Gusty conditions were encountered over the German Bight, whilst over the British Isles they had to contend with a strong north-westerly wind and almost continuous cloud cover that made accurate navigation difficult. Oberleutnant zur See Frankenburg aboard *L21* had experienced severe icing problems as he came inland over Norfolk and Lincolnshire, but set course for Manchester, when at 12.30 p.m. he saw far to the south 'a burning airship held in the beams of a searchlight'.

Kapitänleutnant Koch, commanding *L24*, was bound for the Midlands industrial belt when he realised that he had been driven south by the wind. He boldly resolved to attack London instead. As he approached the capital he witnessed the frightening sight of a falling airship in flames ahead of his ship. Despite this, and with a demonstration of great bravery, Koch held to his course. He flew his imperilled airship across north London, dropping his bombs on Stoke Newington and Hackney before making good his escape.

Mathy had come inland over Lowestoft at 9 p.m. and steered towards Chelmsford, where he followed the shining rails of the Great Northern Railway that led him directly to London. At 10.45 p.m. just north of the capital he was picked up by searchlights, causing him to veer away northwards, only to quickly turn to the north-west in the hope of approaching the city undetected from the west.

At 12.10 p.m. over Hertford, with engines throttled back and drifting silently down wind towards the city, he was suddenly pounced upon by four RFC aircraft. Mathy dropped water ballast and rang the engines up to flank speed, but his fate was already sealed.

Second Lieutenant Wulfstan Tempest had taken off at 11 p.m. to intercept the raider. After an hour he had coaxed his machine to almost 15,000ft and spotted the Zeppelin to the north-east illuminated by searchlights and surrounded by bursting anti-aircraft fire. Tempest dove through the exploding shells, overhauling the airship and raking her with a drum of ZPT ammunition as he passed under her, but without effect.

Banking round, Tempest returned to attack her from the stern and below, distributing a second drum along the hull that caused her to take fire. In his report Tempest stated: 'suddenly she shot up in the air about 200ft, paused and came roaring down straight on to me, I nose-dived my machine into a spin, as she shot past me roaring like a furnace.' The *L31* crashed into a field at Potters Bar, with the front portion impaled on a shattered oak tree and the stern section piled up some distance away. Both portions continued to burn for several hours.

Military and civilians who made their way to the burning wreckage found the horribly mutilated and burnt bodies of the German crew, apart, that is, from Mathy who had jumped from the blazing airship, rather than face being burnt. He lay on his back some distance from his command, his broken body having left a deep impression in the soft wet earth.

So died the most resolute and determined of the Zeppelin commanders, who had led by example and bravery. With his death, much of the offensive spirit that he had engendered went out of the service. Strasser's vision of overcoming England by means of the airship alone now seemed an ever more remote possibility.

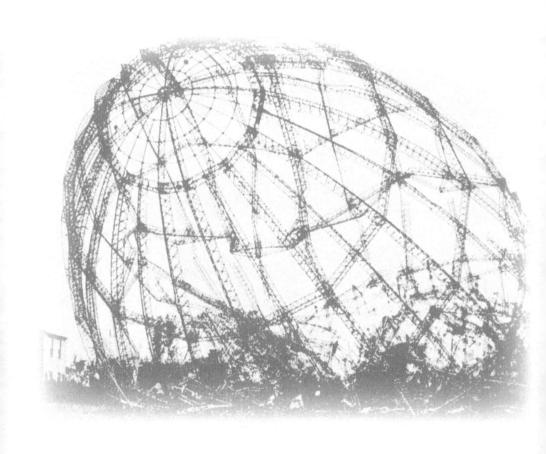

11

Zeppeline Gegen England und Weiter
1915–18

Despite the discouraging losses of four of their most modern airships in action, and two older airships by accident during September and October of 1916, the Imperial German Navy continued to maintain their belief in the offensive capacity of the Zeppelin as a weapon of war.

The leader of naval airships, Korvettenkapitän Peter Strasser, was so convinced that the Zeppelin airship could successfully carry the war against the British Isles that in August

28. Zeppelin *L59*, the *Afrika* ship built to carry supplies to von Lettow-Vorbeck's colonial forces fighting in German East Africa, 1917.

1916 he wrote a memorandum to the commander-in-chief of the High Seas Fleet, Admiral Reinhard Scheer. In the letter he communicated his belief that 'England could be overcome by aerial bombardment using a large fleet of airships', and concluded by submitting that 'airships offer a certain means of victoriously ending the war'. Whether the commander-in-chief may have shared such a dramatic view of the capabilities of the Zeppelin airship is doubtful, he being more interested in their ability to provide a vital early warning capability to the High Seas Fleet.

At high command headquarters the disruption caused to war output in British factories, together with the tying down of large numbers of men, guns and aircraft for home defence that otherwise could be employed at the front in France, was recognised as a major factor for the continuance of such raids. Accordingly, every encouragement was given to Strasser from Berlin to undertake as many bombing raids against the British Isles as practical, whilst the Zeppelin Company continued to receive heavy government support to develop still more efficient airships designed specifically for the dual purpose of scouting and strategic bombing.

Prior to September 1916 both naval and army Zeppelins operating over the British Isles had enjoyed almost complete immunity from attack by either guns or aircraft, although the army had become discouraged by their early losses in 1914 whilst employing early model Zeppelins against frontier fortresses in France and Belgium. With the introduction of the improved L10 class of Zeppelin in April 1915, Strasser felt he had at last a more effective weapon with which to prosecute the war against England, with eleven of the type being supplied to the navy and a further ten to the army by the end of the year.

Throughout 1915 in a series of wide-ranging raids this class of airship caused the highest level of material and monetary damage to be inflicted on the United Kingdom by Zeppelin raiders throughout the course of the war. The spring and summer months were largely free from any incursions by Zeppelin raiders over the mainland, but with the onset of autumn and the shortening days the German navy took advantage of the kaiser's somewhat reluctant agreement made in July 1915 with the high command authorising attacks on the British capital.

Early forays using the new naval Zeppelins were not always successful in reaching their main targets, as on the night of 9/10 August 1915, when five of the latest L10 class Zeppelins set out from the bases at Hage, in east Frisa, and Nordholz to raid London. In thick, squally weather none of the airships reached their targets, with Mathy's *L13* turning back with engine trouble and *L10*, captained by Oberleutnant zur See Wenke, bombing Eastchurch naval air station in the belief he was over central London.

Oberleutnant zur See von Buttlar's *L11*, after coming under anti-aircraft fire, dropped his bombs in the sea off Lowestoft under the impression he was attacking the port of Harwich before turning for home. The older *L9* under Kapitän Loewe was more successful. Having been ordered to the mouth of the Humber where, despite a broken rudder cable, she still managed to attack Goole to the west of Hull, causing some material damage and loss of life before returning safely to base.

Peterson, commanding *L12*, was unsure of his position; mistakenly identifying Dover for Felixstowe he shed his cargo of bombs and incendiaries, and had the misfortune to be hit amidships by a 3in anti-aircraft shell fired by the local defences. Despite her severe damage Peterson effected temporary repairs and, after throwing overboard unnecessary equipment and empty fuel tanks to save weight, made off to the south-east towards the Belgian coast whilst continuously losing gas from the damaged cells.

Finally, at 3.40 a.m., she came gently down in a calm sea off Zeebrugge, where despite repeated attacks by British aircraft from Dunkirk the airship was successfully taken in tow by a German torpedo boat, safely reaching Ostend by noon. Here, once the airship was brought alongside the harbour wall, an attempt to salvage the *L12* was made utilising a large crane, but as the forepart was being lifted from the water it exploded, resulting in the total loss of the ship.

Losses among Zeppelins during this period were high and not always ascribed to enemy action, as with the *L10* in September 1915 under the command of Kapitän Hirsch, which, whilst valving hydrogen over the island of Neuwerk prior to landing, was struck by lightning, and lost with all hands. Equally unfortunate was the loss of the *L18*, destroyed by fire whilst being inflated in the *Toska* shed at Tondern in November 1915, with one of the crew killed and seven injured. The *Toska* shed was to prove particularly unlucky, as a year later in December 1916 the *L24* broke her back whilst attempting to enter the shed and took fire; a fire that also ignited the *L17* that lay on the north side of the shed. Fate, however, had not finished with the *Toska* shed, and eventually five airships were to be destroyed within its confines.

At the beginning of October 1915 Strasser was instructed by Admiral von Pohl, the then commander-in-chief of the High Seas Fleet, to attempt a raid on Liverpool, which was now seen as the main port for the delivery of war material from the United States and where the disruption of such trade would strike a severe blow against the Allied cause. Bad weather held the ships in their sheds for a week and when eventually, on the night of 13/14 October, five airships set out, weather conditions dictated that a raid on Liverpool was impractical and the airships were diverted to targets further south. Despite later attempts to raid Liverpool, German naval airships never managed to reach this prime target throughout the course of the war.

By mid-1917 the army had abandoned the use of airships as a weapon of war, putting their faith in aeroplanes and, in particular for long-range bombardment, employed the Riesenflugzeuge, or 'R' planes; giant machines with several types being of all metal construction spanning up to 140ft and weighing upwards of 10 tons. These huge aircraft were powered by four to six engines, which in some cases were mounted within the fuselage where they could be tended by engineers in flight. They were at the cutting edge of aeronautical development, a type of machine that would have seemed inconceivable ten years before.

In some designs the power of the engines was transmitted to the propellers mounted outboard through complex transmission shafts, up to 15ft long, which in themselves were a major technical development. The development of these extraordinarily advanced aircraft was led, paradoxically, by Count Zeppelin himself, who by early 1916 had come to recognise the limitations of his own invention as an instrument of war.

The count had earlier handed over the reins of Zeppelin development to Dr Ludwig Durr and Karl Arnstein at Friedrichshafen in order to concentrate on the development of these giant multi-engine bombers. Some seventy or so 'R' planes were built at the Zeppelin Staaken works near Berlin, whilst other manufacturers, such as Link Hoffman, Siemens-Schuckert, DFW and Dornier, also produced similar craft. Most of the 'R' planes saw service on the Eastern Front, where again due to the less organised air defence system operating in that theatre they proved effective in their bombing role, and to a lesser extent were also employed in raids on England.

The Dornier and Zarsch companies in turn produced giant flying boats that were capable of flying extended patrols for the navy of over twenty-four hours at speeds of up to 120mph. Had the war continued beyond 1918 these machines would have undoubtedly replaced the airship for maritime reconnaissance work.

On 27 November 1916 Kapitänleutnant Max Dietrich, commander of the *L34*, was celebrating his 46th birthday together with Oberleutnant Frankenburg of the *L21* in the officers mess at Nordholz, when the call came to attack the English Midlands.

29. Zeppelin *L30* flying over SMS *Ostfriesland*.

Soon seven airships were in the air steering west along the Friesian Islands, with *L34* and *L35* setting course for Newcastle, encountering light winds and excellent visibility. However, von Buttlar, aboard *L30*, again experienced engine trouble and was forced to turn back at 8.30 p.m.

L34 reached Hartlepool at 12.30 a.m. where she dropped her bombs from 9,500ft only to be picked up by a searchlight, causing Dietrich to order the airship to turn out to sea. Second Lieutenant I.V. Pyott of 36 Squadron and flying a Sopwith Camel had taken off from Seaton Carew at 11.22 p.m. on receipt of warning of the raid. Now flying at 10,200ft Pyott sighted the *L34* on the coast between Sunderland and Hartlepool, held in the beams of a searchlight some 200ft below him.

Going into a shallow dive, Pyott made a beam attack from starboard without effect. Then, passing under the airship he made a second attack on the *L34*'s port quarter, resulting in the airship bursting in flames and falling into the sea. Her consort the *L35* was at that time flying over Seaham Harbour, where Kapitän Ehrlich witnessed his comrades' destruction, before reversing course for home. Likewise Hollander in *L22*, 70 miles distant over Flamborough Head, and Friemel in *L24* standing 15 miles out to sea, saw the *L34* take fire and hastened off eastward to safety.

The *L36* had attempted to raid Edinburgh, but due to a temperature inversion Schütze was unable to climb higher than 10,000ft causing the commander to abandon the undertaking.

Meanwhile, Frankenburg aboard *L21*, who had so recently been celebrating with Dietrich, found himself under heavy anti-aircraft fire over the coast at Atwick where he bravely, but unwisely, turned inland again towards Leeds. After passing east of Leeds, the *L21* bombed Barnsley at midnight and Macclesfield soon after, before passing over Norwich and wandering slowly across Norfolk.

By now Frankenburg was in great danger having been over England for over eight hours, only passing out to sea again at Great Yarmouth at 6 a.m. Here she experienced engine trouble and had been spotted by RNAS aircraft that were now climbing to intercept her in the clear early morning air. *L21* was only 10 miles out to sea when she was attacked simultaneously by three aircraft, with Lieutenants Fane, Pulling and Cadbury delivering separate attacks, causing the *L21* to fall in into the sea a blazing wreck. Once again Strasser had lost two airships in a single day.

The final loss for 1916 came on 29 December 1916 when the *L38*, newly commissioned in November and commanded by Kapitänleutnant Martin Dietrich (no relation to Max Dietrich of the *L34*), operating from the Baltic base of Wainoden in Courland attempted a raid on the Russian port of Reval. In deteriorating weather the airship rapidly accumulated a heavy load of ice on the envelope, which together with intermittent engine trouble required the helmsman having to fly at an angle of 20° to maintain her in the air.

The *L38* managed to struggle back to the coast in a fierce snowstorm, but by now flying almost 3 tons heavy, the airship finally stalled and came down in the forest at Seemuppen, Russian territory, which fortunately was now in German hands. The impact of the crash was absorbed by the airship's hull being impaled on the pine trees, allowing the crew to escape uninjured, but the ship was a total loss and the *L38* was dismantled where she lay in the snow.

Although the Baltic airship detachment made brave, determined and strenuous efforts to carry the war against their Russian foes under conditions of great difficulty,

30. *L62*, a lightened 15m bay Height Climber fitted with Maybach 245hp Mb IVa altitude motors.

their efforts were largely ineffectual in influencing the outcome of operations in that theatre.

Perhaps the greatest achievement of the German naval airship division in the Baltic was the extended endurance flight carried out by the *LZ-120*, seconded to naval service from the army prior to the disbanding of the army airship service in mid-1917.

The *LZ-120* was an L30 class Super Zeppelin based at the Seerappen base in East Prussia under the command of Hauptmann Ernst Lehmann, who was later to find fame as the captain of both the *Graf Zeppelin* and the *Hindenburg*. The *LZ-120*, with a crew of twenty-eight, loaded with 20 tons of fuel and 6,000lb of bombs, embarked on a long flight to determine the maximum possible duration attainable for this type of airship.

Cooking plates on the engine exhausts provided hot meals for the crew, together with additional hammocks and other comforts in the crew space to ensure more tolerable conditions on such a long flight. The *LZ-120* departed Seerappen on the evening of 30 July in calm weather, cruising westward along the northern German coast as far as Kiel Bay, before turning north towards the vicinity of Copenhagen, then following the southern Swedish coast towards Stockholm. From here the airship flew further into the northern Baltic, then eastward towards the coast of Latvia where greater caution was called for, although no enemy craft were sighted.

By running the engines alternately no major problems were experienced with the machinery, although a broken exhaust pipe required mid-air repair. As the *LZ-120* continued her flight along the southern shores of the Baltic the airship encountered increasingly stronger winds and it was decided to curtail the flight and course was set for home.

The airship landed at Seerappen at 4.40 a.m. on 31 July after 101 hours in the air, having covered 4,000 miles without serious incident and with sufficient petrol on board for a further thirty-six hours' flight. This cruise was a magnificent achievement that ably demonstrated the potential of the airship for long-distance commercial purposes.

Another long-range flight took place four months later that was even more remarkable, not only for its endurance but the fact that it took place largely over the desert regions of Egypt and the Sudan in conditions of searing heat by day and contrasting cold at night; conditions that no airship had previously encountered.

General von Lettow-Vorbeck, commanding German forces in East Africa, had been conducting a successful campaign against Imperial forces in northern Tanganyika but was desperately short of ammunition and medical supplies. In May 1917, the German colonial office were approached with the suggestion that a Zeppelin should be used to transport the urgently needed supplies on a one-way trip to re-supply the German garrison.

The plan involved the lengthening of one of the new L53 class airships by two 15m bays, making it the largest airship in the world; measuring 743ft in length with a capacity of 2.42cu ft, it imparted a gross lift of over 70 tons of which an amazing 51 tons was available as useful load. A shortcoming of this conversion was that due to its immense size the airship would be seriously underpowered, which militated against effective control in flight and an adequate reserve of power in an emergency.

The *L57*, then building at Friedrichshafen, was chosen for conversion. It made its first flight on 25 September but was almost immediately destroyed by fire a few weeks later at Jüterbog. A replacement airship, the *L59* under construction at Staken, was hastily completed to the required specification and placed under the command of Kapitänleutnant Ludwig Bockholt. Following her first flight on 19 October 1917 the *L59* was dispatched to the forward base at Jamboli in Bulgaria from where the proposed African flight was due to start.

News had reached the army command in Berlin that von Lettow-Vorbeck's troops were holding out in the Makonde highlands, and plans were now advanced for the relief flight to depart immediately. After two abortive attempts being frustrated by weather, the *L59* eventually set out on her epic journey at dawn on 21 November 1917 loaded with 14 tons of supplies and a crew of twenty-two.

With the air temperature 0°C and a strong northerly wind, the *L59* was soon over the Sea of Marmara and, by 10.15 p.m., was off the east coast of Crete. Soon after this they encountered violent thunderstorms and had to wind in the wireless aerial, unaware that high command was desperately trying to contact them with the vital information: 'enemy forces now control the greater part of the Makonde Highlands, and German colonial troops are now in retreat.'

As the airship flew on through the storm a spectacular display of St Elmo's fire engulfed the ship, with lurid tongues of blue and green fire dancing on the ships metal structure momentarily causing the crew to imagine the ship was on fire. The *L59* successfully weathered the storm and at 5.15 a.m. the next morning she passed over the African coast at the Gulf of Solum.

From here the *L59* set course relying on celestial navigation across the inhospitable Libyan desert towards the River Nile, experiencing severe buffeting from the scorching columns of hot air rising off the desert. Bockholt had to fly the airship several degrees nose-down in a light condition to avoid going over pressure height and loosing valuable gas.

In the late afternoon the reduction gear housing on the forward port engine cracked and although repaired was not used further, while the remaining four engines were rested alternately for one hour in every four. With the onset of darkness and the cooling of the gas, Bockholt now had to fly at an up angle of 4° to maintain his heavy ship in the air. Even so, at one point the great airship stalled and fell 2,000ft, where only the instant release of over 7,000lb of ballast and ammunition adverted disaster.

At 10 p.m. the *L59* reached the River Nile at Wadi Halfa and followed the great westward curve of the river towards Khartoum. At 12.15 a.m. on 23 November the *L59*, whilst flying 100 miles to the west of Khartoum, finally received a recall signal from the powerful Nauen wireless station near Berlin, informing them that 'the whole of the Makonde Highlands are now in the hands of enemy forces and the undertaking is to be abandoned'.

With heavy hearts the *L59* turned northward and set course for Jamboli (where they again experienced thunderstorms over the Mediterranean) then eastwards towards the Turkish coast at 10,000ft, finally landing at their base at 7.40 a.m. on 25 November, after ninety-five hours in the air. The airship had covered 4,200 miles in extreme conditions of weather and severe temperature variation, with fuel still on board for a further sixty hours' flight.

The question of what to do now with the *L59* exercised the minds of the Admiralty. Strasser wanted the airship returned for use in the North Sea, whilst other councils saw benefit in keeping the craft in the Balkans where it could be used to raid Italian, Russian and British interests. In the end the latter view prevailed, and on 16 March 1918 the *L59* made a successful attack on Naples and its naval base.

This was followed on 20 March by a daring attempt to bomb the installations at Port Said and the Suez Canal. An attempt that had to be abandoned due to contrary winds when only 3 miles from her target. Bockholt then headed for his secondary target, the British naval base a Sudra Bay in Crete, but was again frustrated when thick cloud covered his objective.

On 7 April 1918, the *L59* left Jamboli on her final mission to attack Malta and its important naval base. At 8 p.m. the captain of the U-boat *UB58*, cruising southward on the surface in the Strait of Otranto, was overhauled by a giant airship flying low on the same course, which he identified as a German Zeppelin.

Ninety minutes later the crew of the *UB58* observed to the south, at a distance of perhaps 30 miles, a huge flame in the sky accompanied by a deep and reverberating series of explosions as the blazing airship slowly fell. It was concluded that as the Italian forces failed to claim any part in her destruction, some form of accidental internal explosion had occurred.

During 1917 improved versions of the L30 class were under constant technical development in the race to outpace the British defences. Progressive lightening of structural components of the framework saved weight, increasing the ratio of useful to gross lift from 40 per cent, in the case of the L30 class of 1916, to 63 per cent for the L53 class of 1917.

Other weight-saving measures, to attain ever greater altitude, included the removal of the third engine from the rear gondola together with the suppression of the clumsy bracket-mounted propeller drives; the introduction of smaller, streamlined wing and control cars; and the use of lighter, two-ply gas cells. In the engine department the 240hp HSLu motors were replaced by the newly developed MBIVa altitude motors, designed to give improved power at altitude. By the introduction of these structural changes the Height Climbers class were initially able to achieve altitudes of up to 18,000ft, whilst the L53 class, with a useful lift of 40 tons, was able to rise dynamically to 20,700ft during trials.

The L53 class represented the apogee of development of the airship as a weapon of war. In this class the spacing of the mainframes was increased from 10m to 15m with two light, intermediate frames to further lighten and strengthen the structure. These airships were equipped with five of the newly developed 250hp Maybach MBIVa altitude motors, capable of driving them at almost 70mph. These new motors had been developed to overcome the fall-off in power of the older HSLu motors (which had lost more than half their power at altitude and were unable to successfully fight the winds at those heights).

With these developments in hand, on 16/17 March 1917 Strasser ordered out five of his improved airships to raid 'England south, London if practical'. This was the first raid by his high-flying Height Climbers, which were capable of reaching up to 18,000ft. Yet once again he was let down by an inadequate weather reporting service and at altitude the airships found they were still fighting against an unsuspected gale of great force.

German Naval Airship L70 (LZ 112) - 1918
Zeppelin"X" Type
Luftschiffbau Zeppelin GmbH, Friedrichshafen

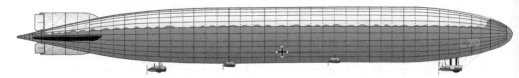

2,196,000 Cubic feet capacity
693' x 79' x 91'
71 tons gross lift
43 tons useful lift
7 x 240 h.p. Mabach MBIVa 6 cylinder in line
Speed 81 m.p.h.
Range at cruising speed 7,400 miles
Crew 24
Ceiling 21,000 feet

100 Feet

Despite the determined efforts of their commanders on this raid no targets of significance were reached, monetary damage was limited to a mere £173 and, fortunately for Britian, no casualties were incurred. Driven before the gale the unfortunate *L39*, commanded by Kapitänleutnant Koch, was blown over the Western Front where he was shot down by anti-aircraft fire with the loss of the entire crew. The other airships managed, with difficulty, to struggle home to safety.

During May six Zeppelins were again dispatched on another high-level raid, and although their commanders claimed to have bombed imagined targets of military importance along the east coast, no bombs were traced over land. Clearly the strategy of high-altitude raids was not succeeding; in the frozen wastes of the icy sub-stratosphere the crews, even with the aid of oxygen, suffered with altitude sickness and general lethargy brought on by the intense cold.

Operating at these extreme altitudes made navigation and the accurate identification of targets even more difficult, whilst in the North Sea the Zeppelins were now forced to patrol at greater heights due to the danger posed by the RNAS flying boats. The new engines still continued to suffer with frozen radiators, iced-up carburettors and congealed oil blocking their pipes, while disappointingly the power output at 20,000ft was reduced from 240hp to 140hp.

On 16 June 1917 the *L48*, under Korvettenkapitän Schütze, Strasser's second-in-command, was ordered to lead a raid of six airships against London and the south of England. Due to the short nights prevailing at this time of year the raid was ill-advised, as the visual conditions would favour the defenders. In the event, two of the airships detailed failed to leave their sheds due to crosswinds and two others turned back with engine trouble, leaving only *L42* and *L48* to cross the coast.

As early as 8.30 p.m. Kapitän Dietrich was in sight of the English coast, but stood off deeming it too light to go on. At 11.30 p.m. Schütze wired the attack signal to all airships, unaware that he had only one consort.

At 2 a.m. Dietrich and the *L42* were over Ramsgate, where he dropped his bombs on an ammunition dump causing £28,000 worth of damage before heading back out to sea as the first glimmers of dawn were seen to the east. The *L48* had meanwhile made landfall at Harwich, where he dropped his bombs ineffectively in the surrounding fields.

From 18,500ft the *L48* dropped down to 13,000ft where, at 3.30 a.m., Schütze asked for radio bearings from the home D/F stations. Several British aircraft were already airborne, having sighted the Zeppelins in the clear air earlier, and Lieutenant L. Watkins of 37 Squadron was closing on *L48* as she sent her wireless request. Closing at the stern, Watkins sent the *L48* blazing to the ground at Holly Tree Farm near Theberton, Suffolk, where amazingly two of the crew survived the inferno.

The loss of the *L48* and Schütze was a bitter blow to Strasser and, although raiding continued through the second half of 1917, such attacks were ineffectual and could only be justified by the retention of troops and guns for defence against them.

Undaunted, Strasser continued to improve and expand his air fleet with the new seven-engine L70 class being introduced in June. These huge craft were of 2.195 million cu ft capacity, capable of 70mph and able to reach 23,000ft carrying a bomb load of 8,000lb.

With these new airships Strasser endeavoured to convince an ever more sceptical high command that his beloved airships could break the will of the British to continue the war, despite the turning tide of evidence on the Western Front. Yet January 1918 opened with an unmitigated disaster for Strasser, when on 5 January at the new headquarters base at Ahlhorn at 5 p.m. a huge orange flame erupted from Shed 1, which housed both *L51* and *L47*.

Walter Dose, the captain of the *L51* which lay on the south side of Shed 1, had earlier ordered crew members to clean out the engine gondolas. In the late afternoon with dusk falling as they worked by the light of electric cable lamps, two of the crew of the *L51* noticed a glow of fire that seemed to be coming from under their ship. Immediately sensing the danger they scrambled out of the gondola and ran under the *L47* and out through a picket door in the side of the shed, which saved their lives.

As they did so two huge explosions engulfed the *L51* and the *L47*. The blast from the explosions rained debris down on Shed 2, which housed the *L58*. She exploded in turn with even greater violence. Flaming debris from this explosion even fell on to Shed 3, which was almost completely destroyed along with the *L46* that had lain inside. Shed 4 was similarly damaged and set on fire by falling wreckage, inside the wooden-framed *SL20*, which was under repair, was incinerated amid the collapsed shed.

The whole incident had taken just four minutes to reduce the proud base to a shapeless mass of twisted and burning metalwork. In this short space of time four sheds and five airships had been destroyed, together with a death toll of fourteen and over 100 personnel injured. The actual cause of the fire was never satisfactorily explained, but naturally sabotage was considered as a possible cause. This now seems the least likely cause, however, and some form of petrol fire, perhaps ignited by a frayed lighting cable, was a more probable culprit.

The destruction of the base must have been a devastating blow for Strasser and the service in general, but despite this terrible catastrophe the naval airship division continued their patrol work over the North Sea and the base itself was rebuilt before the end of the war. In the last months of the war the Zeppelins also made raids on the British coast, although the material damage inflicted was minimal.

Finally, on 5 August 1918, Strasser ordered out five of his most modern airships to 'attack south or middle England, London to order of Leader of Airships'.

Strasser himself was flying aboard the brand-new *L70*. Under the command of Kapitänleutnant von Lossnitzer, the *L70* was a monster seven-engine craft able to reach 25,000ft and capable of a top speed of 81mph.

At this time of year, with its light evenings, the Zeppelins were clearly seen as early as 8.30 p.m. flying in a V-formation 30 miles off the Norfolk coast. Further alerting the defenders to the attack were their wireless transmissions, which had already been intercepted by British listening stations. Within minutes, scores of aeroplanes from air stations along the coast were climbing hard to intercept the German air fleet.

Major Edgar Cadbury, together with Captain Robert Leckie in their de Havilland DH.4, climbed to 16,400ft and at 10.20 p.m. attacked the *L70* bow-on. It immediately took flame and fell burning into the water off Wells-next-Sea on the Norfolk coast.

So perished Fregattenkapitän Peter Strasser, the most resolute and determined leader of naval airships who inspired his men by example against all odds and was admired by his crews and enemy alike.

The remaining Zeppelins, on seeing their leader fall, turned tail with the *L65* (which, flying within 1 mile of the *L70* was lucky not to share her fate) when an attacking aircraft's guns suffered a stoppage, allowing her to escape.

On 11 August Kapitän Prolss, on patrol at 18,000ft over the German Bight, was shot down by Lieutenant S.D. Culley flying a Sopwith Camel that had been launched from a lighter towed by a destroyer. This was the final Zeppelin destroyed by British forces in the war.

Within a few months Germany was brought to its knees and sought an armistice, followed by mutiny within the fleet, the abdication of the Kaiser and revolution. The remaining airships of the fleet were hung up in their sheds whilst the conditions of the peace treaty was hammered out in Versailles.

However, following the scuttling of the High Seas Fleet at Scapa Flow on 21 July 1919, the revolutionary sailors' councils ordered the crews at the naval bases to wreck seven of the remaining airships. At Nordholz and Wittmundhafen crews disabled seven of the airships by dropping them on to the concrete floors of the sheds and smashing their framework. At other bases the order was ignored. This action resulted in the call for the immediate surrender of all other existing airships to the Allied Control Commission.

12

The R23 & 23X Classes

Due to the stop-go policy that delayed the completion of *R9* from its proposed entry into naval service in mid-1914 until late 1916 (by which time it was hopelessly behind current Zeppelin practice), the Admiralty found themselves in a dilemma as to which line of development to follow.

From their intelligence sources they were aware of the rapid advances in airship development taking place in Germany, with building times being measured in weeks as opposed to the years required in the British case to complete a rigid airship. The Admiralty had already cancelled the smaller *No. 14* and *No. 15*, each of 22 tons gross lift and so inadequate for the purpose in hand, but were under pressure to supply some form of airship to work with the fleet irrespective of their performance capabilities.

In May 1915, while *R9* was still fitfully under construction with the work hindered by constant changes of specification and the slow delivery of materials and parts, it was proposed at a defence committee meeting that ten improved *R9*'s should be built with all possible speed. This urgent decision was taken following the German government's declaration that the waters around the British Isles were now regarded as a war zone, while the increasing losses of naval and merchant vessels to submarine attack also focused attention on this requirement.

Later the same month the Vickers design team under H.B. Pratt began work on an improved design, the structure of which was based largely on that of *R9*: while retaining the same hull diameter of 53ft it was longer by 12ft and contained an additional gas cell. Vickers were contracted to build two ships, while simultaneous orders were placed with Armstrong Whitworth and Beardmore to provide two ships apiece. The new craft had a deeper radius of the bow and stern sections, which increased the capacity to 910,000cu ft with the gross lift rising to 26.5 tons, providing an anticipated useful lift of 8 tons.

Early in the design procedure the Rigid Airship Committee, in the person of Sir Tennyson d'Eyncourt, insisted that the factor of safety of the keel structure be increased to a value of two, representing twice the normal anticipated forces acting on it under standard flight conditions. This recommendation required the introduction of steel tubing in the keel and lower hull structure, which while achieving the desired result increased the weight penalty.

Engine power was also increased to a total 1,000hp by the installation of four 250hp Rolls-Royce Eagle1 engines, mounted in three cars arranged on the centre line of the hull. Once again swivelling propellers together with their attendant transmission gear were employed, manufactured by the Wolseley motor company and fitted to the bow and stern cars to aid control and descent. In this design the complex multiple rudder and elevator arrangements were abandoned in favour of a simple cruciform tail configuration, while the deep-V external keel was retained, from which the three power cars were suspended.

Work on the first ship, *No. 23*, began at Vickers in March 1916, which in itself seems an unpardonable delay of ten months from the placing of the order, especially considering that the greater part of the hull structure was a direct copy of that of *No. 9*, and that to a large extent the necessary jigs and materials were already available.

Naval Airship HMA R23 - 1916
Messrs Vickers Ltd Barrow-in-Furness

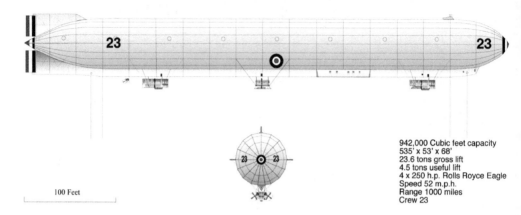

942,000 Cubic feet capacity
535' x 53' x 68'
23.6 tons gross lift
4.5 tons useful lift
4 x 250 h.p. Rolls Royce Eagle
Speed 52 m.p.h.
Range 1000 miles
Crew 23

100 Feet

The building schedule for *No. 23* was a much extended period, covering as it did some twenty months, with the airship not making its first flight until September 1917, almost a year later than the proposed completion date. This delay can be explained by the fact that although the greater part of the hull structure, such as girder work, main transverse frames, gas bags and most of the outer cover, were complete by the end of 1916, much trouble was experienced with every other aspect of fitting out the ship, including the electrical, ballast, control and petrol installations, which were necessary to complete the ship. The work was further hampered by the non-delivery of essential materials, strikes and two periods when work ceased completely for two weeks on each occasion for 'Works Holidays'.

The later ships *No. 24*, *No. 25* and *R26* were all completed within a nine-month building period, benefiting from the experience gained in constructing of *No. 23*.

As in *No. 9* the main transverse frames were of a wire-braced triangular section without kingpost bracing, set at 30ft intervals with two light steel intermediate frames between. The deep-V keel section was rigidly attached to the underside of the hull and also made mainly from steel tube section for additional strength. The main petrol tanks were fitted athwart ships and dispersed along the keel above the walkway, while each of the engine cars had a 20 gallon gravity feed service tank fitted in the keel above.

Upon completion, and following lift and trim trials, it was found that *No. 23* had substandard conditions: the disposable lift was only 5.7 tons as against the 8 tons envisaged in the original design. In order to meet the Admiralty requirements similar measures to those taken with *R9* were employed, involving the removal of much of the naval equipment, including the heavy swivelling propeller gear, and the substitution of lighter gas bags.

Additionally, the existing stern car and engine were replaced by the wing car from the downed Zeppelin *L33* containing a single Maybach 245hp MBVa engine, which drove a single 16ft diameter propeller. The combined engine arrangement giving a top speed of 53mph.

Following these changes the disposable lift was increased to 6.2 tons which was deemed acceptable to the Admiralty inspectors.

Of the remaining ships of this class *No. 24* and *No. 25* were both completed in October 1917, with *R26* following in March 1918. *No. 25* and *R26* were also modified in the same way as *No. 23*, by having the stern power car removed and the substitution of a wing car from a downed Zeppelin installed in its place.

As *L33* had only two wing cars it must be assumed that these replacements came from the salvaged parts of the *L32*, which had been shot down in October 1916 at Great Burstead, Essex. Although presumably, as this airship crashed from a great height in flames, unlike the *L33*'s relatively gentle denouement, some considerable rebuilding of the engines and cars must have been necessary.

In October 1915 the responsibility for the naval rigid airship construction programme came solely under the Admiralty's control, with all design matters being centralised under the direction of Constructor-Commander C.I.R. Campbell and his staff.

No. 23 was delivered to Pulham via Howden in October 1917, flying over London in December to demonstrate to MPs and the public alike the 'success' of the new rigid airship programme. *No. 23* flew forty-four hours in 1917 and a further 278 hours during 1918 on training and patrol work. One of her longest flights in service was made under the command

31. Rigid *R26* in flight, 1918; the first British airship to carry the 'R' designation.

of Major Little in May 1918, when she stayed out from Pulham for over forty hours over the North Sea under war patrol conditions.

Later, following the Armistice, she also flew a series of victory flights over several cities including London and Liverpool, and was present to oversee the surrender of German U-boats at Harwich on 20 November 1918, commanded by Flight Lieutenant G.M. Thomas DFC.

Among those who received training aboard *No. 23* was Lieutenant Commander Zachary Lansdowne USN, who was to command the first American rigid ZR1, *Shenandoah*, meeting his death when she crashed in a storm at Ava, Ohio in September 1925.

Various experimental armaments were fitted to *No. 23* including the Davis 2lb recoil-less quick-firing gun mounted on the upper platform. In this weapon the recoil was absorbed by a large plug of lard at the breach end, which while fairly effective had the disadvantage of spraying the gunners and the outer cover with copious quantities of pig fat.

During November 1918 an unpiloted Sopwith Camel fighter was taken aloft, suspended on a release mechanism under the midship's cabin and launched over the sea near Great Yarmouth to test the feasibility of carrying aircraft for protective purposes. Later the same month a Camel piloted by Lt R.E. Keys of the Royal Air Force was successfully dropped over Pulham air station, with the pilot descending safely to the ground.

The German navy had some months earlier carried out similar experiments when an Albatross fighter had been dropped from *L35*, together with a steerable, guided, winged aerial torpedo.

During March 1919 *No. 23* was taken in hand to have a strengthened bow section installed, to enable tests to be carried out with the 'three wire' method of mooring. This system allowed an airship to be temporarily moored floating some 30ft above the ground using the minimum of mechanical infrastructure and ground crew, and could be readily accessible for servicing by rope ladders to the cars. Although this system had some distinct advantages, it was basically suitable only as a temporary method of mooring to be used primarily in an emergency.

Following these trials *No. 23* did not fly again, her flying career had been, not unnaturally, devoted to the area of flight and ground crew training. Her performance, although leaving much to be desired, was the best that could be expected at that stage of development in Britain. In September 1919 she was deleted and broken up along with her sister ships, due both to their poor performance and as an economy measure.

No. 24, the second of these ships, was built by Beardmore at Inchinnan, first flying on 18 October 1917. Like *No. 23* she was deficient in useful lift with only 5.1 tons disposable against the projected 8 tons for her design. She was also heavier than the other ships of her class due to the use of heavier gauge materials and heavier rivets of a different specification.

Upon completion *No. 24* was flown from Inchinnan to East Fortune, her war station, on 28 October 1917 as the Beardmore building shed needed to be clear for the imminent laying down of the new *R34* based on the captured Super Zeppelin *L33*.

Before this flight and due to her poor lift, drastic measures were required to ensure she could gain sufficient altitude to safely over-fly the Pentland hill that lay between Inchinnan and her new base. This included the removal of all unnecessary equipment, auxiliary machinery and the engine from the rear car, which reduced her top speed to 38mph on the three remaining engines. Once at East Fortune the engine, propeller and other equipment were reinstalled, increasing her speed to 50mph. This, together with further modification, increased her disposable lift to 6.1 tons, but in all other respects she was a most unsatisfactory airship.

A major problem was the gas bag wiring which caused excessive surging and chaffing. While common to all ships of this class, it was particularly exaggerated in *No. 24*. Despite these shortcomings she managed ten hours in 1917, and 154 hours in 1918 including war patrols. Later, in 1918, she was equipped with a mooring socket in the bow and was employed in testing the new high mast at Pulham.

During these tests *No. 24* added a further twenty-eight hours of flying to her logbook and remained moored to the mast for sixty-three days, being alternately winched up and down in all weather conditions until finally, in December 1919, she was finally hauled down and scrapped.

No. 25, the Armstrong Whitworth ship, was flown from Barlow to Howden in October 1917, undertaking acceptance trials in December of that year. Like her predecessors, *No. 25* suffered a catalogue of failings with serious problems of gas bag surging due to the ineffective design of the retaining netting, either to contain the cells in position or effectively transmit the lift to the best advantage. The surging problems also resulted in constant changes in the centre of lift and gravity, affecting the trim and controllability of the ship in the air.

She logged six hours in 1917 and a total of 214 hours in 1918, being under the command of Captain G.H. Scott in July 1918 when she performed some limited patrol work. Later she was moved to Cranwell for experimental work and training, undertaking little or no flying duties during 1919 before being deleted in September 1919.

The last of the quartet of these ships to be completed, the *R26*, was the first British rigid to bear the 'R' designation. Certain modifications were incorporated in her construction increasing her useful lift of 6.5 tons, an improvement which enabled her to operate with a degree of efficiency. At her full speed of 54.5mph she was capable of ten hours' endurance and could reach the design ceiling of 3,500ft.

The drag coefficient on all these ships was high; in tests with all engines simultaneously declutched the speed dropped from 53mph to 20mph in under two minutes. As previously mentioned, a lighter rear car from a L33 class Zeppelin had been fitted to *R26*, driving a single propeller which also improved her performance.

She was delivered to Pulham in May 1918. Here, during July under the command of Captain T. Elmsley, she made a cruise under war conditions of forty hours forty minutes over the North Sea. After further patrol work and training flights *R26* was sent on a demonstration flight over London in October 1918 in support of a war bonds drive. In total the *R26* logged 208 hours, mostly on war patrols, and could be considered to be the first successful British rigid, insofar as she was able to undertake patrol work of a reasonable duration and radius of action carrying a useful war load.

She was present together with *No. 23* and *SR1* at the surrender of German U-boats at Harwich in November 1918, and further performed several scientific flights on behalf of the NPL studying aerodynamic measurements. Later, in January 1919 during mooring mast experiments at Cranwell, she was severely damaged in a snowstorm, the load of snow and rain causing her to be forced to the ground. Following this accident *R26* was taken down and shedded for inspection, which determined repair was undesirable. She was finally dismantled in March 1919.

The 23X Class

Even before the 23 class had commenced building, Constructor-Commander Campbell and his team were aware that these ships would not achieve the desired performance figures required.

Their ceiling was limited to 3,500ft, while their endurance (apart from *R26*) was only eight hours at full speed, a performance that was hardly better than the cheaper Sea Scout non-rigids already in service. Accordingly, in January 1916 the Rigid Airship Committee under the auspices of the Director of Naval Construction Sir Tennyson d'Eyncourt, with Constructor-Commander Campbell in attendance, was convened to produce an improved version of what was in essence a failed design.

Several design variants were evaluated by the committee, with the selected design representing the best all-round compromise in terms of useful lift, range and speed. The most promising design, the 23X as the new class was known, was estimated to have a disposable load of 35 per cent of the total lift as opposed to the 24 per cent achieved by *No. 23*.

This increase was a commendable improvement, were it not for the fact that at this time comparable Zeppelins were already reaching a ratio of 40 per cent useful lift over gross lift. This figure was to go as high as 65 per cent when the Height Climber class appeared in 1917, where the structural weight was reduced to a dangerous extent so that their much lightened structures were in danger of collapse under certain conditions in order to obtain maximum altitude.

A new design of girder, of deeper section and of greater strength, was employed in the 23X. It was still of triangular latticed form but with the apex of the girder pointing outwards as opposed to the base in the proceeding class. In this design all the main weights were suspended directly from the mainframe girders or the radial wiring of the frames.

The 23X class offered no great advance in design thinking over the existing class, having the same general layout as the earlier ships, but benefitted from the removal of the drag-producing external keel structure, which the committee estimated contributed less than 5 per cent to the hull strength.

With the removal of the external keel, access to the inside of the ship was provided by an internal corridor composed of inverted U-shaped girders attached to the mainframes forming an internal walkway between the gas bags the full length of the hull. The suppression of the external keel greatly improved the turning coefficient from the 11.2 of *No. 23* to 9.8 in the 23X class, giving them a tighter turning radius at the same engine speed.

The bow and stern radius were again increased, together with a further lengthening of the hull resulting in an increase of gas volume of 48,000cu ft, or an extra 1.5 tons additional lift. The same disposition of cars was again employed, but the rear engine car now drove a single propeller replacing the heavy swivelling propeller unit fitted initially to the previous class.

The improved design was approved in June 1916, with a very optimistic delivery date set for the end of the year – in reality the two ships of this class completed took almost two years to build. Orders for two ships, *R27* and *R28*, were placed with Beardmore, while *R29* and *R30* were placed with Armstrong Whitworth at Selby.

R27 was the first to fly on 6 June 1918. Initial trials showed that the ship handled well, was stable and achieved a useful lift of 7 tons, which together with a reasonable top speed of 55mph and a range of 1,400 miles carrying a crew of sixteen made her, in part, able to fulfil her designed purpose. One of the more undesirable features of the airship was its ability to absorb up to 1 ton of additional load in heavy rain due to the absorbent nature of the outer cover and the ineffective doping medium applied. A US navy inspection team reviewing the British programme in 1919 reported adversely on her performance, citing the problems with her outer cover and inefficient power utilisation.

Following acceptance on 29 June 1918, *R27* was delivered immediately to her war station at Howden and employed on regular war patrols. She made a scouting flight over the North Sea of twenty-three hours twenty minutes with Major Ommany in command in July, and by the end of August had flown eighty-nine hours forty minutes. Unfortunately, her promising commission was cut short after just six weeks when she was accidentally destroyed by fire in her shed.

The circumstances of her spectacular denouement was the result of a US navy crew who, whilst rigging a spare SS Zero envelope and car in the shed, were in the process of testing a wireless transmitter. As the set was switched on, sparks from the transmitter set fire to petrol that had accumulated in the bottom of the car, the resulting fire quickly engulfed the Zero and spread to two other Zeros, *SSZ38* and *SSZ54*.

R27 was also completely destroyed in the conflagration, the only survivor being the Vickers-built Parseval *P6* which escaped destruction with some scorching to her envelope; her survival being attributed to her upwind position in the shed and her highly streamlined form, which allowed the searing blast to pass rapidly by her without obstruction.

In contrast, the *R29* had a career of some sixteen months. She was first flown shortly after the *R27* on 20 June 1918 and delivered from Armstrong Whitworth at Selby to Pulham. *R29* was considered the first really effective British rigid airship, with a disposable load of 8.64 tons; over twice that of *No. 23*.

As previously mentioned, her improved turning radius allowed her to be used effectively against U-boats operating in the North Sea. Here, operating out of Pulham, she was put to work on convoy protection, mine-spotting and patrol duties, during the course of which she engaged three U-boats: the first escaping underwater; the second destroyed when it ran into a mine under pursuit; and later, under the command of Major G.M. Thomas DFC, a third U-boat, the *UB115*, received a direct hit with a 250lb bomb and was sunk.

From commissioning, the *R29* flew a total of 337 hours in 1918, traversing 8,300 miles on patrol and training missions; this was followed by a further 100 hours in 1919, with her longest continuous cruise being thirty-two hours twenty minutes in July 1918.

R29 was damaged whilst being shedded at Pulham in January 1919, requiring reconstruction to the lower bow section of girder work. This work resulted in her not being able to fly again until March of that year, when she undertook a wireless telegraphy trial, followed by some local training flights and finally making two fairly long flights in company with the new *R34* in June 1919, before being taken into her shed where she was decommissioned and deleted in October 1919.

Of the remaining two ships of the class, *R28* and *R30* were cancelled in September 1917 whilst in an advanced stage of construction in favour of more advanced types. *R28*, for instance, had 60 per cent of her hull erected and her engine and control cars attached before she was demolished.

Naval Airship HMA R29 - 1918
Messrs Armstrong-Whitworth & Co. Engineers , Barlow, Yorkshire

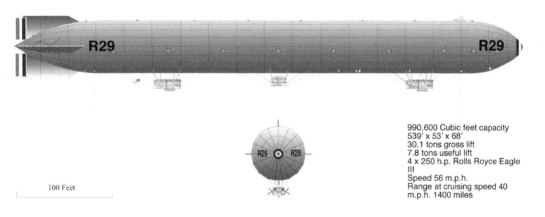

100 Feet

990,600 Cubic feet capacity
539' x 53' x 68'
30.1 tons gross lift
7.8 tons useful lift
4 x 250 h.p. Rolls Royce Eagle III
Speed 56 m.p.h.
Range at cruising speed 40 m.p.h. 1400 miles

By mid-1917 the superior Short's wooden-framed *R 31* and *R 32* were well in hand, as was the construction of the even more advanced copies of the captured L33 class Super Zeppelins. The cancellation of *R28* and *R 30* was, therefore, an urgent necessity to allow the *R 33* and *R 34* to be erected at their berths at Inchinnan and Barlow in their place.

Once again, with more advanced designs swiftly succeeding each and other incorporating greater technical advances with such rapidity, earlier models were rendered obsolete almost before they left the drawing board. Because of the extended development times of British airships the 23X class of 1918 were, at best, only the equal in performance to the German L3 class of four years' previous.

The construction of the 23 and 23X classes involved a considerable investment in terms of money and resources, and the programme was handicapped by trying to extrapolate from an inefficient design dating from 1913, which, consequently, failed to produce an efficient and useful airship suitable to work with the fleet.

In hindsight, the effort and money expended could possibly have been better employed in developing the non-rigid airships of the North Seas class and the larger, projected 500,000cu ft capacity K class airships, both of which showed great promise and could have been developed to adequately fulfil the patrol and scouting requirements of the fleet.

13

Short Brothers' Wooden-Framed Y Type: *R31* & *R32*

In mid-June 1915, while *R9* was still in an early stage of construction, the board of the Admiralty met to discuss the ordering of four larger airships of an improved design, which were subsequently to emerge as the 23 class.

These ships were again designed by Vickers, with the construction and development being overseen by Constructor-Commander Campbell and a team from the Royal Corps of Naval Constructors, who were now to assume official responsibility for the design of all future airships to be built to Admiralty orders.

In January 1915 the Admiralty had approached Messrs Short Brothers as one of the potential builders to tender for the construction of two of the approved 23 class craft, which were to be built in a new large double shed erected on land at Cardington in Bedfordshire. The projected cost of the shed, some £110,000, was to be provided in the form of a loan from the Treasury to each of the four participating constructors.

Sir Tennyson d'Eyncourt, a famed and respected naval architect who had designed many of the battleships of the fleet, chaired a meeting at the Admiralty in early 1916 to discuss the supply of airships to the Grand Fleet. At this meeting he declared that it was imperative to provide the Grand Fleet with ten of the 23 class ships at the earliest possible date, in order to improve the range of the fleet's protective screen. This total was to be further increased to twenty airships once suitable new double sheds could be provided to house them; this large number being indicative of the fleet's urgent requirements in this area.

The pressing need for scouting craft to serve the Grand Fleet had influenced this bold decision, although even at this stage doubts were being expressed as to the capabilities and expected performance of this untried class of ship. With these anticipated shortcomings in mind, various studies were already in hand prior to their construction in an effort to improve their disposable lift, range and speed.

As a more accurate appraisal became available through various sources on the performance of the latest German airships of the L10 class, it became apparent to the committee that the British ships would be greatly inferior in all respects to the new German craft. This assessment was indeed to be accurate, as the L10 class were to prove to be the most successful type of Zeppelin used in the Great War in terms of efficiency, general handiness and all-round performance. The L10 class were to achieve the highest level of utilisation of any of the Zeppelin variants used throughout the war, although in part their success during 1915 was aided by the lack of an effective gun or aeroplane defence at that time.

In consequence of the uncertainty regarding the performance of the 23 class, the original order to Shorts was countermanded by the Admiralty later the same month, instructing them to disregard the existing plans in their possession. Subsequently, they were to receive new drawings for two of the improved 23X class ships, which were refined and lightened versions of the earlier model and judged to be the best of the three variants of the original design that had been considered by the committee.

Short Brothers, unlike Vickers, had little experience in working with metal structures, particularly in the use of Duralumin for aircraft design. Consequently, the requirement to construct a metal airship was a radical departure for them, which would involve assembling a new workforce with specialist metalworking skills.

Prior to receiving the Admiralty orders Short Brothers' own design team, led by Oswald Short and C.P. Lipscombe, a designer of repute, had produced their own design during mid-1915 for a large airship utilising a wooden structure – a material they were familiar with due to their pioneering work on aircraft. Their own design was no doubt influenced by the designs of the Schütte-Lanz concern of which they were aware, although there is no evidence that they were in receipt of any technical information from that source.

On 14 March 1916 in a meeting at the Admiralty in London, Short Brothers submitted their own proposal for a wooden-framed rigid, which incorporated many advanced features that, at least on paper, were clearly superior in performance to the 23X design.

The proposed craft was of 2,077,000cu ft capacity, measuring 700ft by 67ft with a disposable lift of 20 tons, this being more than twice that of the Admiralty 23X class. This powerful airship was estimated to have an endurance of twenty hours at a full speed of 70mph, giving a still air range of 1,400 miles and capable of working far out to sea carrying a strong defensive and offensive armament.

The design represented a major technical advance in airship construction and such a ship would be ideal for scouting operations with the battle fleet. Its anticipated performance would allow it to operate in all but the very worst conditions in the North Sea and the Western Approaches.

Short Brothers stated to the board that they were confident of building such a craft within ten months, and that subsequent ships could be produced in a shorter time frame to allow for the rapid build-up of a fleet of airships that would satisfy Admiralty requirements in this area.

The Admiralty board, however, rejected the Short Brothers' design, indicating that in future all design work would be under the direct control of Commander Campbell and the Royal Corps of Naval Constructors, reiterating the injunction previously issued that the government would not sanction further private ventures in this area. The Admiralty already had its own design in hand, albeit in an early stage of progress, and Short Brothers were offered, by way of compensation, to be allowed to tender for two wooden-framed ships of the as yet uncompleted Admiralty design.

Anxious to procure the loan for the new double shed, Short Brothers agreed to accept the tender as the first of four companies so contracted. Similar sheds were to be erected under the same scheme for Vickers at Barrow, Messrs Beardmore at Inchinnan near Glasgow, and Armstrong Whitworth at Barlow in Yorkshire. These buildings would be the largest clear-span steel structures in the country with internal dimensions of 700ft by 180ft by 110ft, each using sufficient steel in their construction to build a light cruiser.

At the time it was felt that the acquisition of the shed would put Short Brothers in a strong position to obtain future Admiralty work. In the event, however, and following the wars' end, not only did they have to repay the full amount of the loan but the shed itself was taken over by the government in 1919 to be operated as the Royal Airship Works, in return for which Short Brothers received only £40,000 in compensation.

Short Brothers accepted the terms of the Admiralty contract and were informed that the dimensions of the new ships would be 594ft long by 64ft beam and were to be designated *R31* and *R32*.

Perhaps the most interesting development during early 1916 was the defection to the British secret service of Hermann Muller, a shadowy figure of Swiss or German origin who

Naval Airship HMA R32 - 1919
Messrs Short Brothers, Cardington, Bedfordshire

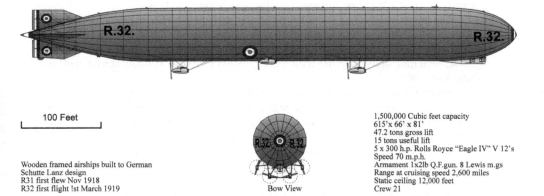

100 Feet

Wooden framed airships built to German
Schutte Lanz design
R31 first flew Nov 1918
R32 first flight 1st March 1919

Bow View

1,500,000 Cubic feet capacity
615'x 66' x 81'
47.2 tons gross lift
15 tons useful lift
5 x 300 h.p. Rolls Royce "Eagle IV" V 12's
Speed 70 m.p.h.
Armament 1x2lb Q.F.gun. 8 Lewis m.gs
Range at cruising speed 2,600 miles
Static ceiling 12,000 feet
Crew 21

claimed to be a former designer-draughtsman at the Schütte-Lanz factory in Mannheim.
Muller either supplied or produced a full set of drawings representing the then latest practice
in airship design at Schütte-Lanz, based most probably on the *SL6* type-D craft which first flew
in September 1915.

By March 1916 Muller was able to instruct Short Brothers in the method of construction
of the 10in by 10in triangular- and square-section slotted spruce ply girders, manufacture
of which started at the Rochester works mid-year. Unfortunately, the highly effective German
Casein glue was unavailable at that time and it was not employed until the latter stages of
construction of *R31*.

In its place the majority of the girder work was completed using the traditional cabinet-
makers Scotch glue. Yet the use of this animal-based glue had unfortunate results on the
integrity of the structure as the glue was subject to absorbing moisture, which affected
its efficacy as an adhesive. Muller is reputed to have brought with him the formula for
the patented Casein glue, known as *Kaltleim*, which possessed these superior adhesive
qualities and was subsequently utilised in successfully bonding the girder work of the *R32*'s
structure.

Short Brothers possessed previous experience as balloon manufactures and were Britain's
earliest aeroplane constructors, at that time supplying the Royal Navy with a range of highly
effective seaplanes that were to perform sterling work throughout the war period.

Work on the constructional facility at Cardington went ahead later that year. The new shed
took only four months to complete, being finished in December 1916 together with attached
fabricating shops, gas production plant and a housing development, Shortstown, to the south
of the airfield to house the workers. Immediately after this date the first of the completed
wooden girders began arriving at Cardington from Rochester in quantity and construction got
under way.

The new ships were to have a proposed gross lift of 43 tons and a disposable lift of 18 tons
– over 10 tons more than the yet-to-be-built 23X-type *R29*. This great improvement in
performance represented the leap forward in design necessary to allow Britain the chance to
catch up with Germany in airship development.

The two projected craft designated by the Admiralty as the Y class were each of 1.5 million cu ft capacity, with a length of 615ft and a diameter of 66ft, the engine power being provided by six Rolls-Royce Eagle III, twelve-cylinder vee engines of 300hp each. This made them the most powerful airships yet built in Britain, able to drive through the air at over 70mph.

As originally designed with the intention of employing them on distant anti-submarine patrol work in the Western Approaches, the Y type were to be given an exceptionally heavy armament, comprising two 2lb quick-firing shell guns on the upper platform for use against other airships, with a further two 2lb guns mounted in casements on either beam for anti-shipping work. This heavy weapons fit was completed by a defensive armament of up to ten machine guns, together with a heavy bomb load of 4,300lb in the form of anti-submarine bombs.

The general design was an extrapolation of the Schütte-Lanz *SL6* type D, the design of which had begun in January 1915 and first flew in September 1915. It is also likely that Muller had not only smuggled out plans and other technical details of the type D, but also had access to information on the later and more advanced *SL8* type E, the design work for which had commenced in Germany in early March 1915.

In the British copy the twenty gas bags were made of rubberised cotton fabric lined with goldbeaters' skins, and equipped with manoeuvring and automatic valves designed by Messrs Short Brothers, which discharged through trunking to hoods on the top of the ship. An axial cable ran the length of the ship connected to the bulkhead radial wiring that retained the gas cells in place, this being the first use of this feature on a British airship. The purpose of the cable, which consisted of individual segments joined at the bulkhead wire bracing, was intended to spread the shear loads on the hull and reduce the lateral strain on the structure should a gas cell become deflated.

The triangular, slotted-section wooden mainframes were set at 40ft intervals, with two slotted-I-section light intermediate frames between. Whilst over the engine cars the frame spacing was reduced to 20ft with one intermediate frame in order to better distribute the load in the immediate area. The *R31*'s mainframe spacing was 10ft greater than that of *SL6* from which the design derived, and it would later transpire that both *R31* and *R32* demonstrated an extreme flexibility of the hull structure in flight – to the extent that a crewman standing aft in the keel corridor would see a fellow crew member disappear from view during a turn in the same manner one can observe while looking into adjoining coaches on a tube train as it follows a curving section of track.

Detail design had started in May 1916, while the fabrication of girders was already under way. The actual construction and erection of *R31* took twenty-seven months, this at a time when the Zeppelin Company were capable of turning out a new ship on a three-month basis.

Work proceeded at a leisurely pace, hampered by material shortages and constant changes to the specification required by the Admiralty, together with problems caused by the use of Scotch glue. This last problem resulted in many girders being rejected by the A.I.D. inspectors as unfit for use, due to the rapid deterioration of its adhesive qualities leading to girders un-sticking when exposed to moisture. Towards the end of construction *Kaltleim* glue, a product of the German chemists, was substituted for the girder work, which quickly demonstrated its superior qualities.

Had this class of ship, with its relatively high useful lift, speed and a range of 2,000 miles, been produced more rapidly, and the structural problems associated with the glue been solved sooner, it could almost certainly have provided the solution for both the long-distance convoy protection requirement in the Western Approaches and the fleet scouting role.

Following a short maiden flight in the vicinity of Cardington, which revealed a degree of flexibility in the hull structure but otherwise indicated that the ship was satisfactory and handled well, the Admiralty expressed their satisfaction with the design.

The *R31* was prepared for her second flight in mid-August, on what was to have been a forty-hour test over the North Sea. Immediately after take-off the test programme was initiated

32. *R32* flying over Amsterdam in 1919, photographed from the *R33*.

and the engines were tested at various speeds, fuel consumption monitored and turning trials undertaken. These initially indicated satisfactory performance. However, after less than two hours in the air and following a series of high-speed turns the wire-braced, flat-surfaced upper fin and rudder collapsed together with the starboard elevator, which jammed at a down angle.

The captain, Flight-Lieutenant E.H. Sparling, ordered the engines stopped and sent riggers on to the top of the ship to secure the damaged control surfaces and to detach the control wires so that the lower rudder could operate freely. Eventually, with temporary repairs complete and after being blown as far south-west as Reading, the crippled airship was safely nursed back to Cardington.

After inspection of the damaged fin, all four surfaces were strengthened and re-fitted with additional modification to the ring girders supporting them, together with improved wire bracing to support the vertical and horizontal planes. Additionally, it was decided that the *R31* was overpowered and that this factor had contributed to the damage to the structure. Consequently, it was proposed that one of the Rolls-Royce Eagles should be removed, with the No. 5 engine being relocated on the centre line in an effort to reduce the strain on the structure and minimise the degree of hull flexure.

Following these changes *R31* was then accepted by the Admiralty, and on 6 November left Cardington for East Fortune which was to be her war station. Yet after three hours' flying a number of girders failed and she put into the Howden shed in Yorkshire for immediate repair.

After six days of repairs, which included strengthening the lower longitudinal girders, she again left for East Fortune, but once again after two hours' flying further hull girders began to fail and her captain considered it unsafe to continue and returned to Howden. Subsequent examination showed that, as expected, the Scotch glue had been adversely affected by dampness and was fast losing its adhesive qualities.

On shedding the dirigible the problem was exacerbated by the fact that the Howden shed had been damaged in August 1918 when the *R29* caught fire; the fire damaging the roof causing it to leak badly, a situation which in turn accelerated the deterioration of the *R31*'s framework. No further remedial work was carried out on either the *R31* or the shed during the winter and, following a further Admiralty inspection in the summer, she was condemned and finally dismantled in late 1919, having flown a total of less than twenty hours.

Construction on the *R32* also proceeded at a slow pace but, due to the delay, benefited from the experience gained with *R31*. All girders were now joined using the superior *Kaltleim* glue that was resistant to the effects of moisture, and as with *R31* the engine arrangement was again reduced to five engines producing a total of 1,500hp. The design of the fins was also modified by shortening their length and introducing a sharply angled leading edge, together with an improved wire-bracing system. In addition, all the mainframes and longitudinals were further strengthened as a protection against the powerful aerodynamic bending moments imposed in flight, and again to reduce the flexure of the hull.

R32 conducted her trial flight on 3 September 1919 and displayed satisfactory flight characteristics: her turning coefficient was 7.5 compared to *R26*'s 11.2; with her turning circle being tighter she could also turn faster than the earlier ship; her speed on her original six engines was 70mph and when reduced to five was still capable of 65mph; while her static ceiling was estimated to be 12,000ft, a major improvement on the 3,500ft of the previous class.

Once preliminary trials were completed and she had been accepted by the Admiralty she was flown to her permanent operational base at Pulham in Norfolk. From here, on 10 September 1919, *R32* together with *R33* crossed the North Sea to over-fly the International Aeronautical Exhibition taking place in Amsterdam, from where both ships flew over Brussels and Antwerp before visiting the battlefields of Flanders.

Following the Armistice in November 1918 all the armed services had been swiftly rundown, with tens of thousands of men being demobilised as fast as the administration would allow.

In the navy hundreds of ships were decommissioned or placed in reserve and their crews dispersed, while the RAF (who by the end of 1918 had maintained ninety-five squadrons in France, with a further fifty-five operational squadrons at home) were reduced by mid-1920 to a mere twenty squadrons abroad and only two home defence squadrons.

Contracts for aircraft and equipment were cancelled, and existing aircraft and engines were sent to the scrap heap; in the space of eighteen months the Royal Air Force had almost ceased to exist as a fighting service, the overriding priority being economy and retrenchment. The Chief of the Air Staff Lord Trenchard, when presenting the air estimates in March 1920, was fighting for the very existence of the RAF as an independent service, a task which required every conceivable economy to be made both in terms of manpower and equipment in order to survive.

In October 1919 the navy handed over all its airships to RAF control; this expensive arm of air power was regarded by Trenchard as an expensive luxury of dubious value and superfluous to the requirements of the Royal Air Force. It was, therefore, one of the first areas to be affected by financial stringency. Almost all the blimps were immediately decommissioned and deleted together with the older rigids, whilst work on rigids under construction was suspended pending a decision on their disposal. Despite the success in July of the *R34*'s Atlantic crossing, both the new *R33* and *R34* were shedded, deflated and slung from tackles on a care and maintenance basis. Here they remained for most of 1920.

Among the exceptions to these cuts were the *R38*, which was being completed for sale to the US navy, and the *R32*, which was scheduled to be employed by the National Physical Laboratory for a range of aerodynamic experiments to be flown under the control of an RAF crew. The *R32* subsequently conducted a series of successful trials with the NPL, many of which centred on recording the stresses imposed on her structure under differing flight conditions and with particular reference to the powerful forces acting on her fins and control surfaces.

Although this data must have been made available to the Royal Airship Works technicians who at that time were working on the *R38*, Constructor-Commander Campbell, the designer of the *R38*, seems to have made little or no use of these vitally important statistics. Later analysis of Campbell's workbooks following the loss of the *R38* indicate that all the stress calculations for the giant new craft were based solely on static conditions, with no reference to aerodynamic loadings being included.

Following the NPL tests *R32* was also scheduled for demolition under Treasury restrictions, but the airship was reprieved in order to provide an initial instructional airframe for the American naval detachment that had arrived at Howden in April 1920 for training in rigid airship operation in preparation for the proposed transatlantic flight of the Admiralty A type *R38*, which had been sold to the US navy as the *ZR2*. The *R32* was thus flown from Pulham to Howden on 20 March 1920, where she was hung and deflated awaiting the arrival of the Americans.

The US navy contingent was divided into two groups: engineers and riggers. The engineer group were instructed on the Sunbeam Maori engines on the *R33*, which shared the shed with *R32*, and visited the Sunbeam works in Wolverhampton where *R38*'s 350hp Cossack engines were under test. The rigger section were given practical training on *R32* and the metal framed *R33*, while both groups made regular visits to Cardington where *R38* was under construction.

The original plan had been for the Americans to receive flying training on *R34*, but at this time both *R34* and *R33* were, as previously mentioned, out of service for reasons of economy, the *R34* additionally being overhauled following its Atlantic flight of the previous year. Because of this situation the initial flight training of the US crew was carried out on the pressure airships *SSE3* and the larger North Seas class *NS7* at Howden.

The Americans were set to work overhauling the *R32*; stripping off and replacing most of the outer cover and inspecting the gas cells where examination showed them to be in poor condition, being by then some eighteen months old and in need of repair where the goldbeaters' skin had cracked or separated from the cotton backing material, requiring laborious re-cementing with rubber solution. The refitting proceeded with all speed, the gas cells were repaired and replaced, engines serviced and petrol and electrical systems overhauled. The cost of repairs and the later flying training was borne by the US government. This work was completed by the end of June, allowing the ship to be ready for re-inflation by the eager American crew.

During this period a shortage of coke for the production of gas prevented any of the rigids being available for flying duties, however by the end of May sufficient gas had been produced to allow R33, which had priority, to be inflated and put into flying condition. By the end of July a further supply of hydrogen was available for R32 to be inflated to capacity, after which she was ballasted and loaded with fuel to allow flying training to begin. Once she was airworthy the R32, although still under Air Ministry control and furnished with a British crew, was in fact in all but name an American airship for the duration of the training period.

On 8 June 1920 R33 made her first training flight under the agreement with the US government, with an impressive cruise of seventy hours thirty minutes ranging over 2,273 miles and a number of US navy crew members on board. This was followed by three further flights of long duration in July, also with a large proportion of US navy personnel on board. On 11 August, the refurbished R32 (still under the command of Flight Lieutenant Wann) commenced her first training flight with Commander Maxfield and a largely American crew.

In July 1920, the Air Ministry had informed the RAF of its decision either to sell or give away the existing rigids to any consortium that might be interested in operating the ships in a civil commercial role. An interesting proposal was made in late 1920 by Sir Hubert Wilkins, the famed Australian Arctic explorer, who had previously flown aboard the airship as a guest, to use R32 for an exploratory flight over the North Pole. He saw the airship as the ideal vehicle to map these still unknown areas of the high Arctic.

Had the money for such a venture been forthcoming it is entirely possible that, given the political will, favourable conditions and planning, such a flight could have been accomplished operating from a forward base in Spitsbergen. However, the scheme failed to impress the Air Ministry as either practical or viable.

As no serious contenders could be found for the civilian use of R32 or the other rigids, and in view of the grave economic situation facing the country, in January 1921, following the accident to the R34 when she was severely damaged on a training flight, the Air Ministry took the decision to disband the airship service temporarily.

At the Royal Airship Works at Cardington a partial closure of the works was instituted with the suspension of all work of the R37, although work was allowed to continue on R38. At Howden the training programme for the US navy detachment continued with two further flights in August aboard the R32, totalling thirty-four hours, followed in September by four flights amounting to fifty-six hours. On the last of these flights Commander Maxfield was in command for the first time.

Seven flights took place in October totalling sixty-three hours, including an extended cruise of twenty-one hours on 11 October over the North Sea. In the majority of these flights Commander Maxfield and his crew occupied key stations onboard under the watchful eye of Flight-Lieutenant Wann.

Even with the onset of winter weather the American crew continued to acquire and improve their airship handling skills, managing two flights each in November, December and January. The last series of training flights in the programme consisted of five trips made in February 1921, with the US crew having recorded an aggregate total of 203 hours out of an estimated total flying time of 252 hours – this figure could in fact be even higher as the records of most of these flights are incomplete.

In April 1921 *R32* was housed in the single shed at Howden, where her outer cover was stripped off and her engine cars cut away, the framework being slung from tackles in the roof structure. On 27 April, following the installation of strain gauges in the hull, the frame was tested to destruction with cell 18 being filled to 100 per cent capacity. Subsequently the cell pressure was increased first to 20mm water pressure, where the structure began to fail, and then to 35mm pressure, where several longitudinal girders carried away together with the shear wires and mainframe failures. Following these tests the whole framework was dismantled over the following month, with some examples of the framework later being preserved at the Science Museum in London.

Although the *R32* never fulfilled its designed role as a distant scouting and anti-submarine airship, it was a task for which the class, had it been produced more rapidly, could have made a valuable contribution by virtue of it being basically a sound design. *R32* did, however, provide a useful source of training to the American detachment at Howden, where aboard *R32* Maxfield and his crew had received the best possible instruction from Wann and his colleagues in the complicated art of flying a rigid airship.

This very complete training from the second most experienced airship service in the world included all the aspects of airship operation: loading, trim, determination of lift, meteorology, landing, docking in gusty conditions, mooring and navigation. All essential knowledge if they were to successfully fly the giant *ZR2* across the Atlantic in a few months' time.

14

R34 Across the Atlantic

During the autumn of 1916 the spectre of fear that had for so long lain over the British Isles had been dealt a decisive blow when, within a few short weeks, three German airships were seen to fall blazing from the night sky before the astonished gaze of millions of Londoners. These awesome events were a very public manifestation of Great Britain's ability to protect her citizens from aerial attack, and evidence that at long last the authorities had the measure of the Zeppelin menace.

The vision of these feared predators illuminating the night sky as they plunged to their fiery fate provided a much needed boost to public morale, at a time when the news from the Western Front was anything but encouraging. The feeling had been growing for some time that with

33. Rigid *R33*, shortly after commission into the Royal Navy, on the mast at Pulham, Suffolk, 1919.

the unprecedented advances being made in aeroplane design, together with its ability to operate more effectively at night, albeit still in conditions of great danger, the future of the war in the air lay with the aeroplane.

The majority of military analysts had concluded by this stage of the war that the airship as a military weapon had run its course and was doomed to obsolescence, at least as a strategic bomber. From now on the airship would be relegated solely to a scouting role at sea, operating only where it was safe from fast, high-flying aircraft. Indeed, for some time prior to these events supporters of the heavier-than-air machine had claimed that the aeroplane offered the only practical solution to aerial navigation for either military or passenger transport purposes due to its greater speed, practicality and safety.

With the onset of the autumn raiding period, on the night of 23 September 1916 twelve naval Zeppelins left in favourable weather conditions from their North Sea bases to raid London and the Midlands. Of the airships involved in this raid, four were of the newly delivered L30 Super Zeppelin type of almost 2 million cu ft capacity, with a gross lift of 35 tons and able to rise to an altitude in excess of 13,000ft.

Korvettenkapitän Peter Strasser, commander of the naval airship division, felt confident that with these new airships he at last had the ultimate air weapon with which he could alter the course of the war. The recent loss on 3 September of the army's Schütte-Lanz wooden-framed airship to the incendiary ammunition of Leefe-Robinson's BE2c did little to dent his confidence as they set out into the gathering night.

Eight of the older airships were detailed to attack the English Midlands, where they found they were unable to climb higher than 10,000ft due to the relatively high temperatures. The four L30s took a southerly course, with von Buttlar's *L30* cruising along the Norfolk coast dropping her bombs on supposed targets in the mistaken belief he was over London; such gross errors in position being not uncommon at night in these early days of aerial navigation.

L31 and *L32* made landfall together in the vicinity of Dungeness at 11.45 p.m. and flew on in company until they reached the vicinity of Sevenoaks at 12.10 a.m. From here the *L31* turned in a north-westerly direction towards London, crossing the Thames and distributing her bomb load in a line of destruction across the city.

Meanwhile, Peterson's *L32* had turned north-east at Sevenoaks to cross the Thames north of Dartford. Here she was illuminated by searchlights and soon after this attacked by Second Lieutenant Fredrick Sowery flying a BE2c, who fired three drums of ZPT into the airship, immediately causing her to take fire and explode in a massive ball of flame. She fell to earth at Snails Farm, South Green near Billericay at 1.10 a.m. with all those on board perishing.

The loss of *L32* was witnessed by Strasser who was flying with Mathy in *L31* some 15 miles to the north, and by Bocker in *L33* who had troubles of his own at that time. The *L33* had approached the capital over Essex reaching as far into London as Bow and Stratford, dropping a string of bombs as she went.

Turning eastward, *L33* came under sustained anti-aircraft fire over Bromley where a shell exploded inside cell 14 throwing shell splinters throughout the length of the ship, causing the axial cable to be severed and ripping adjacent cells open resulting in a severe loss of gas. Other pieces of shrapnel smashed girders, cut wiring and cables, and caused major structural damage to the framework.

While the crew worked frantically to repair the torn cells and secure the damaged girders with rope, all surplus equipment, including the remaining bombs, fuel tanks and the wireless station, were thrown overboard in a desperate effort to lighten ship.

At 12.15 a.m. near South Ockendon, Lieutenant de Bath Brandon saw *L33* picked out by searchlights to the north and immediately attacked, emptying two drums of ZPT into her hull without any apparent effect, but undoubtedly further damaging the stricken craft, before losing the airship in cloud.

Bocker struggled on to the coast, with the ship becoming heavier by the minute. He took *L33* some 5 miles out to sea from Mersea Island, leaking gas from her damaged cells. Any hope of

34. *R34* landing at Pulham after the transatlantic flight of June 1919.

reaching Belgium was now out of the question as it became increasingly impossible to keep the *L33* in the air, and the commander realised that to continue would mean the loss of his crew.

Bocker turned the floundering airship slowly back to land and on crossing the coast again he ordered the remaining gas to be valved off, allowing the exhausted ship to come gently to earth at 1.10 a.m. beside a group of cottages at Little Wigborough, 5 miles south of Colchester. Luckily, there was so little hydrogen left in the cells that the ship did not catch fire on impact and there were no casualties amongst the crew.

After warning the terrified occupants of a nearby cottage Bocker set fire to the wreckage, with some difficulty, using signal pistols to fire the fabric outer cover and almost empty gas cells. He then assembled his men and marched off towards the nearby coast, with the intention of seizing a boat and sailing across the North Sea.

However, after a short time as they marched along the country lanes in the darkness they were stopped by a police constable on a bicycle, who, after cautioning them in the standard manner, accepted the surrender of the demoralised crew.

Whilst the loss of two of his most modern ships was a serious setback for Strasser, the fortuitous lack of serious damage to the hull and gondolas of the wrecked *L33* was a gift to the British. The towering wreck, laying incongruously in the fields beside a sleepy English village, was soon to be visited by the Minister for War Lloyd George, together with other important members of the naval and army hierarchy.

Following the realisation that good fortune had presented them with a virtually undamaged example of the latest product of the Zeppelin works, plans were immediately set in train to copy the design. Under the direction of Commander Colemore and Constructor-Commander Campbell from the Admiralty a team of draughtsmen was organised to make detailed drawings of the structure, which was so vast that it was necessary to employ an ordnance survey team to map the disposition of the complicated girder work.

At an Admiralty conference in November 1916, authorisation was given to build three ships (*R33*, *R34* and *R35*), which were to be based on an amended design of the *L33*.

R33 and *R35* were ordered from Armstrong Whitworth to be built at Barlow, where it should be noted that due to the internal dimensions of the shed (150ft in width) it was impractical to build two 78ft-diameter ships at the same time.

R34 was to be erected by Beardmore at Inchinnan, where similar restrictions also applied and immediate enlargement of the shed was required for the purpose.

In the preparation of the drawings by the Admiralty draughtsmen the original metric measurements of the structural members were converted into imperial units, presumably as all concerned in the construction process would be more familiar with this system. Actual fabrication of girders and erection began in June 1917 for *R33* and in December 1917 for *R34*.

Although the design was almost a replica of *L33*, the capture of the improved and undamaged *L49* in France in October 1917 demonstrated certain structural and engineering advances that were thought to be advantageous and should be included in the new airships. Improved engine and power transmission arrangements resulted in the suppression of the three-engine after car, with its cumbersome bracket-mounted, shaft-driven wing propellers used in the *L33* in favour of a smaller twin-engine coupled unit driving a single 19ft-diameter propeller.

Construction proceeded at rapid pace with the *R33* being the first to fly on 6 March 1919 at Barlow, closely followed by *R34* on 14 March from Inchinnan. Initial results from early test flights were generally favourable but indicated a degree of instability to the airship's flight path, particularly in the lateral plane. Both ship were also hunting over a vertical gradient and following a corkscrew motion in flight, which led to an erratic flight path. These ship also gave poor responses to the helm, it being considered that the area of the fins and rudders were too small and ineffectual.

As completed, the ships had a top speed of around 55mph, making them slightly slower than the original, and had a still air range of 4,500 miles carrying a crew of twenty-three.

Civil Airship HMA R33 - 1919
Messrs Armstrong Whitworth & Co. Engineers, Barlow, Yorkshire

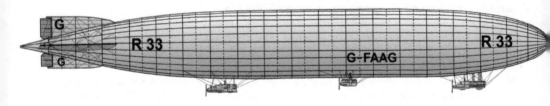

|_____ 100 Feet _____|

1,960,000 Cubic feet capacity
643'x 79' x 92'
59.5 tons gross lift
25 tons useful lift
5 x 240 h.p. Sunbeam "Maori 4" V 12's
Speed 62 m.p.h.
Range at cruising speed 4,600 miles
Static ceiling 13,000 feet
Crew 22

Cost £350,000
1st flight 1st March 1919

NAVAL AIRSHIP HMA R34 - 1919
Messrs William Beardmore Engineers, Inchinnan, Glasgow

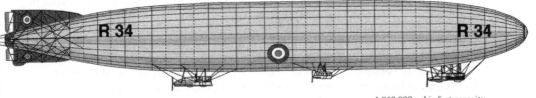

1,960,000 cubic feet capacity
643' x 79' x 92'
59 tons gross lift
26 tons useful lift
5 x 240 h.p Sunbeam "Maori 4" V 12's
Speed 62 m.p.h.
Range at operating speed 4,400 miles
Crew 22

100 FEET

Disposable lift was 26 tons out of a gross of 60 tons, with 3–6 tons being accounted for by water ballast. *R34* had a maximum fuel tank of 6,000 gallons, or 19.2 tons, which was sufficient for 100 hours, whilst the fuel consumption of the five Sunbeam Maori engines was estimated to be 400lb per hour.

The 643ft by 78ft diameter, 1.95 million cu ft ships had cost £350,000 to build and could be considered, despite certain problems, the first and indeed the most successful British airships.

R34 was dispatched in May 1919 to her war station at East Fortune on the Firth of Forth, where, after several test flights, the government ordered the ship to make a prolonged trial along the north German Baltic coast. This voyage was not only to fully test the ship for the proposed Atlantic flight, but in the expectation that the appearance of a British rigid airship cruising off their coast would encourage the Germans to finally sign the Versailles Peace Treaty.

The trip was not without hazard as the two countries were technically still at war. With this in mind the airship was armed with depth charges and machine guns were mounted in the cars and on the top platform, with the crew operating under war conditions.

R34 departed from East Fortune with Major Scott in command on the evening of 17 June, crossing the North Sea, passing Heligoland, Wilhelmshaven, over the Kiel Canal to Hamburg and along the Baltic coast. On several occasions during the course of the flight German aeroplanes closed on the airship and the crew went to action stations, but nothing came of the encounters.

Major Scott headed north over the Baltic Sea, over-flying the coasts of Sweden and Norway – infringing their neutrality as he did. After radioing base, Scott turned *R34* westward across the North Sea against adverse winds, where the craft experienced some engine trouble before finally landing at her base after a flight of fifty-six hours in the early morning of 20 June.

In late 1918 meetings were held at the Air Ministry involving the main constructional firms, financial institutions, the Post Office and the Cunard shipping company for the express purpose of planning a commercial airship service.

Following discussions in December 1918 between the Air Council and the board of the Admiralty concerning civil usage of existing airships, the Admiralty offered to lend the yet-to-be-completed *R 34* to the Air Ministry to undertake long-distance commercial demonstration flights, including a proposed transatlantic flight.

After complicated deliberation, the decision was taken that the *R 34* should undertake the Atlantic crossing at the earliest date, specifically to gather operational and meteorological information for a transatlantic passenger service.

The necessary arrangements were made in America for the US navy to provide facilities for receiving *R 34*, with the servicing of fuel and hydrogen for the airship on Long Island. With only two weeks until the proposed flight the *R 34* was speedily repaired and overhauled, the troublesome after engine was replaced and she was gassed up and loaded with 1 ton of oil, 3 tons of water ballast and 16 tons, or 4,900 gallons, of fuel. Finally, with Major Scott again in command of a crew of thirty, the heavily laden airship lifted off from East Fortune at 2 a.m. on 2 July 1919. She turned her nose north-westward and climbed slowly into an overcast sky, bound for New York.

At the outset of the voyage the *R 34* was so heavily laden with fuel that it had only just enough buoyancy to lift her own weight, and relied on the dynamic lift provided by the airflow over the hull to be able to fly almost 1 ton heavy. Flying via the Clyde and over Rathlin Island

35. *R 34* at Mineola Field, Long Island, 6 July 1919.

in heavy weather, the *R34* pressed on into the Atlantic relying on dead reckoning to plot their course on this epic flight.

By the end of the first day the airship had travelled 1,000 miles over the ocean and, apart from some minor problems with the engines and the discovery of a stowaway (a crew member who had not been selected for the flight), all was well.

During the second day a weather system developing to the south-west was utilised by Major Scott to assist the progress of the airship westward, whilst a rare glimpse of both the sun and the surface of the ocean allowed the navigating officer to fix their position with some degree of accuracy. By carefully resting the engines in turn, the engineering staff could carry out maintenance and minor repairs while husbanding the fuel reserves as they fought the strength of the westerly headwinds.

On the morning of the third day Newfoundland lay only 250 miles to the west, but the *R34* required a further two days to cover the remaining distance to New York. During this period the *R34* had to endure almost constant headwinds and extremely rough weather, and at one time Major Scott was sufficiently worried by the fuel situation to ask the US navy to send a ground crew to the northern point of Long Island should an emergency landing be necessary.

However, after her trials and tribulations the *R34* arrived safely at Mineola outside New York on the morning of Saturday 6 July 1919, after being in the air for 108 hours, and having covered 3,130 sea miles. Large crowds of enthusiastic Americans were on hand to welcome the arrival of the *R34*. After the airship had been secured by being moored in the open on the three-wire system assisted by a large ground crew, the officers and crew were taken off to a round of receptions and speeches and to enjoy the hospitality of their hosts.

Anchoring an airship in the open without the security of a protective shed, relying on the ground crew to hold her safely over a period of days, was naturally a far from satisfactory situation. After being successfully held in this way for four days, Major Scott heard news of an approaching storm front and so hurriedly departed from Mineola at midnight on the 10 July. After circling New York where the airship was lit by searchlights, she turned eastward heading for home.

The homeward voyage was by comparison almost without incident, except that the airship was instructed by the Air Ministry to divert to Pulham instead of flying direct to their home station of East Fortune. Accordingly, their arrival after this epoch-making flight was somewhat muted when the *R34* dropped her lines over Pulham in the early morning of 13 July after a flight of seventy-five hours.

Whatever the reasons behind the playing down of publicity associated with the flight by the Air Ministry, nothing can detract from the magnificent achievement it represented both technically and in the skill and bravery demonstrated by those involved.

Upon her return from the Atlantic flight, and after some essential servicing including again replacing the troublesome rear engine, *R34* left Pulham to return to her base at East Fortune via London, where she circled St Paul's and the Houses of Parliament at low altitude in order to show her off to MPs and the public alike. At East Fortune *R34* was taken into her shed where she spent August–January 1920 in a much needed refit.

In February 1920 she was flown to Pulham again in a flight of seven hours, remaining at that station for six weeks performing a few local flights to test her equipment. In mid-March *R34* was assigned to Howden where she had her bow strengthened in order to be fitted with bow-mooring gear. This was never installed, however, the assigned coupling later being fitted to *R33* instead.

Following the government decision to disband the airship service in November 1919 the *R33* and *R34* were both deflated and hung in the giant No. 2 shed at Howden, with the smaller wooden-framed *R32* occupying No. 1 shed and also empty of gas. All three ships were made available for instruction to the American airship contingent who were to take over the *R38/ZR2*.

Due to shortages of coke for gas production it was not until May 1920 that sufficient gas was made available to fill the *R33*. In June, with US naval personnel included in the crew, the *R33* made a flight of over seventy hours covering 2,300 miles. This was followed later that month with a flight in co-operation with the Atlantic Fleet. During July a further three flights, all of over 1,000 miles, were made with the US personnel on board, giving them invaluable training for their intended Atlantic flight.

In January 1921 it was decided that the recently overhauled and refilled *R34* should be put at the disposal of the Americans, as *R33* was required for scientific tests to be carried out by the NPL and to operate with the navy. As discussed, under Treasury restrictions on expenditure the *R32* was due for demolition, whilst *R34* was to remain in commission for the benefit of the US navy airship detachment at Howden in preparation for the proposed transatlantic flight of the *R38*.

These plans were thrown into confusion when, at noon on 27 January 1921, *R34* with *R32* in company departed Howden on training flights, with *R32* carrying an American crew under Commander Maxfield USN. The two ships separated off the Norfolk coast and cruised independently over the North Sea for several hours until *R32* received a recall signal in the late afternoon forecasting bad weather. Responding to the signal Maxfield set course for Howden, landing safely two hours later in deteriorating weather conditions.

The *R34* meanwhile, due to defective wireless communications, failed to receive the recall signal and flew on through the worsening weather over the sea. In the late evening Flight Lieutenant H. Drew turned inland, but due to navigational errors her four navigators placed the *R34* 40 miles north of her actual position as she crossed the coast. Unable to determine her true position, shortly before midnight she ran into the rising ground of the Cleveland hills and scraped an unseen high moorland ridge. This damaged the control car, smashed the forward propeller and caused more serious damage to the after engine car, wrecking its two engines.

Flight Lieutenant Drew quickly dropped ballast aft to re-trim the ship, stopped the damaged engines on the centre line and increased power on her two wing engines to regain control. Under her reduced power the *R34* limped back to Howden twelve hours later where, due to rising crosswind, she was unable to be housed in her shed and had to be moored out on the three-wire system.

During the following hours, with the wind increasing in violence, the famous airship was battered to the ground and reduced to a total wreck. Over the next few days she was dismantled where she lay. This was a tragic end for this historic craft, which had accumulated 800 hours flying time in its short but eventful life.

Whilst her sister ship, the *R33*, was not to undertake a transoceanic voyage, she did have the distinction of becoming Britain's longest lived, and arguably most successful, rigid airship.

Following her trials at Barlow, *R33* was commissioned into the Royal Navy and dispatched to Pulham air station where, between 6 March 1919 and October 1919, she made twenty-three flights totalling 337 hours.

One of these flights in July 1919 was of thirty-one hours' duration, flying over the Midlands, Liverpool, North Wales, the Isle of Man and down the Irish coast before returning to Pulham. A long flight in September 1919 was made to the aeronautical exhibition held at Amsterdam in the Netherlands, accompanied by the wooden-framed *R32*, where both airships cruised over the exhibition grounds before visiting the battlefields of the Western Front.

In April 1920, having survived the government economy cuts, *R33* was registered as a civil ship with the registration G-FAAG and put at the disposal of the National Physical Laboratory in order to accumulate aerodynamic data for future airship development.

During these trials the *R33*, although a civil ship, was crewed by RAF personnel and moored at the high mast at Pulham where she successfully rode out winds gusting at up to 80mph.

Experiments were also carried out with the releasing of aircraft when, during May 1920, an unmanned Sopwith Camel was dropped over the North Sea as the first stage in a drop-and-recovery programme.

In April 1921 *R33* was employed checking the night lighting for the London–Paris air service, operating from a wooden mobile mast at Croydon, and later the same month was occupying the No. 1 (single) shed at Pulham. When the *R36* was badly damaged at the Pulham mast during deteriorating weather conditions, which required her to be housed immediately, the surrendered *L64* had to be dragged out of the No. 2 shed and was demolished on the field to make room for her.

Between February and June 1921 the *R33* made a total of fifty flights totalling 172 hours, and during June she made an appearance at the Hendon air display. From Hendon she was dispatched to occupy the vacant *R38* berth at Cardington where she was initially housed in flying trim, but was deflated and decommissioned following the loss of the *R38* in August of that year.

The *R33* remained in her hangar for the next three years until, in 1924, the Labour government led by Ramsay MacDonald proposed to undertake the establishment of an airship service on imperial air routes linking the empire. This decision led to the part adoption of the Burney civil airship development scheme, which resulted in the building of the privately funded *R100* and state-controlled *R101*.

In order to provide much needed data on airship operations the *R33* was reconditioned, fitted with new gas bags and engines, and by 2 April 1925 had made several test flights.

On 17 April, whilst moored at the Pulham high mast in a 50mph wind, the mooring bracket fractured, forcing the airship to collide with the mast and wrenching the mooring cone and part of the bow out of the airship. This damaged girders and bow frames and deflated No. 1 gas bag. The *R33*, with Flight Lieutenant Booth in command and eighteen men on board, drifted downwind, narrowly missing the shed as the airship, with a gaping hole in her wrecked bow and exposed gas bags, was driven before the wind.

Flight Lieutenant Booth quickly dropped ballast, bringing the crippled airship into trim. Rising to 4,000ft, he sent men forward to secure the damage by sheeting in the torn outer fabric to protect the fragile gas bags. This work was successfully carried out at great risk by the riggers who, after several hours of struggle, managed to secure the damaged girders and jury rig an effective covering at the bow of the vessel.

Whilst these emergency repairs were in progress Booth and his crew flew the damaged craft as a free balloon, where, under the force of the wind, the *R33* was blown across the North Sea as far as Holland. Booth and his crew managed to get two of the wing engines running and, cautiously, the airship made her way back across the North Sea. After a flight of twenty-nine hours she landed safely at Pulham.

The incident was a major news story, followed avidly by the newspapers who were on hand at the airfield to record the damaged airship dropping into the hands of a hastily assembled ground crew augmented by a large number of local women and children. For their devotion to duty and outstanding airmanship Booth was presented with the Air Force Cross while Coxwain Hunt and other crew members received the Air Force Medal.

Despite the severe damage *R33* had received she was again repaired and had her bow strengthened in order to continue with her experimental work, which included the dropping and recovery of a de Havilland Hummingbird light aircraft in December 1925. Further experiments with releasing aircraft were carried out with manned Gloster Grebe fighters being dropped on two occasions in October and November 1926.

Plans were put forward for the *R33* and the *R36* to carry out a series of experimental flights to Egypt in support of the imperial airship service scheme, but in neither case was this proceeded with. By then the structure of the *R33* was showing signs of its age and metal fatigue was evident. Standing orders in 1926 recognising her fragility by stating 'that no two men should be allowed aft of frame 33 at any time'.

Accordingly, and conscious that another ill-timed accident would be politically disastrous to the development of the planned empire airship service then in hand, the *R33* was shedded at Cardington and eventually dismantled during 1928.

The *R33* and *R34* were the most effective rigid airships built by the British. They demonstrated great structural strength and the ability to survive severe weather conditions and major structural damage, which they owed to their Zeppelin lineage in copying a basically sound design. The *R33* flew for a total of 735 hours, whilst the *R34* flew for 500 hours, the combined figures representing more than half the total hours flown by all other British rigid airships.

15

The Last of the Aerial Cruisers:
R36 & Shenandoah

On the night of 19 October 1917 eleven naval airships set out from their North Sea bases to attack the industrial regions of the British Midlands; targeting Sheffield, Manchester, Leeds and, if weather conditions allowed, the great seaport of Liverpool on the west coast.

Liverpool in particular was seen as a target of great importance to the enemy as the main destination for war material and food coming from the United States, which was now itself involved in the war. The Irish Sea and the approaches to Liverpool were the subject of continuous and intense predation from the U-boat campaign, and a successful aerial attack on the great port would underline the ever tightening stranglehold the Germans could bring to bear on our trade.

British intelligence based in Room 40 at the Admiralty were aware of the impending raid by picking up the wireless message as the airships left their bases, stating that they had 'Only HVB on board'. The Admiralty cryptographers recognised this abbreviation for the *Handelsschiffverkehrsbuch*, the German merchant service code book that had long been compromised and indicated by its presence that as the airships, having left their more secret naval codes behind, were destined for an attack on the British Isles.

This was to be the first large-scale raid by the Naval Air Service for some time and Peter Strasser, the leader of naval airships, hoped to achieve a decisive blow by employing his new Height Climbers. With their ability to rise to 20,000ft whilst carrying a heavy bomb load he confidently expected that these superior airships would have little to fear from either anti-aircraft fire or the home defence squadron aircraft, none of which, in his estimation, were capable of rising to that altitude.

The weather conditions were forecast as being 'exceptionally favourable'; the wind for the German Bight being light south to south-west and over the British Isles moderate west to north-west later, with tellingly increasing stronger currents at higher altitude. The German weather service, unlike their British counterparts, received no weather information from stations further west than occupied Belgium, which severely affected their ability to accurately forecast conditions over the British Isles during raids.

In hindsight, their U-boats patrolling in the Channel and the Western Approaches could have been employed to record wireless local meteorological data from the direction of the prevailing weather systems to augment their weather maps.

After setting off in the early afternoon from their bases the airships steered west into the sunset along the Friesian Islands against a light wind, but on approaching the British coast in the early evening, as they began to climb to their attack altitude of around 16,000ft, the first signs of trouble came. At this altitude the Zeppelin commanders found the wind backing to the north and increasing in strength as they climbed; soon the air fleet found itself fighting in the teeth of a full gale.

Kapitänleutnant von Buttlar commanding L54 was alerted to the increasing strength of the wind and realised that it would be impossible to reach the intended targets against the force of the storm. Bringing L54 down to a lower altitude into less turbulent air at around 5,000ft, he crossed the coast near Happisburgh then dropped his bombs in a haphazard manner in the Colchester area before turning out to sea near Clacton and setting course for home.

L54 managed to elude a solitary BE2c from Great Yarmouth Air Station sent up to intercept him, making out to sea to return safely to his base at Tondern after a flight of some twenty hours.

Of the other ships, L41, L46, L47, L53 and L55 – after wandering across the Midlands dropping their deadly cargoes of high explosive on whatever targets they could discern (or imagined they saw through the cloud wrack at heights of between 16,000 to 20,000ft) – eventually safely returned to their respective bases.

During the course of these wanderings L55, with Kapitänleutnant Hans Flemming in command, set an altitude record reaching 25,000ft, this being the greatest height ever attained by any rigid airship. Kapitänleutnant Flemming managed to struggle home, but after a flight of over twenty-eight hours the exhausted ship made a crash landing in a forest clearing near the River Werra in Thuringia. Fortunately without loss of life, although the ship itself became a total loss.

Under these high-altitude conditions the cold and the lack of oxygen affected the efficiency of both the ship and the crew, who, despite the provision of oxygen apparatus, became lethargic. Their responses were slowed, they found it difficult to perform their duties and severe altitude sickness was induced.

Due to the unexpected change of wind direction and its ferocity, together with the extreme altitude at which they were flying, the airships now became hopelessly lost over the darkened countryside and were having to rely on dead reckoning, with the occasional wireless bearings from Germany, to help them home.

In the control car as rime ice covered all the metal surfaces, half-frozen crew stood at their posts peering through frost covered windows, the alcohol frozen in the compass and differential contraction of metal and wiring causing the rudder and elevator cables to jump off their guiding sheaves. The crews fought against the wind to bring the storm-battled ships back to their bases.

Meanwhile, in the exposed wing cars the engines were also struggling, working at the very limits of their mechanical ability, starved of oxygen, plagued by congealing lubricants and frozen supply pipes as the exhausted mechanics sought to keep them running.

Four of the airships in the fleet were never to see Germany again, the first of these being the L44 with Kapitän Stabbert in command. After making landfall over the Wash at 8.30 p.m. Stabbert realised he was far to the north of the capital and so turned south and raced before the gale as he bore down on London at high speed. Due to their altitude and the layered cloud below that only occasionally allowed a view of the ground, Stabbert was unable to determine what lay below him and dropped his bombs in a random manner over rural Hertfordshire, some of which fell on Leighton Buzzard.

Speeding southward, L44 then crossed the Thames at Erith before turning out to sea at Folkestone. Here, she fell in with L52 and in company they headed off on an east-south-easterly course, battling all the time in an effort to counter the leeward drift to the south.

At first light the L44 found herself and her exhausted crew drifting over the trenches of the Western Front with two engines frozen to a standstill. At once she came under heavy and accurate French anti-aircraft fire. It was only a matter of time before the L44 was hit by an incendiary shell amidships, exploding immediately with great violence to fall blazing to the ground with the loss of her commander and the entire crew.

The L52 viewed the loss of her consort at a distance to the north-east where, alarmed at the sight, Kapitänleutnant Friemel dropped his remaining ballast and made off eastwards at flank

36. US naval airship *ZR1 Shenandoah*; the first rigid designed and built in the United States.

speed. His half-frozen crew was relieved to eventually land at the Alhorn base with only a few gallons remaining in her tanks after a nineteen-hour flight.

The *L45*, under Kapitänleutnant Kolle, struggled against the wind at 19,000ft as far as Northampton. Although Kolle was unaware of his true position he dropped salvos of bombs at supposed targets below, before being swept by the gale southward far from his intended course. Towards 11.30 p.m. the crew were astonished to see the lights of a huge city below them and realised that they were over London.

The majority of the *L45*'s bombs fell on north-west London, whilst her main cargo of three 660lb high-explosive bombs fell in rapid succession near Piccadilly Circus, Camberwell and Hither Green, causing considerable damage.

A peculiar feature of the raid was that, due to the gale and the altitude, the airships engines were not heard as they passed overhead, neither were the defences initially aware of their presence over the city. Subsequently, this raid became known as the Silent Raid.

Crossing the Thames and the Medway the *L45* made out to sea, where a BE2c briefly pursued it, but was unable to make contact. With most of the crew suffering from altitude sickness the airship drifted over France with three engines stopped, unable to make any headway to the east and safety as she passed over Amiens and Compiègne.

By daylight the *L45* was over Lyons still being driven south by the gale, until finally, at Sisteron and almost out of fuel with only two engines running, Kapitänleutnant Kolle brought the ship down in a broad river valley. Here, after slashing the cells the crew staggered away from the wreck, firing the ship with flares from a signal gun, which effectively destroyed the great airship. The unharmed but exhausted crew surrendered to French soldiers who had

The "Silent Raid" on the night of the 19th-20th October 1917 Showing tracks of the attacking airships

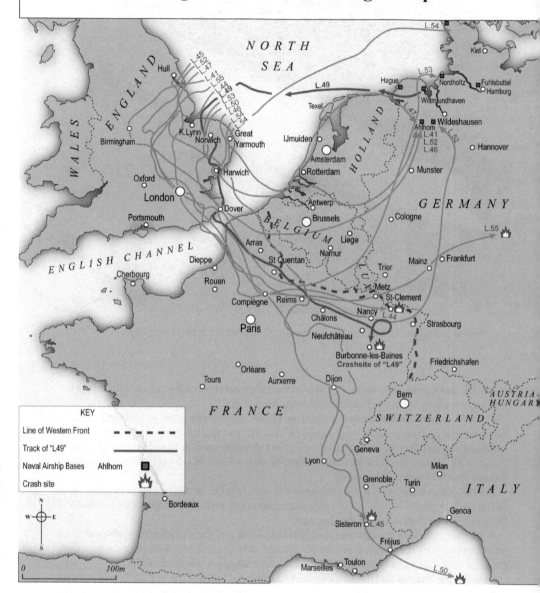

hurried to the scene, safe in the knowledge that at least they had not allowed their ship to be captured.

The third of the raid's casualties was the *L50* under Kapitän Schwonder. He came inland over Norfolk around 8.40 p.m., dropping his bombs in open countryside from 16,000ft before leaving the coast in the vicinity of Harwich an hour later. The *L50* was also carried far to the south by the fierce gale blowing out of the north-east. Soon she lost power on two of her engines and, being unable to fight the wind, her course also took her over the trenches to the east of Reims.

Flying now at only 7,000ft they saw the *L44* flying close aboard to the north-east and observed her fiery destruction by the French gunners. Horrified by what he had witnessed, Schwonder turned *L50* away to the south-west in an effort to escape a similar fate, and later passed over the *L49* at Bourbonne-les-Baines lying seemingly undamaged on the ground.

In the early afternoon Schwonder brought the *L50* lower, to within a few hundred feet of the ground, with the intention of destroying her by flying into a hillside. Misjudging his approach the *L50* hit a small hill, tearing off the control car, and as the ship hit the majority of the crew jumped from the engine cars and the keel. Relieved of this load the hull of the *L50* rose rapidly into the air with four of her crew still left on board, whilst on the ground the remaining sixteen crew members were safe apart from minor injuries.

The *L50* rose to 23,000ft, far above the attacking French aircraft that tried to bring it down. Here, despite their best efforts, the *L50* escaped southward and was last seen heading far out over the Mediterranean. Its ultimate fate, and that of the four crew members, is unrecorded.

The most important victim of this raid from the entente cordial point of view was the *L49* under the command of Kapitänleutnant Gayer. This most modern of the Height Climbers was destined to fall into their hands in an almost undamaged condition.

German Naval Airship L49 (LZ 96) - 1917
Zepplin "U" Type
Luftschiffbau Zeppelin GmbH, Lowenthal

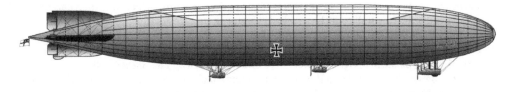

100 FEET

The L49 took part in the "Silent Raid" of the 19th October 1917 over London, during the return flight the airship made a forced landing at Bourbonne-les Baines in the Haute Marne. She was captured intact, her design serving as the basis for the British R36 and the American ZR-1 "Shenandoah".

1,970,500 Cubic feet capacity
645'x 79' x 92'
62 tons gross lift
37 tons useful lift
5 x 240 h.p. Mabach HSLu 6 cylinder in line
Speed 66 m.p.h.
Range at cruising speed 4,400 miles
Crew 19
Ceiling 19,000 feet

Frustrated from reaching her chosen objectives by the gale, the *L49*, like her compatriots, wandered over Norfolk unaware of her position, depositing her bombs in a random pattern without causing any military damage before going out to sea about 10.30 p.m. Dawn found *L49* still lost over France, where far to the north the crew witnessed the *L44* falling in flames, which further depressed their flagging morale.

Gayer had no idea he had been driven so far south, believing he was in fact over Holland until he was suddenly attacked by French scouts, from whom he escaped by dropping his remaining ballast and climbing to a safe altitude. Realising there was no possibility of making it back to Germany, Gayer resolved to bring his ship down and destroy it. He was now flying over the rolling French countryside of the Haute Marne, and began to valve gas to descend.

There were no aircraft in sight so he selected a wide, wooded valley at Bourbonne-les-Baines. Sailing up into the wind at low altitude he gently set the giant airship down into a small wood where the trees helped to cushion its fall and hold it safely in their branches. The exhausted Germans stumbled from their ship, glad to have survived their ordeal, but were quickly captured by French troops before they could destroy the intact airship.

During the descent Gayer had ordered his maps and codebooks to be destroyed by tearing them up and throwing the pieces overboard. Later, an observant American army officer searched through the woods along the track of the airship and managed to recover the torn portions and piece together much of this confidential data.

From Strasser's viewpoint this raid, meant to be a decisive blow against the enemy, was in fact an unmitigated disaster with the loss of five of his latest ships. These losses only served to convince the high command of the vulnerability of the airship as a practical offensive weapon, and whilst the navy continued to use Zeppelins for scouting their role as strategic bombers was taken over by the army's 'Giant' bombing aeroplanes from 1917 onwards.

Of the raid an official British historian commented:

> the Germans have very little to show for their appreciable losses, although things could have gone very differently as the defence system had very little to do with the disastrous outcome of the raid, with guns and searchlights being of little use, whilst the aeroplanes that did go up were incapable of reaching the ceiling at which the Zeppelins were flying.

The intact hull of the *L49* was rapidly surveyed by a party of French army engineers sent from Paris, with detailed drawings being made of the entire structure, which was then carefully dismantled before the hull could be wrecked by a storm.

The entente were delighted to have this example of the latest type of Height Climber, the completed plans were quickly copied and distributed to the entente, where later they were to form the basis of the plans for the British *R36* and the American *ZR1 Shenandoah*.

It was naturally assumed that the *L49* represented a technical advance over the earlier *L33*, with the entente noting that the *L49*'s structure had been considerably lightened compared to *L33* in order to increase the useful load. What they didn't appreciate was that this increase in load had only been achieved at the expense of so lightening the structure that the *L49* was proportionally weaker than the earlier airship, and was specifically designed to operate safely only at high altitude,

Both the British and American designers failed to recognise, or chose to overlook, these shortcomings that were inherent in the design, they being anxious to incorporate these advanced features in ships that were meant to fly in a maritime reconnaissance role in the denser air at lower altitudes.

Manoeuvring an airship at speed in denser air at low altitude induced a greater strain on these lightened frameworks; the US navy's *Shenandoah* was later lost when she suffered a major structural failure flying in a squall over Ohio in 1925, while the *R36*, which incurred some serious failure of girders in her short flying life of only some two months, was lucky not to have suffered a similar fate.

By September 1917 the War Cabinet were sufficiently confident of the progress being made on the Super Zeppelin copies to approve a new airship programme that was to comprise of sixteen R33 class rigids, at a total cost of £2 million. This programme also called for twelve North Sea tri-lobes and thirty-eight SS Zero non-rigids for an additional cost of £1.169 million, with this part of the plan being successfully accomplished over the next eighteen months.

This ambitious programme required that five rigids were to be immediately laid down: R33 and R34 (already in hand), followed by R35, R36 and R37. The remaining eleven ships were to be ordered as soon as new sheds to house them became available.

Following the capture of the more advanced Height Climber L49 and an evaluation of its performance and structure, improved versions of the two prototypes, R33 and R34, were sought by the Admiralty. The Admiralty immediately issued a revised specification, calling for a minimum operational requirement which included an endurance of ninety-six hours (or four days) at full speed.

It was proposed that the structure of the new craft should be lightened in accordance to L49's scantlings. Additionally, the new class would have their bows strengthened for towing by destroyer to increase endurance – this last measure being considered necessary to counter the increased range of the new U-boats operating to the west.

In this advanced class a crew of four officers and twenty-six men in two watches were to be carried, and the hull was to have a fully waterproof outer cover together with swivelling propellers on the forward engines, which were to be either 260hp Sunbeams or 240hp Fiats. A wireless apparatus with a range of 600 miles was to be fitted, together with an offensive bomb load of 2,300lb.

Naval Airship HMA R37 - 1919
Admiralty "Z" Class-Anti-Submarine Patrol Airship
Messrs Short Brothers Ltd, / RAW Cardington, Bedfordshire

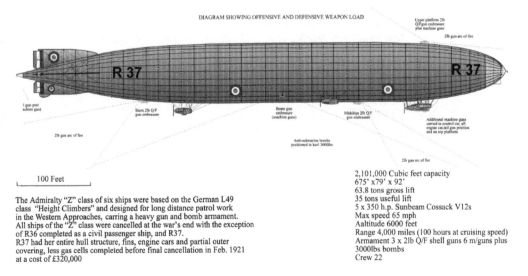

DIAGRAM SHOWING OFFENSIVE AND DEFENSIVE WEAPON LOAD

100 Feet

The Admiralty "Z" class of six ships were based on the German L49 class "Height Climbers" and designed for long distance patrol work in the Western Approaches, carring a heavy gun and bomb armament. All ships of the "Z" class were cancelled at the war's end with the exception of R36 completed as a civil passenger ship, and R37. R37 had her entire hull structure, fins, engine cars and partial outer covering, less gas cells completed before final cancellation in Feb. 1921 at a cost of £320,000

2,101,000 Cubic feet capacity
675' x79' x 92'
63.8 tons gross lift
35 tons useful lift
5 x 350 h.p. Sunbeam Cossack V12s
Max speed 65 mph
Aaltitude 6000 feet
Range 4,000 miles (100 hours at cruising speed)
Armament 3 x 2lb Q/F shell guns 6 m/guns plus
3000lbs bombs
Crew 22

The *R35* was ordered at the same time as *R33* and *R34*, but was enlarged by one bay (as in the *L49* design) to give a capacity of 2,101cu ft and a gross lift of 62 tons, with power being provided by four 360hp Sunbeams. Erection of *R35* had to await the completion of *R33*, as the Barlow shed with an internal span of 150ft made it impractical to build two ships at the same time.

Fabrication of the mainframe diamond truss and longitudinal girders, however, began in March 1918 in the adjacent assembly shops, continuing until work was suspended in November 1918. The completed girders were stored along the walls of the shed ready for assembly, and by this method the greater part of the units comprising the hull structure were completed awaiting assembly.

Once the *R33* had departed the berth at Barlow in March 1919 erection of the framework commenced, with a considerable portion of the frame being assembled (possibly up to 60 per cent complete) before the contract was finally cancelled in July 1919. The completed spare girders, and in all probability girders taken from the dismantled *R35* together with those of the *R40* (which was a cancelled Admiralty A class design) were delivered to Short Brothers at Cardington to be incorporated in the *R38*.

Following the cancellation of the *R35* a more advanced type, the Admiralty Z class of six airships designed for long-distance patrol work in the Western Approaches carrying a heavy gun and bomb armament, were ordered. The first of this class, the *R36*, was authorised in January 1917 with the contract being awarded to Beardmore, and was laid down at Inchinnan in March 1919 following completion of *R34*.

Her sister ship *R37* was laid down at the same time at Shorts, Cardington with both these ships, like *R35*, being lengthened by an additional 32ft (10m) bay.

The *R36* was the only one of the class to be completed and to fly; the *R37* was abandoned when 85 per cent complete at the time of the Armistice, with the remaining airships also being cancelled at the same time. *R36* and her sister ship *R37* had a capacity of 2,101cu ft, giving them a gross lift of 62 tons and a disposable of 32.5 tons.

The original proposed power layout was four 350hp Sunbeam Cossack engines – one less engine than *R34* – in order to achieve the higher ceiling of 17,000ft and which were

Civil Airship HMA R36 -1921
Messrs William Beardmore & Co. Engineers, Inchinnan, Glasgow

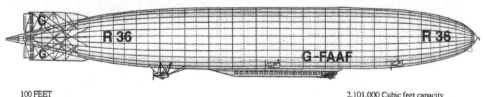

100 FEET

Cost £350,000
First flight 1 April 1921
Last flight 21 June 1921
Total hours flown 97
Demolished June 1926

2,101,000 Cubic feet capacity
675' x79' x 92'
63.8 tons gross lift
16 tons useful lift
2 x 240 h.p. Mabach MbIV
and 3 x 350 h.p. Sunbeam Cossack V12s
Max speed 65 mph
Range 7700 miles
Crew 23
Passengers 50

Passenger accomodation aboard the airship R36 - G-FAAF
Messrs William Beardmore & Co Engineers Inchinnan Glasgow

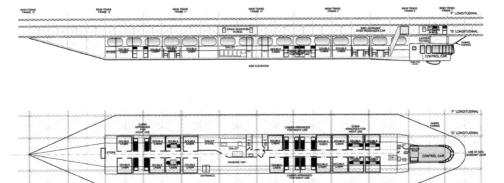

R36 LIFT AND LOAD DETAILS
Gross lift : 63.8 tons
Disposable lift : (prior to car being fixed) 35 tons
Disposable lift : (after attachment of car) 16 tons
Fuel comsumption at cruising speed : 0.065 tons per mile
Fuel requirement for London - Cairo flight :13.65 tons
Remaining disposable available for ballast, crew, passengers,
 mails etc. : 2.35 tons!

R36 PASSENGER CAR DETAILS
Dimensions Length 131' x width 8.5' x height 7.5'
Weight 16.5 tons
Passenger capacity 50 (Day and night accomodation)
Crew 26
Airship considered by Air Ministry to be capable of carrying
30 passengers plus one ton of mail between London and
Cairo (2,100 miles) in 72 hours.

optimistically expected to provide a top speed of 65mph. As completed, *R36* carried three of the Sunbeam engines, supplemented by two 260hp Maybach's complete with their original gondolas taken from the surrendered *L71*, which by this date was in the process of being dismantled.

During 1919 the government, in the form of a committee headed by Field Marshal Jan Smuts, were considering the development and establishment of empire and European air links once the peace treaty with Germany was signed. As part of this plan they ordered that *R36* should be completed as an experimental passenger-carrying airship, and it was assigned the civil registration G-FAAF.

In view of its original role as a high-altitude, structurally lightened naval scout, the decision to employ her in a low-altitude civil capacity can only be regarded as a serious error of collective judgement. It would seem to have been a rather desperate measure on the part of the government in order to recoup some of the massive unfulfilled investment that had been made in rigid airships up to that time.

Naturally, airships designed for war purposes were entirely unsuited to fulfil a civilian passenger-carrying role, but even Vickers made a proposal to convert the private venture *R80* to a passenger transport in an effort to salvage some financial return from their efforts.

At Inchinnan the Beardmore-built *R36* was completed as a passenger ship by attaching a rather narrow 131ft-long cabin to the G longitudinals on the underside of the hull. The cabin contained accommodation for up to fifty passengers, complete with sleeping berths, galley, toilets, wicker chairs and tables that afforded a modest degree of comfort to passengers if one overlooked the lack of heating.

The cabin and its fixtures weighed in excess of 12 tons, and this of course affected the airship's available disposable lift adversely, which in turn reduced her fuel capacity and range. The *R36* nevertheless made her first flight on 1 April 1921 from Inchinnan, a short local flight of three hours' duration during which the engines and controls appeared to operate satisfactorily.

On the following day, under the command of Flight Lietnant A.H. Wann together with General Maitland and Air Marshal Robert Brooke-Popham, director of research at the Air Ministry, the R 36 left on its delivery flight. Again the airship handled well, appearing to be stable and responsive to control movements. The R 36 arrived at Pulham after an uneventful night flight of just over twelve hours, where she was moored for the first time to the high mast, riding comfortably in the breeze and taking on additional hydrogen and water ballast via the services incorporated in the tower.

Two days later, on 5 April, R 36 was readied for her third flight with Major Scott in command and accompanied by a group of officials from the Air Ministry and representatives of the press on board. Slipping from the high mast and departing Pulham at 8 a.m., R 36 set course for London in good weather. After passing over St Paul's and the Houses of Parliament, Major Scott set course for Bristol with the intention of gradually running up the engines to full power to test the behaviour of the control surfaces at varying air speeds.

This programme was followed over the next two hours with the rudder and elevator controls being thoroughly tested over a range of speeds, at which they appeared to be performing satisfactorily. Later, however, while flying at 4,000ft at 50 knots in the vicinity of Bath, the ship suddenly pitched up, the height and steering coxswains immediately loosing all steering control, before the ship plunged into a steep dive dropping down to 2,000ft under power.

In response to the emergency the navigating officer in charge of the control car, Flight Lieutenant T. Elmhirst, instantly rang the telegraphs to stop all engines and simultaneously ordered 1,000lb of water ballast to be dropped from the bow to bring the ship level. Major Scott, who had been dining in the lounge with the guests, hurried to the control car and ordered crew members into the keel to balance the crippled ship.

An assessment of the damage was carried out by the coxswains who, on climbing to the top of the hull, saw that all four of the horizontal and vertical stabilisers had either buckled or collapsed. As they drifted over Somerset temporary repairs were effected to the lower rudder while the elevators were fixed in a neutral position. Once repairs were completed the R 36 set course for Pulham at 30 knots, arriving safely at 9 p.m. with Scott having used crewmembers in the keel to adjust the ships attitude and trim when ascending or descending.

Once again, as with R 31, the strength of the girder work and wire bracing of the flat section fins had proved inadequate; a problem that would not be solved until the introduction of the thicker section cantilever fins of later craft.

The R 36 remained in her shed for most of May as repairs were made with strengthening plates and reinforcing channels being fitted to the fin girders (a similar expedient was applied a few months later to damaged girders on the R 38), emerging on 8 June for a short test flight. On 9 June a longer flight of eleven hours was undertaken from Pulham via Nottingham and Liverpool, which was adjudged to be most satisfactory.

The next day the R 36 slipped the Pulham mast at 10 p.m. heading for London then on to Portsmouth, passing out to sea en route for the Channel Islands where she over-flew Sark and St Helier. From here the airship turned westward down the English Channel crossing the English coast again at the Lizard, then followed a course through Devon and Somerset, over Oxford and Aylesbury, eventually dropping her lines at Pulham after a successful and trouble-free voyage of thirty hours.

The airship was out again on 14 June for the Ascot race meeting where she assisted the police in a traffic control exercise, sending her reports by wireless. The editor of the *Aeroplane*, W.H. Sayer, who had been on board for the flight, complained about the absence of heating in the cabin and the lack of food in what he considered an uncomfortable twelve-hour flight.

U.S. Naval Airship ZR.1 U.S.S. "SHENANDOAH" - 1923
Naval Aircraft Factory Philladelphia - Constructed at Lakehurst N.J.

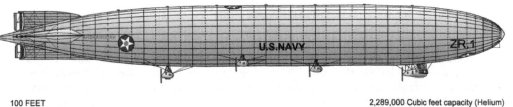

100 FEET

Cost $1,500.000
First flight September1923
Last flight 3rd September 1925
Total hours flown 740

2,289,000 Cubic feet capacity (Helium)
680' x79' x 93'
63.8 tons gross lift 129,000lb
16 tons useful lift 48,000lb
5 x 300 h.p. Packard 1As
Max speed 51kts
Range 3,980 miles
Crew 54

Following this she was again in the air on 17 June when she carried forty-nine MPs on a three-hour trip, mooring again at the new Cardington mast.

On 21 June, flying from Cardington, the R36 was again damaged whilst attempting to moor at Pulham. Here she had collided with the mast as her mooring cable fouled in the winch, causing severe damage to the lower longitudinals and frames, and also deflating gas bags No. 1 and 2. The mooring cable held so the ground crew were able to haul the damaged ship down to the ground, where in the face of a rising wind it was deemed imperative to house Britain's most modern airship.

As the No. 1 shed housed both R33 and R34, and the No. 2 shed contained the surrendered L64 and L71, it was decided that the only solution to the dilemma was to sacrifice the unfortunate L64, which was unceremoniously dragged out on to the field and summarily demolished to make room for R36.

The R36 was again repaired and had her bow rebuilt and strengthened, remaining in the Pulham shed until 1925 when it was planned to re-commission her together with R33 in order to undertake a further series of experimental flights for the new airship programme. It was proposed that R36 could be employed on a trial flight to Egypt and in preparation for this flight she had her outer cover stripped off for replacement, and it was also planned to replace her gas bags.

The Air Ministry rated the R36, which had originally cost £350,000, as being capable of carrying thirty passengers plus 1 ton of mail between England and Egypt in seventy-two hours. The R36 had a gross lift of 63.8 tons and a disposable of 32.5 tons, but this was reduced to only 16 tons when the heavy passenger car was fitted. The fuel consumption was estimated to be 0.65 tons per 100 miles; with the distance between London and Cairo being 2,100 miles this would require 13.65 tons of petrol and oil, leaving only 2.35 tons for ballast, passengers and mail etc, making the whole undertaking quite impractical.

The Air Ministry partially refurbished the R36 during 1924–25 at a cost of £15,000, before realising that the airship was totally unsuitable for such a long flight and the scheme was quietly dropped. The airship was scrapped during 1926 after only eighty hours' flying time.

The R36 was a structurally weak airship with her factor of safety being only half that of the earlier R33 class, as little consideration had been given by her designers to adapting her scantlings to provide a stronger structure, better suited to the conditions that she would be expected to operate under on overseas flights. In view of this weakness it could be considered fortunate that this airship flew so little and only in benign conditions, as she would have almost certainly have been lost in the first adverse weather she encountered on a long flight.

One other airship was constructed to the design of the *L49*, this being the *ZR1 Shenandoah* in July 1919 when the US Treasury approved a bill for the purchase of two large rigid airships for the US navy, allocating the sum of $4 million for the purpose.

One airship, the *R38*, was to be purchased from Great Britain, whilst the other was to be designed and built in the United States. The Bureau of Construction and Repair were responsible for the design of the American-built airship, which was to be based on the plans of the Zeppelin *L49*. Starr Truscott was designated as chief engineer with C.P. Burgess as chief calculator.

It should be noted that, unlike the *R36* and *R37*, from the outset the Americans devoted a great deal of effort into undertaking a full scientific analysis of both the aerodynamic and static loads and stresses imposed on an airship under all conditions. The strength of all structural members were also carefully considered and at the time of her construction she was the strongest airship ever built.

The structure of the airship was to be fabricated at the National Aircraft Factory in Philadelphia, with the completed components to be assembled at the yet uncompleted airship hangar at Lakehurst, New Jersey. During the early design process seven design variants of the original *L49* were considered, with the final approved design being one bay (10m) longer, giving an overall length of 680ft and a capacity of 2,289,000cu ft of helium contained in twenty gas cells.

The hull structure, as previously mentioned, was far stronger than in *L49*, with heavier gauge longitudinals installed between frames 10 and 40, while the lower three longitudinals on either side (H, J and L) were of deeper cross-section than the standard girders and constructed of thicker gauge material. The lower fin and the keel girders were similarly strengthened, while the fins were of a strong, thick cantilever section of an aerofoil shape, with the minimum of external bracing.

The surface area of the rudders and elevators were reduced in area from the original in order to mitigate the bending moment imposed on the control surfaces during high speed turns, with all surfaces limited to 15° of movement each way for the same reason. Design work occupied the period between August 1919 to October 1921, with actual construction commencing with the arrival of fabricated girders from the naval aircraft factory at Lakehurst in April 1922.

When the *ZR1* had been first conceived it was expected that the finished ship would fly with hydrogen in her cells. However, following the public outcry at the loss of life in the US army airship *Roma* disaster in 1922, and the *R38/ZR2* in 1921, these plans had to be revised drastically.

A considerable amount of work had already been done concerning the usage of helium, with the first flight of a helium-filled airship, the navy *C5*, being accomplished in December 1921. While at the Fort Worth plant sizeable quantities of the rare and expensive gas were then being produced. Although the use of helium had undoubted advantages from the viewpoint of safety, it came at a price in terms of an 8 per cent reduction in gross lift as compared to a hydrogen-filled airship of similar volume.

In addition, the substitution of helium with a lift of only 60lb per 1,000cu ft, compared to the 68lb per 1,000cu ft of hydrogen in the *ZR1*, would reduce the cruising range from 3,700 miles at 40 knots to only 1,800 miles at the same speed on helium. Helium at that time was also very expensive, costing approximately $120 per 1,000cu ft compared to $3 per 1,000cu ft for hydrogen.

A leading and enthusiastic proponent for the use of rigid airships in the US navy was the chief of the Bureau of Aeronautics, Admiral William A. Moffett. This officer was single-minded in belief that airships employed in a scouting role would be invaluable to the fleet operating in the vast reaches of the Pacific. Admiral Moffett had been instrumental in the ordering of both the ZR2 and the ZR1, and following the loss of forty-nine lives aboard the ZR2 Admiral Moffett was undaunted, stating: 'We will carry on, building and operating as many large rigid airships as deemed necessary by the Navy.'

Admiral Moffett was under pressure from Washington to demonstrate the new airship to US taxpayers, and encouraged the use the ZR1 *Shenandoah* for publicity purposes, sending her on demonstration trips to large cities across the nation. This course of action interfered with the well-planned flight test programme that had been jointly devised by Commander McCary, Commander Weyerbacher and the civilian test pilot Captain Heinen, a former employee of the Zeppelin Company. Later in her career, Admiral Moffett's enthusiasm for demonstrating her in public also made her unavailable to take part in several planned fleet co-operation exercises, the very purpose for which she had been designed.

The original engine layout called for the installation of six de-rated Liberty 12A engines with a nominal output of 250hp. However, this was subsequently changed to six of the lighter and more fuel efficient Packard 1A-1551 six-cylinder, water-cooled engines developing 300hp.

The design was finally approved by the NACA on 4 December 1921 and, parallel to this decision, the fleet was asked to supply volunteers for rigid airship training, supplementing the survivors of the ZR2 Howden detachment to form a ground school at Lakehurst. The first of the group underwent training on free balloons in March, and later in May on the non-rigid *J1*.

Construction began in April 1922 and proceeded at a steady rate. The parallel centre section was erected first, whilst construction of the stern section and the cantilever fins were more cautiously approached due to their more advanced design compared to the original. By November 1922 the hull was 75 per cent complete, and by June 1923 the hull structure was entirety complete. All gas cells were fitted and the outer cover laced on, with engines, petrol, ballast and electrical systems installed by 12 July, allowing inflation to commence during early August.

On 20 August ZR1 was officially complete. She was weighed-off and airborne for the first time, and walked to the north side of the shed for engine test runs and systems testing. Finally, on 4 September 1923 with Captain Anton Heinen in command, the ZR1 was walked out on to the field by a ground crew of 420. At 5.20 p.m. she lifted off for the first flight carrying twenty-nine persons, cruising in the vicinity of Lakehurst for just over an hour.

Over the next month the new airship made a series of flights to New York, Philadelphia, Washington and other cities, showing herself off to the taxpayers who had stumped up $1.5 million for her completion. In September it was proposed that the ZR1, with just twenty-five hours' flying time to her credit and only a fraction of her trials programme completed, should make a publicity flight to St Louis to coincide with the International Aeronautical congress being held in that city.

The flight commenced on 1 October carrying forty souls. During the flight some damage occurred to the port fin girder work, requiring temporary repairs to be effected with ropes and planks, but despite this mishap the airship successfully reached Lambert Field safely.

The ZR1 made several other flights before the end of 1923, which included mooring mast trials, and in a ceremony on 24 September the wife of the secretary of the navy, Denby, at Lakehurst had named the airship *Shenandoah*, with Commander McCary being appointed as her first captain.

Early in 1923, even before the ZR1 was complete, a plan was conceived to send the new airship on a polar flight to the North Pole and back. This plan was initially approved by President Calvin Coolidge and Denby appointed Admiral Moffett as expedition leader. Over the next nine months detailed plans were made for the expedition, with support ships, mobile masts and advanced bases established on the west coast and in Alaska.

The flight was planned to depart from Lakehurst in June 1924, flying via Fort Worth to Camp Lewis, Washington and on to the base at Nome Alaska. From here the *Shenandoah* was to fly over the Pole to Spitsbergen, returning by either the same route or going on to England and completing the journey with an Atlantic crossing. In order for the airship to accomplish this long flight, consideration was briefly given to lengthening the *Shenandoah* by an additional bay and refilling her with hydrogen. In the event, neither of these measures were implemented.

Before this could happen, however, the *Shenandoah* suffered an accident in January 1924 when she was torn away from the Lakehurst mast in a gale. The fierce winds damaged her bow by wrenching out the mooring cone, deflating the forward two cells and causing much structural damage to the bow girders.

Captain Heinen dropped water ballast and ordered the engines started, sending the watch crew to secure the fabric at the bow in order to protect the gas cells. After fighting the wind for nine hours, Heinen brought the damaged airship back Lakehurst for a safe landing; a feat which earned him scant praise from the authorities who were anxious to see the back of him. His contract was not to be renewed at the year's end.

Following the accident, McCary was relieved of command and Lieutenant Commander Zachary Lansdowne became the new captain. Lansdowne was possibly the US navy's most experienced airshipman, having been seconded to the RNAS during the Great War. He had served on both pressure airships and rigids. Additionally, he had been the US navy official observer on the westbound section of the transatlantic flight of the *R34* in July 1919.

As previously described, plans for the polar flight required alterations to be made to the *ZR1* in order to increase her range. These changes were duly carried out in early 1924, however by then enthusiasm for the polar flight had waned due to opposition in congress at the estimated cost of the expedition at $350,000, together with the of fear of failure. Following this, President Coolidge cancelled the scheme in February 1924.

The *ZR1* was re-inflated in May and performed a series of publicity flights, as well as fleet exercises in the North Atlantic and mooring trials on the fleet oiler *Patoka* that produced favourable results. Based on this success the navy board proposed sending the *ZR1* on an extended deployment to Pearl Harbor, Hawaii in order to operate with the Pacific Fleet.

Lansdowne wrote a report stating that such a flight, involving as it did the necessity of the airship operating 6,000 miles from her base for up to six months without either shed of proper facilities, was a impossible task for an experimental airship. In place of this, a further proposal was made that required the *Shenandoah* to fly to the west coast to co-operate with fleet manoeuvres. This scheme being judged viable, plans were put in place.

On 1 October 1924, the *Shenandoah* left Lakehurst and flew via Fort Worth to San Diego through the mountains in dangerous conditions – she nearly came to grief in a mountain pass. After a five-day layover she flew up the coast to Oregon to operate with the fleet, returning to San Diego after a flight of forty-seven hours. The return flight to Lakehurst by way of Camp Lewis, Washington, Arkansas, Indiana and Ohio was without incident. She was away for nineteen days and covered a total of 9,300 miles.

The *Shenandoah* had carried out her tasks well, but in the environment of the continental United States and performing duties that were largely removed from its primary function of scouting for the fleet at sea.

With the arrival from Germany of the *ZR3 Los Angeles* in October 1924 the *Shenandoah* was taken out of commission and into the Lakehurst shed so that her helium could be transferred to the new airship.

While the *Shenandoah* was laid up the weight-saving measures that had been proposed to increase her range for the polar flight were implemented. These included the removal of the forward No. 6 engine, the substitution of a lighter control car and other measures that resulted in the saving of 7,800lb in weight. Additionally, the number of automatic valves were reduced both in number and size; a measure that was to have unforeseen dangers for the future.

The *Shenandoah* was re-commissioned in July 1925, and in turn filled with helium from the *Los Angeles*, as the navy did not posses sufficient helium to fly both airships at the same time. The airship continued working with the fleet throughout July and August operating from the mast on the *Patoka*.

During July 1925 the explorer Roald Amundsen, together with his wealthy backer Lincoln Ellsworth, fitted out two Dornier Wal flying boats in an attempt to fly from Kings Bay, Spitsbergen to the North Pole and back. Some 160 miles short of the Pole one of the flying boats developed engine trouble and both craft put down on a temporary open lead of clear water, which rapidly began to freeze over. Amundsen and his three companions beached the undamaged Wal on the firm ice and began to construct a runway on the ice flow.

At Lakehurst, when news of Amundsen's disappearance was announced, Lansdowne asked for permission to launch a rescue mission using the *Shenandoah*, but this was turned down by secretary of the navy, Denby, as being too dangerous – finally ending any possibility of polar flight.

Meanwhile, after an ordeal of twenty-five days both crews crammed into the lone Dornier and managed, with the narrowest of margins, to take off successfully from the ice. They returned to Kings Bay in triumph, having being given up for dead.

Their survival was a tribute both to Amundsen's survival skills and knowledge of Arctic conditions, and to the exceptionally rugged construction of the Dornier flying boat.

Earlier in the year a flight to Minneapolis in the Midwest had been scheduled for Los Angeles, but excessive leakage of her gas cells had caused her to be taken out of service and the re-inflated. *Shenandoah* was instructed to take her place.

The proposed flight during late August included visiting twenty-three cities and covering over 3,000 miles of the Midwest. In response, Lieutenant Commander Lansdowne expressed his concern regarding weather conditions in this area during the well-known 'thunderstorm season'. However, political pressure from Washington insisted that the flight go ahead at the earliest possible time and, accordingly, a departure date of 2 September was set.

The *Shenandoah* left the mast at 3 p.m. carrying a crew of forty-one and two passengers, loaded with 7.5 tons of fuel and 4.4 tons of water ballast. The first part of the flight went smoothly, *Shenandoah* finding herself over Philadelphia at 4 p.m. and later passing over the Alleghenies and on to Wheeling, West Virginia by 1.45 a.m. next morning.

At 4 a.m., flying at 1,800ft above Cambridge, Ohio, a long dark cloud denoting turbulence was observed to the north and west, building rapidly and violently illuminated by lightning. The meteorological officer Lieutenant Anderson recommended turning south away from the threatening cloud, but at 4.25 a.m. the airship started to rise out of control on a fierce vertical air current at a rate of 1,000ft per minute.

Control was regained at 3,200ft with the engines running at flank speed, but a second rise commenced some ten minutes later at over 2,100ft per minute to 5,000ft. The airship was now above pressure height and the gas pressure, unable to vent quickly enough from the reduced number of automatic valves, built up to a dangerous level. This put an intolerable strain on the structure, causing several longitudinal members in the lower part of the hull to carry away.

At this juncture the elevator man, Allen, reported that he could not hold the ship down and a third, even more abrupt rise to 6,200ft at a rate of 1,000ft per minute commenced. Lansdowne ordered the manoeuvring valves open to aid the automatic valves and halt the rise, and noted that despite the rate of rise the airship was some 3.5 tons heavy. He further ordered 1 ton of water ballast dropped in anticipation of the inevitable downward plunge to follow.

From her peak altitude the airship momentarily halted, then with a shudder accompanied by the sounds of breaking girders the *Shenandoah* pitched nose-down and began her earthward plunge at over 1,000ft per second.

Lansdowne quickly released an additional 2 tons of water ballast, but the ship was still 1.5 tons heavy and at this time the struts supporting the control car began to break away. Immediately following this a violent blast of air wrenched the nose upwards, causing the ship to crack open at frame 125.

Lieutenant Rosendahl had been sent from the control car into the keel to oversee the possible dropping of fuel tanks, should there be a further rise and descent – an action that was to save his life. Suddenly, in rapid succession, the No. 4 engine car broke away and with a tortured roar of breaking girders the bottom of the ship opened up like a cracked egg.

In the control car the army observer Colonel Hall, hearing the sound of breaking girders, shouted: 'She's gone, everybody out now'. He then sprung to the ladder and was followed by Lietenant Anderson. They were the only survivors from the control car; Lansdowne and six other crew members plunged to their deaths seconds later as the control car ripped free of the hull.

After drifting for several minutes the two sections of hull finally ripped apart, with a further break occurring at frame 110. The tail section now fell relatively slowly, coming to earth with eighteen survivors half a mile from the wreck of the control car. The smaller centre section fell rapidly until both engine cars broke away, killing their crews. Relieved of the weight this section then came to earth containing four further survivors.

Meanwhile, the larger bow section with seven men aboard, including Lieutenant Rosendahl, had risen to 10,000ft and was being driven south by the now moderating wind. Rosendahal took control, having both a manoeuvring valve and water ballast and petrol on board, and he was able to fly the wrecked section as a free balloon. After an hour he landed the bow section on a hillside at Sharon, Ohio 12 miles south of the main wreckage at Ava.

As well as Lieutenant Commander Zachary Lansdowne thirteen other crewmembers had died, some of them being survivors from the ZR2. Lieutenant Charles Bauch, as senior survivor in the stern section, took control of the survivors and collected together the bodies of the dead, assisted by then by a few sheriff's deputies. He next telephoned details of the disaster to Lakehurst at 7 a.m. from Caldwell, the nearest town to the crash site.

A county fair was in progress in Caldwell and following the crash hoards of sightseers descended on the wreck, where the few sheriff's deputies were powerless to stop the systematic looting of the wreckage. By late afternoon a greater portion of the ship had been stripped with the crowd making off with portions of girder work, gas cell fabric, outer cover, instruments and personal effects; a truly disgraceful episode.

The court of inquiry concluded that the *Shenandoah* had been lost due to:

> ... large unbalanced external aerodynamic forces arising from high velocity air currents.
> Whether the ship if entirely intact and undamaged would have broken under these forces, or prior minor damage due to gas pressure was the determining factor in the final break up are matters which this Court is unable to determine ...

The court also severely criticised Admiral Moffett for using what was patently a naval vessel on high-profile publicity flights over land, particularly in areas where weather conditions were unfavourable to airship operations, such as the Midwest of the United States.

The blame for her loss must ultimately devolve on both the navy department and the governmental agencies in Washington, who failed to understand the performance limitations of the experimental craft, or even its primary purpose. The authorities were more interested in showing off their airship to the public than allowing her to operate at sea with the fleet.

The attempt by the entente to cash in on the technological advances contained within the *L49* Height Climber design failed to achieve the desired results, with the British recklessly extrapolating from a proven but an exceptionally complex design while only paying lip service to the calculations to ensure structural integrity. Against this background the British were fortunate that their only copy, the *R 36*, enjoyed a brief flying career, for she would certainly have been lost had she encountered adverse conditions.

The *Shenandoah* on the other hand was constructed on sound scientific principles with close consideration given to the mathematical stressing of the structure. Frame for frame the *Shenandoah* was superior in strength to the original L49 class, and the *R 36*, and was subsequently lost more to political expediency rather than to the extreme weather she encountered on her last voyage.

16

The Vickers *R80*, Airship Sheds & Mooring Masts

The Vickers *R80*

During 1916 Barnes Wallis, who had by this time been promoted to chief airship designer at Vickers, proposed together with Hartley Pratt and J.E. Temple the design of a streamlined ship based on their own experience obtained from the construction of the highly streamlined Vickers Parseval.

This accumulated data showed that a correctly streamlined hull offered a considerable reduction in the drag coefficient over that of a hull shape based on the calculations propounded by the American scientist Professor Zahm, and which had been employed in the design of the R23 and 23X classes.

Zahm's theory contended that the optimum form of an airship's hull could be determined by employing a ratio that increased the bow radius to twice that of the diameter of the parallel hull section, whilst the stern radius should be three times that value. Wallis and his team correctly considered these calculations to be in error, and were confident that by applying a true streamlined form to the hull they could both increase the range and speed over that of a comparable airship designed on Professor Zahm's principles.

Wallis was able to demonstrate that the drag coefficient of the streamlined hull of the proposed craft was only 3 per cent of that compared to a flat plate of similar cross-section, as opposed to the calculated 16 per cent head resistance of the Zahm-formed hull of the *R26*, currently completing at Barrow. A similar streamlined shape without any parallel cross-section had originally been proposed for the *Mayfly* as far back as 1908, being based on Vickers hydrodynamic hull testing experience working on submarines, but had been rejected on the grounds of complexity and expense.

The design of *R80* closely followed the current Zeppelin practice with few major innovations, apart from the application of the continuous streamlined form and the incorporation of all the most advanced data acquired from downed German airships.

The Wallis team applied for and received permission from the Admiralty to build the private venture Vickers design; a move that was contrary to the Admiralty's own policy and possibly indicative of their recognition of the superiority of the design over their own efforts. So impressed were they, in fact, by the Vickers design that the Admiralty did at one stage consider cancelling all the remaining ships of the 23X class included in the ten-ship programme, in preference for allowing other firms to build *R80* type airships in their place.

In a further reversal of Admiralty policy, Vickers were to be given a free hand in design matters with only occasional visits from the Admiralty inspectors. This freedom of action allowed them to incorporate improved features of their own and relieved them of employing the excessive strength of structural elements, a requirement set out by the Admiralty for

NAVAL AIRSHIP H.M.A. R 80 -1920
Messrs Vickers Ltd - Barrow in Furness

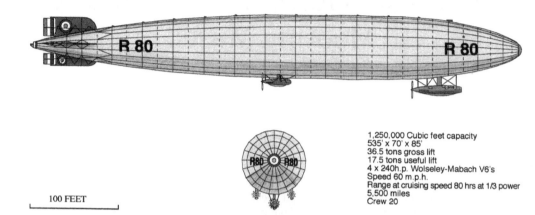

1,250,000 Cubic feet capacity
535' x 70' x 85'
36.5 tons gross lift
17.5 tons useful lift
4 x 240h.p. Wolseley-Mabach V6's
Speed 60 m.p.h.
Range at cruising speed 80 hrs at 1/3 power
5,500 miles
Crew 20

100 FEET

earlier ships. These requirements, although laudable and suitable for airships being handled by novice flight and ground crews, had resulted in them possessing low useful lift characteristics, which militated against their useful service employment.

The hull of the *R80* was a twenty-one-sided polygon, 535ft in length with a maximum diameter of 70ft. Eleven main and ten intermediate longitudinals ran from nose to tail, with the whole hull being of a continuous streamlined shape without any parallel portion.

The actual design of the hull framework closely followed Zeppelin practice, with king post, braced diamond trusses supported by diagonal wiring in the plane of the frames forming the main transverse frames which were set at 10m intervals. Alternate flat, un-braced intermediate frames were set between the mainframes, without any form of radial wiring to allow for the installation of the gas cells.

Tensioned, diagonal steel shear wiring braced the hull structure, with continuous mesh wiring covering the hull to support the stout doped canvas outer cover, which was applied in vertical sections between each set of mainframes. Here it was laced together and taped over to provide a smooth, friction-free surface.

Contained within the hull were fifteen gas cells with a total capacity of 1.26 million cu ft, which at 95 per cent full imparted a total lift of 36.5 tons. In the design, due to the careful attention given to saving weight, the *R80* attained a respectable disposable lift of 17.5 tons, or a ratio of 47 per cent net to gross; this figure was the highest so far achieved by any British airship.

R80 was designed to carry 10 tons of fuel and 8 tons of water ballast, giving a range of 3,900 miles at 50mph with a top speed of 60mph and a ceiling of 11,000ft. As with previous designs, under war conditions the range could be greatly increased by substituting petrol for water ballast.

Motive power was provided by four 240hp MbIV six-cylinder, in-line engines contained in three streamlined cars; two equipped as single engine wing cars, with the forward control car containing two coupled engines driving a single, larger diameter airscrew. The MbIV engines were copies produced by Wolseley Motors, who had procured licensing rights before the war from Maybach for the engines installed in *Mayfly*.

It is to be assumed that these copies were closely modelled on examples recovered from Zeppelins downed over Britain and that, despite the fact that the two countries were at war, arrangements would have been concluded through intermediaries in Switzerland to ensure that the terms of the licensing agreement and payments were still met.

The design of the control car, together with its attached engine car, was also of streamline form, without any protuberances apart from the sliding radiators. It was attached to the underside of the hull by streamlined struts and served by a fully enclosed ladder trunk from the keel gangway. The design of the flat section, wire-braced fins were a distinctive feature of the airship, being of deep low-aspect ratio design some 20m long by 6m or so deep with balanced elevator and rudder surfaces.

The structure of the fins were strongly supported by two sets of cruciform girders running across the front and rear of the No. 15 gas bag, making them powerful anchoring points. The shape of these fins was to prove highly effective in comparison to the longer, shallower fins of the preceding classes that had been shown liable to collapse under high aerodynamic loading. It is attested that throughout the *R80*'s short flying career no troubles were recorded with her directional control, nor was there any failure of her control surfaces such as had occurred on the *R31*.

Design on the airship began in early 1917, with the team later incorporating preliminary data that had been obtained from the *L49*, brought down intact in France in October of that year. Construction commenced following submission and approval of final plans in November 1917, with rapid progress being made in the erection of the framework. Fabrication of girders began early in the new year, with actual construction starting in April 1918. Rapid progress was made in the erection of the framework, just as *R26* was completing in the adjacent bay of the Walney Island shed.

Vickers had hoped to acquire a large shed some 700ft by 150ft that was being erected at Flookburgh where they were to build the *R37*, having been awarded the Admiralty contract for one of the new 'improved' *R33*s, lengthened by one bay. However, for reasons that are not completely clear but are associated with the shortage of steel and problems with the foundations, the partially completed shed was cancelled in late 1917. This resulted in the Short Brothers being awarded the contract for *R37* instead.

Proposed Trans-Atlantic Airship R 85 - 1919
Messrs Vickers Ltd - Barrow in Furness

100 FEET

A model of this proposed passenger airship which included accomodation for 100 passengers, and an observation lounge on top of the hull was displayed at the Amsterdam Aeronautical Exhibition in September 1919,
The design concept was impractical and too ambitious and was not proceeded with

3,500,000 Cubic feet capacity
750' x 98' x 105'
105 tons gross lift
60 tons useful lift
6 x 300h.p. Sunbeam Cossack V12's
Speed 80 m.p.h.
Range at cruising speed 180 hrs at 70% power. 10,000 miles
Crew 30
Passengers 100

It can be assumed that the original intention of the Vickers team was to build a much larger airship of streamlined form with a greater gross lift, but the eventual product of the Vickers team, the *R80*, was a compromise imposed by the confines of the exiting 531ft-long Walney Island shed. Not for the last time was the size of an airship dictated by the dimensions of the shed.

As *R26* was being fitted out for trials Vickers had approached the Admiralty with the suggestion that a further example of the evidently superior *R80* design should be ordered. This suggestion was surprisingly accepted by the Admiralty, who in the past had been wary of allowing Vickers too much freedom of action where airship construction was concerned. They were, however, sufficiently impressed with the design to approve a second ship, the *R81*, in March 1918. A proportion of the structure was put in hand in the bay vacated by the *R26* before the contract was cancelled at the time of the Armistice.

Prior to cancellation further consideration had been given in Admiralty circles to allowing Vickers to produce a series of these ships, as a solution to the requirement for a distant scouting type to work in the Western Approaches against U-boats, where their comparatively low ceiling was considered acceptable.

The proposed heavy shell gun armament, which was to be provided by the Admiralty but never installed, was to consist of three 2lb shell guns mounted in the gondolas together with two quick-firing rifle guns, one mounted on top of the hull and the other positioned in a gun pit at the stern. Additionally, a heavy anti-submarine bomb load was carried in the keel. Had these ships become operational there is little doubt that their smaller size and all-round handiness would have rendered them highly effective weapons in the war against the U-boat.

Between September 1918 and March 1919 extensive changes to the *R80* engine specifications and electrical systems delayed construction, with the work being further held up by the lack of skilled labour and the extremely cold winter working conditions, aggravated by the necessity of the shed doors being left open to provide adequate light.

Her fuel was carried in thirty-four aluminium tanks arranged on the starboard side of the keel gangway, giving a total capacity of 10 tons of petrol. While eight water ballast bags, each with a capacity of 1 ton of water, plus eight 500lb emergency water sacks were distributed along the port side. For convenience of servicing and easy recognition in emergency all control, water piping and electrical systems were individually colour coded.

Following the Armistice, work proceeded slowly with the gas cells and outer cover eventually being fitted in April 1920, and the ship finally being walked out of her shed in July 1920. Upon completion it was shown that the *R80* had a disposable lift that was 2 tons in excess of the contract figure, which was a major success and to the credit of Wallis and his team in designing the first really successful British airship.

Unfortunately, the first trial flight took place on a particularly hot day, where trouble with starting one of the engines resulted in the *R80* laying out on the field for some considerable time in the sun which resulted in the superheating of the gas in her cells. This rapid expansion of gas caused the ship to become excessively light and it became increasingly difficult for the ground crew to handle and hold her down.

As the gas pressure built up in the cells, efforts were made to valve off the excess gas to stop her rising prematurely. Despite these measures the ship was excessively light when the lines were finally released, resulting in the airship rising rapidly to 4,000ft at a rate of 1,500ft per minute. This rapid rate of ascent caused severe damage to the framework with over sixty longitudinal and transverse girders either breaking or being distorted, together with damage to the gas cells, cell exhaust trunking and the breakage of fin bracing wires.

The *R80* was immediately returned to her shed for repairs and for the installation of bow mooring gear, the installation of which required an additional 1 ton of water ballast to be

37. *R80*, designed by Barnes Wallis, taking her first flight at Walney Island, 1920.

carried aft to counteract the additional weight at the bow. After repairs the *R80* finally emerged to be commissioned in January 1921 at Barrow, making a short trial flight that proved to be satisfactory in all respects, after which she was dispatched to Howden the following month.

On her arrival it was determined that there was no useful purpose she could be put to, and accordingly she was ordered to be deflated and stored. This instruction was rescinded within a few days, however, in order for her to be available for possible use by the US navy. This decision was prompted by the accident to the *R34* in January of that year, on which airship the American aircrew had been receiving training.

The loss of *R34*, and later by the demolition of *R32* in April 1921, left the US naval detachment at Howden in the position where no rigid airship was now available to them. The navy department in Washington, desirous to maintain the terms of the original agreement with the Air Ministry, had made a request in May that the surrendered *L64* be reconditioned in order to meet the further training requirements of the group.

The *L64* had been handed over by the Germans at Pulham in July 1920, by which time she was already over two years old and, due to damp and deterioration, would require a complete set of new gas bags to be installed to make it fit for service. Paradoxically, in the light of the existing economic situation, the Air Ministry informed the US navy that such a refurbishment could be effected and even indicated that the ship could be made serviceable as early as August 1921.

Doubtless this date would have been acceptable to the Americans, but in truth the Air Ministry had no serious intention of restoring the *L64* to flying condition. Apart from the cost they were only too aware of the specialist nature of this high-altitude craft, and appreciated that the fragility of its lightened structure made it unsuitable for low-altitude operation or for training purposes by novice crews. The *R80*, however, presented a suitable alternative for the Americans to continue their training on.

Lieutenant Commander V. Bieg, the chief engineering officer of the US contingent, visited Barrow on 8 February at the behest of Commander Maxfield to report on the *R80*'s suitability. After considerable discussion with the Air Ministry an agreement for the use by the US navy of the *R80* was finally signed, and on 30 March 1921 the Vickers ship was inflated, put into flying trim and placed at the disposal of the Americans.

Flight Lieutenant Little of the RAF was appointed as commander and acting senior instructor, and between then and July 1921 *R80* made at least seven flights (as evidenced by the American crew logbooks). This included a night flight on 6/7 July of twenty-nine hours over the North Sea with the US crew on board.

At this distance in time there is of course some uncertainty concerning the total number of hours flown by this airship, but it is generally credited with a flying career of seventy-three hours. It is stated that during these flights Little was nominally in command, but it is certain that Commander Maxfield was allowed control of the airship for the greater part of the time under Little's supervision.

Following the loss of *R38* in August 1921 the Air Ministry ordered the cessation of all further airship flights, with the exception of allowing the *R80* to make her final flight in September when she flew from Howden to Pulham. On arrival she was deflated and laid up.

With the inauguration of the civil airship programme in 1924 the *R80* was briefly considered for refurbishment by the government in order to acquire aerodynamic data, particularly as she was of the streamlined shape that was envisaged for the new civil airships. However, as with similar plans that had been put forward for *R36*, who had her damaged bow repaired and her outer cover replaced, neither ship was fully restored to flying condition.

As part of the government civil airship programme the *R80* was demolished during structural strength tests in 1925. The *R80* was a very promising design and although of small size could possibly have been developed into a successful class of scouting airship, while the basic design was also utilised for other projected craft that were of interest.

The US navy was familiar with the *R80* design as earlier in 1919 they had shown considerable interest in the possible acquisition of an *R80*-type airship. In response to this interest Vickers produced a series of specifications for two separate airships: one being of 1.25 million cu ft to a modified *R80* design and a second, much larger ship designed in September 1918 of 3.5 million cu ft capacity for the US government.

This larger ship, *R85*, had a projected disposable load of 68 tons carrying a crew of twenty-four and was to have a net to gross of 60 per cent. This was comparable to that of *L70*, however in retrospect the enlarged design was impractical and would have proved structurally weak. Neither of the two Vickers designs were proceeded with.

The *R85* design was reworked as a possible transatlantic passenger craft of 3.5 million cu ft capacity. It was to carry 100 passengers, be powered by ten engines delivering 3,500hp and have a promenade deck built into the top of the hull. Yet this design was an impractical concept.

A model of the airship was shown at the Amsterdam Aeronautical Exhibition in September 1919 where it caused much interest, as did the visit of the *R31* and *R33* over-flying the exhibition during a North Sea cruise. The publicity attending the exhibition, which following so soon after the successful double Atlantic crossing by the *R34* created the impression in the public's mind that regular transatlantic flights by fleets of passenger carrying airships would soon become an everyday event.

The *R85* progressed no further than the drawing board, however, not least because Vickers still lacked a shed of sufficient proportions in which to build such huge airships.

During the same period Short Brothers, Armstrong Whitworth and Beardmore produced designs for ocean-going airships to meet the US government specification, with all four companies coming up with proposals that they hoped could fulfil both naval and civil requirements.

During late 1919, and in response to the Air Ministry's edict to explore the possibility of using existing airships for civil use, Vickers prepared drawings for the conversion of *R80* as a passenger carrier for use on proposed European routes. Plans exist at the National Archive that show a separate streamlined external cabin suspended below the hull between the fore control car and the two wing engine cars with accommodation for thirty passengers, together with sleeping cabins situated in the keel above the car.

The plan was to employ the *R80* on a London (Cardington) to Rome service flying via Paris (four hours), thence down the Rhone Valley to Marseilles and across the Ligurian Sea to Rome, in a total flying time of seventeen hours. In view of *R80*'s relatively small size and limited useful lift, such expectations of performance were ambitious to say the least, and no such modifications were undertaken to convert the ship.

The *R80* design was well conceived and showed considerable promise; at the time of its launch it represented the high point of British airship constructional development. But it suffered from a lack of structural integrity in the girder work of the otherwise excellent hull design.

Vickers had somewhat naively hoped at the time of construction of the *Mayfly* that they could expect to receive a ten-year government monopoly for the building of airships, but no such undertaking was forthcoming. As mentioned earlier, Vickers were further handicapped by the small dimensions of the Barrow shed that limited the size of airship they could construct, while the planned 800ft-long double shed being built at Flookburgh during 1917 had to be abandoned.

Had a suitable large facility existed earlier the Vickers Company would no doubt have been able to carry through some of their more innovative ideas in airship design under the guiding hand of Pratt and Barnes Wallis. It is interesting to speculate that, had Vickers been given a free hand earlier together with a shed of suitable dimensions, they could have produced effective scouting rigid airships for the fleet, and at the war's end would possibly have been in a position to build practical passenger airships.

Airship Sheds & Mooring Masts

The problem of housing airships, both during construction and when in service, was a perennial headache to their builders and operators, as were the difficulties of handling these huge machines on the ground. Airship sheds were in themselves remarkable and impressive structures, with some of the later steel-framed being some of the largest clear span buildings of the period. They could be compared to medieval cathedrals in the vast soaring spaces of their interiors, with some being of such a size that water vapour could condense under the roof spaces to form clouds.

The earliest surviving airship shed is the famed Hangar Y at the French aeronautical establishment at Chalais Meudon near Paris, in which the electrically propelled airship *La France* was constructed in 1884. This building, which has recently been restored, is an iron-framed structure 200ft in length with a clear height of 60ft: it is a worthy memorial to the early French aviation pioneers.

Other early French airship sheds were built of wood and canvas, which proved to be an economical and durable method of construction. This type was suitable for housing the *Lebaudy* and *Clément-Bayard* airships, which were in the order of 200–300ft in length.

Count Zeppelin's first six airships were built in two floating sheds at Manzell on Lake Constance, which had the advantage that the shed could be turned to align with the prevailing wind making entering or leaving easier. Yet Count Zeppelin abandoned the use of the floating shed, with a double shed being built on land at Friedrichshafen in 1909, followed by two larger sheds in 1914 and 1916.

Vickers also built an over-water constructional shed at Barrow for the 515ft-long *Mayfly* in 1908. This, however, was a fixed structure built on piles over the Cavendish Dock immediately alongside the dock wall. It proved to be an inefficient arrangement as she had to be hauled out by a complicated system of ropes and capstans anchored in the dock while gradually being turned through 90°. This system ultimately contributed to the loss of the airship whilst it was being re-housed in 1911.

The Siemens-Schuckert company built a 446ft long by 82ft wide revolving shed of mixed brick, steel and wooden construction, weighing 1,200 tons, to house their 529,720cu ft airship in 1910. The shed revolved on eight four-wheeled bogies running on two concentric sets of rails, with each bogie being driven by two 10hp electric motors.

A much larger double revolving shed was erected at the Nordholz Air Station, headquarters of the naval airship service, during 1914. Originally 597ft in length, it was subsequently lengthened to 656ft in 1917 and was able to accommodate two of the largest type of wartime Zeppelins. The *Nobel* shed weighed over 4,000 tons and its powerful electric motors could rotate the structure through 180° within ten minutes, enabling airships to be launched in all but the strongest of winds.

During the Fist World War the German army and navy built over forty-five large sheds in Germany and occupied territories, representing a huge investment in both manpower and strategic materials. It was often stated that a single, large rigid shed contained sufficient steel to build a heavy cruiser. These huge structures were also very vulnerable to attack from the air themselves, with the sheds at Friedrichshafen, Tondern, Cologne and the army base at Gontrode in Belgium all successfully attacked by aircraft in the opening months of the war, the airships within them being either damaged or destroyed.

In Great Britain three small airship sheds were erected at Farnborough before the outbreak of war, from which the early British army airships operated. With the commencement of hostilities and the expansion of the naval airship service, airship stations sprang up around the coast, eventually resulting in twelve main stations, each fully equipped with sheds, machine shops, accommodation and gas production plants, providing convoy protection and fleet scouting duties.

Additionally, a further sixteen sub-stations and mooring-out facilities were established, which often included simple shelters cut into woods that could house two SS Zero, protected by the screen of trees from where they could operate with minimum facilities.

During the course of the war the RNAS operated 230 non-rigids and a handful of rigid airships from these facilities, together with two stations across the Channel in France. While in the eastern Mediterranean five Sea Scout airships were shipped out to a base and shed on the island of Murdos in the Aegean in support of the Airship Expeditionary Force in support of the Dardanelles campaign.

The RNAS constructed a series of standardised sheds, initially in wood and canvas, to house the smaller Sea Scout airships. These were later built in steel and corrugated iron for the larger Coastal class. The sheds were provided with protective mobile windscreens at either end of the shed, in an attempt to make the withdrawal and re-entering of airships easier in a crosswind. As an aid, large ground crews were required to handle the airships in and out under such conditions.

Although the use of windscreens continued after the First World War, in practice they were often found to produce turbulent eddies that created disturbed air conditions on the leeward side of the screens.

The larger British rigid sheds, such as those at Pulham, Inchinnan, Howden and Cardington, were 820ft in length with a clear height of 180ft, whilst the later German sheds built to house the *Hindenburg* at Frankfurt am Main in Germany and Recife in Brazil were even larger.

In the United States a large shed over 912ft long (and that is still standing) was erected at Lakehurst, New Jersey in which the US navy's *ZR1 Shenandoah* was built. It was also to house the *Los Angeles*, *Graf Zeppelin*, *Akron* and the *Hindenburg* in its time.

The Akron Airdock (where the two giant 6.85 million cu ft naval airships were built in the early 1930s) and the similar shed at Sunnyvale in California were 1,174ft long by 324ft wide by 200ft high, and the vast clamshell doors weighing 1,500 tons could be opened in five minutes by electric motors. Again, both these huge structures are still in use.

The Germans favoured the use of large ground crews for handling and re-housing their airships during the First World War with few mechanical aids, apart from docking rails that ran out from the shed some distance out on to the field. Mooring ropes were attached to trolleys fore and aft, which allowed a degree of control during the tricky process of an airship entering or leaving its shed.

The British were the first to employ a mooring mast when, in 1911, the army airship *Gamma* was attached to and operated from a 60ft mast at Larkhill. This proved to be a satisfactory method of mooring in the field during army manoeuvres.

At Farnborough the Astra-Torres naval airship *No. 3* and other army airships were frequently moored on the 50ft wooden mast, allowing the services to accumulate experience in this area of airship handling. Tests were also conducted on the *R23* and *R26* during the First World War with the three wire method of mooring. Here, three wires were attached near the bow and laid out at 120° under tension, allowing the airship to ride anchored some 50ft in the air, capable of yawing with the wind.

This method was considered as a useful temporary system when re-housing was impractical, and was employed on the damaged *R34* in January 1921. Unfortunately, the violence of the wind proved too strong for this method of control and resulted in the wrecking of the airship. Experiments were also carried out during 1918 on the *R23* when a small mooring cone was attached to a tank, but this was more as an aid to moving the airship in and out of its shed.

Later, under the direction of Commander Masterman and Major G.H. Scott, a more advanced 180ft-high mooring mast was erected at Pulham air station during 1918. From here the *R33* and *R36* were successfully operated, although it must be said that the *R36* did overrun the mast whilst mooring in April 1921 damaging her bow. In addition, the *R33* spectacularly broke away from the same tower in a storm in April 1926, also severely damaging her bow. She drifted over the North Sea before being brought under control and taken safely back to the station after a flight of twenty-nine hours.

Whilst moored on the British style High Mast, an airship was free to vane with the wind but had to be maintained in neutral equilibrium by continuous adjustment to trim and correct ballasting; effectively being 'flown' at the mast.

The Americans initially erected similar high masts, with the mast at Lakehurst being used for mooring the *Shenandoah* and the *Los Angeles*. On 25 August 1927 the *Los Angeles* had the unusual and alarming experience of being stood on its nose to an angle of 80° and rotated through 180°, whilst still attached to the mast, by a sudden change of wind direction combined with a temperature inversion. As the airship swung round to face the opposite direction it gradually settled back to its normal position; amazingly the airship suffered no major damage.

The US navy also experimented with a mast fitted to the stern of a fleet oiler, the USS *Patoka*, which proved to be a very effective mobile platform. It was to here the *Shenandoah* was first moored in tests during August 1924 at Newport, Rhode Island. Later, during 1925 and 1926, the *Los Angeles* used the *Patoka* as a mobile base during several deployments to Bermuda and Puerto Rico, with the airship spending as much as three days secured to the mast in perfect safety.

The British, meanwhile, continued to develop the High Mast concept, culminating with the sophisticated 200ft-high towers erected at Cardington, Montreal, Ismailia and Karachi built to receive the *R100* and *R101*.

The Americans, however, became increasingly dissatisfied with the drawbacks associated with this system and developed the mobile Low or Stub Mast, which had the advantage of allowing the airship to be accessible for servicing at ground level.

The *Los Angeles* was used for early trials in October 1927, utilising an old 60ft radio mast with the stern engine gondola fixed to a simple wheeled car that ran on a circular track. The airship was carefully ballasted down to ensure the stern would not lift in gusty conditions, and the scheme was considered to be a more practical and safer method of mooring.

The mobile Low Mast subsequently became the standard method for mooring airships in the United States. More advanced versions were constructed for the 6.5 million cu ft airships *Akron* and *Macon* in the mid-1930s, and were even belatedly adopted by the Zeppelin Company for the *Hindenburg* and *Graf Zeppelin II*. During the Second World War the US navy airship patrol squadrons operated exclusively from this type of mast, which is still the most practical method of mooring.

17

The Admiralty A Type: *R38/ZR2*

In June 1918 the Admiralty convened a conference to draw up a specification for a rigid airship of an advanced design. It was to mark a departure from current Zeppelin constructional practice by relying less on copying existing German designs. The new airship was also to be significantly larger than contemporary Zeppelins, combining a far greater lifting capacity with high speed, a much extended range and a high operational ceiling.

The Admiralty specified that the airship should be capable of operating in the North Sea, over 300 miles from its base without support for up to six days and possess a ceiling of 25,000ft. This new class of airship were also required to undertake long ocean patrols in the Western Approaches in the protection of convoys, whilst carrying a heavy offensive war load. This consisted of 4,000lb of bombs or depth charges plus twelve pairs of machine guns and a 1lb Q/F semi-automatic gun on the upper platform. Constructor-Commander Campbell and his design team concluded that such a ship would need to be of at least 3 million cu ft capacity and up to 750ft in length in order to comply with the specification.

As no suitable constructional facility existed at that time, Campbell submitted a revised proposal for an airship of 2.7 million cu ft capacity with a length of 700ft that he was confident would fulfil the bulk of the Admiralty requirements. This design was to become the Admiralty A type, four of which were to be ordered: *R38*, *R39*, *R40* and *R41*. This resulted in a further rescheduling of airship designations, however, as these numbers had already been provisionally allocated to a batch of the improved R33 class.

The intention of Campbell's team was to build a radically new type of airship of British design, incorporating many new and innovative features. The use of internal engine rooms, for instance, were briefly considered. Yet in the event, apart from the unique method of the gas bag wiring system, the completed structure basically followed existing Zeppelin practice.

Design work began in August 1918, where the requirement was to obtain the highest net to gross lift ratio necessary to achieve the exceptionally high service ceiling and high useful load dictated the adoption of the 15m mainframe spacing then being employed in the latest Zeppelin Height Climbers.

Campbell in his design had to satisfy two conflicting requirements: first was the need to build a ship whose hull was sufficiently light that it could ascend to a great altitude in order to avoid attacking aircraft, second the craft was required to carry a heavy offensive load and to be capable of manoeuvring rapidly at low altitude on convoy work.

Every effort was made in the *R38*'s construction to reduce the weight of the structural elements, to the extent that girder for girder those in the *R38* were proportionally weaker than those in the framework of the proceeding and smaller R33 class.

Fortuitously for the Admiralty team, or so it appeared at the time, almost immediately following the decision to commence the design large quantities of the wreckage of the latest Zeppelin, *L70* which had been shot down off Great Yarmouth on 5 August, had been recovered

38. The Cardington shed with *R38* on the left completing as US navy *ZR2* and uncompleted *R37* awaiting disposal on the right.

from the sea. With so much material at their disposal from this, the most advanced of Strasser's Zeppelins, Campbell and his team must have felt that they had been presented with a blueprint for their new airship.

The Admiralty team were very much influenced by the data and material they had obtained, leading them to assume it could be successfully replicated in the same way that *R33* had been based on the *L33*. Yet they were unaware of the dangers that were inherent in the German design.

From the very outset the premise on which the new design was based was flawed and contradictory. The *L70* had been built as a specially lightened, high-altitude bombing craft, and the Germans were well aware of the strong aerodynamic forces that could be imposed on such a lightly built craft, should it be subjected to the forceful application of controls in the denser, lower air.

It is tempting and possibly correct to speculate that, due to the short eight-week period in which the Campbell team had to produce the finished design of *R38*, the team had merely copied the girder work from the German airship and then scaled up the design to suit.

In her construction the *R38* required some 2,170 girders to complete the hull framework, with many of these being fabricated in the adjoining workshops to the required specification. However, it should be noted that apart from purpose-made girders she was also supplied with numerous second-hand girders that had been manufactured for earlier airships (such as *R35* and *R40* of the improved *R33* class whose contracts had been cancelled), which had been designed to an earlier specification.

A 15m bay spacing with two light, intermediate frames was employed, with an almost identical form of girder structure with metal of similar gauge to that of *L70* being used throughout; the transverse frames being scaled up from 78ft to 85ft diameter. One of the few departures from Zeppelin practice in the design was of the keel structure, where the normal triangular section keel gangway running the length of the lower portion of the hull was replaced by one of a trapezoidal section.

Whilst the Admiralty Airship Design Department under Campbell were responsible for the overall planning, the main design detail fell to Short Brothers' chief designer T.P. Lipscomb, who had earlier in 1916 prepared Short's own plans for a 700ft wooden-framed craft and had subsequently rendered the designs for *R31* and *R32* in conjunction with Muller.

The official order for *R38* was placed in September 1918, but following the Armistice in November 1918 work on all existing rigids then under construction was either temporarily suspended or slowed down, as there was an urgent need to make all possible economies due to the serious post-war financial crisis. A series of measures were considered regarding airships, ranging from the total abandonment of the airship service to the Treasury insisting that all airships be sold off for commercial passenger-carrying use.

The dual control of the airship service by the Royal Navy and the Royal Air Force was also terminated in October 1919, when the Air Ministry became solely responsible for all aspects of airship design and operation.

The *R38* order was initially cancelled on 31 January 1919, on the grounds that with the ending of hostilities the *R38* and her sisters would no longer be required by the navy in her now defunct specialised role of distant scouting, or for any other useful purpose that could be foreseen. Almost immediately, however, the original order was reinstated on 17 February 1919 due to the interest being shown by the American Naval Aeronautical Department for the purchase of a large airship from Europe, and the possibility of recouping some of the government's investment in airships from such a sale.

The US navy department had hoped to acquire one or more of the latest type of German naval airships that had been immobilised and deflated at their bases on the orders of the Allied

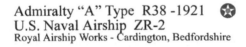

Admiralty "A" Type R38 -1921
U.S. Naval Airship ZR-2
Royal Airship Works - Cardington, Bedfordshire

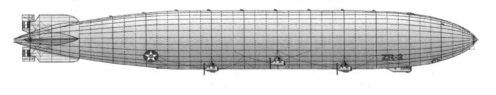

100 Feet

Bow View

2,740,500 Cubic feet capacity
695' x85.5' x 98'
81 tons gross lift
44.5 tons useful lift
6 x 350 h.p. Sunbeam Cossack V12's
Speed 71m.p.h.
Range at cruising speed 180 hrs at 1/3
power. 6400 miles
Static ceiling 20,000 feet
Crew 30

H.M.A. R 38 / U.S. Navy ZR-2 - 1921
Hull structure and wiring detail

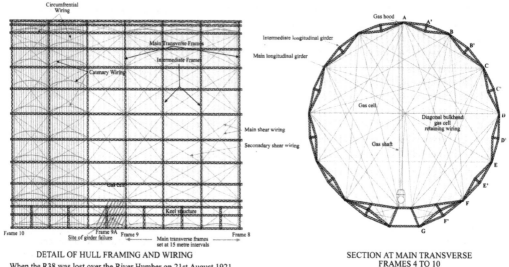

DETAIL OF HULL FRAMING AND WIRING

When the R38 was lost over the River Humber on 21st August 1921 due to structural failure, the framework failed between main transverse frame 9 and intermediate frame 9A due to excessive compressive stress on the lower hull by strong aerodynamic forces exerted by the control surfaces whilst manouvering at full power.

SECTION AT MAIN TRANSVERSE
FRAMES 4 TO 10

Control Commission pending their disposal. The Americans had further hopes of receiving training in Germany from the former Zeppelin crews, and flying two of the most modern airships direct to the United States as adjuncts to the fleet. These plans were to come to nothing, however, following the scuttling of the ships of the High Seas Fleet at Scapa Flow on 21 June 1919.

In response to this action, and to avoid their airships falling into Allied hands, the naval crews at the bases of Nordholz and Wittmundhafen on the morning of 23 June destroyed seven of the remaining airships by dropping them from their tackles in the roof, crushing their fragile frames on the shed floors. Immediately following the wrecking of the German ships, the US naval attaché was instructed by Admiral Knapp, head of the Bureau of Aeronautics, to make investigations with the Air Ministry as to the possibility of obtaining a comparable modern type of British rigid airship.

Negotiations between the US navy's Department of Aeronautics and the Admiralty and Treasury for the sale of *R38* were conducted at the British end by General Maitland. Beginning in June 1919 they were concluded by October when a contract for the purchase of the British airship for $2 million was signed. Once the order had been placed for the *R38*, no efforts were made to modify the structure to suit a very different set of operating conditions required by the US navy. In consequence, no changes were undertaken to the original design that would have strengthened the hull, making it more suitable for low-level operating conditions.

The designers of the *R38* had hoped to create a ship where the stresses on the hull structure would not exceed that of the earlier R33 class, for which they had acquired some data from the NPL. In her construction the scantlings of the *R38* were, as mentioned earlier, appreciably lighter than those used in *R33*, and, as it later transpired, her factor of safety was only half that of the previous design.

In this her designers were either unaware of the risks they were running by extrapolating from the design of the earlier stronger airship, or they chose to ignore what little scientific data that was available to them. With regard to the strength of her hull, turning tests (of comparable severity to those which later caused *R38* to collapse) had been conducted on the wooden-framed *R32* without causing any structural damage to the hull of the wooden ship, although fin failure did occur.

As a general rule it can be concluded that all the earlier British airships were on average twice as strong as *R38*. It is also interesting to note that the design had never been officially approved by the Admiralty, as would have been the case had the *R38* been a surface vessel. In addition, throughout her construction no official technical notebooks were kept, with Campbell relying on giving instructions for modifications to the original design verbally. The methods employed for the determination bending moments and stress calculations generally were most unsatisfactory, and were in most cases not checked by any independent source or, for that matter, recorded.

Admiral Taylor of the aeronautical division of the Bureau of Construction had, as early as March 1920, requested Commander Land and Commander Dyer (who were at Cardington overseeing the construction for the US government) to obtain from the British calculations relating to bending moments acting on the framework under various conditions, together with detailed strength determinations of the hull structure. The Americans failed to obtain this information and were critical of the methods employed by the British team in establishing the airship's factor of safety.

39. *R38* leaving the Cardington shed for trials, June 1921.

The system of gas cell wiring marked a departure from Zeppelin practice, as did the external shear wiring designed to transfer the loading and lift from the cells to the framework. The American Bureau of Aeronautics' representative, C.A. Burgess, was also critical of the external frame wiring with its catenary system that encircled the frame at right angles to the axis of flight. The proven diagonal wiring system employed in Zeppelins was undoubtedly superior, and in his opinion the British method failed to effectively fulfil its requirements to strengthen the framework.

The design of the gas bag wiring was an area where the designers had abandoned the standard and well-tried method of retaining the gas bags and effectively transmitting lift to the structure used in Zeppelins. In so doing substituting it for an inferior and unsatisfactory system, with the that result that gas bag surging and excessive chafing of the fabric cells was commonplace in the *R38*.

Fabrication of girders and construction of *R38* began at Short Brothers' Cardington works in February 1919, while at the same time *R39* was laid down at Armstrong Whitworth. Here, construction proceeded at a rapid pace until the ship was cancelled on economic grounds in August 1919.

Of the other two ships of the Admiralty A class, *R40* was also assigned to Armstrong Whitworth while *R41* went to Beardmore's. Both these ships were marginally smaller and with a reduced ceiling of 20,000ft, but with a greater endurance in accordance with a further Admiralty requirement. Both ships had originally been cancelled in January 1919 but were re-ordered in March, and a sufficient quantity of girders was delivered to allow work to start on them. However, like *R39* both ships were also finally cancelled in August 1919, with their girder work being distributed between *R38* and *R37*.

The actual dimensions of the *R38* were 699ft by 85.5ft by 93ft in height, with a volume of 2,740,500cu ft, imparting a total lift of 82.5 tons and a disposable of 45.5 tons. Six engines producing a total 2,100hp gave a maximum speed of 72mph, with the ship being capable of rising to 22,000ft dynamically and with an endurance of sixty-five hours at full speed.

Construction commenced in February 1919 at Cardington with a pair of the 85ft diameter main transverse frames being fabricated on wooden jigs on the floor. Once completed, radial bracing and chord wires were installed across the face of the frames and tensioned; the second mainframe was lifted above the first on tackles and both sections joined at the corners by thirteen triangular longitudinal girders, with two un-braced lighter intermediate frames incorporated at 5m spacing to form a 15m-high tower.

This procedure was then repeated for the next set of mainframes, after which both sections were rotated horizontally and placed on timber cradles to support the framework separated by a 15m gap. Two further intermediate frames were swung in place and both sections joined together by a further set of longitudinal girders to form three complete bays. Diagonal and shear wiring was then applied and correctly tensioned to the outer framework, and the process repeated for the next section.

By April 1920 work on the hull framework had proceeded rapidly, with the structure between frame 3 and frame 14 being completed. However, much work was still required to complete the fins and rudders, engine cars, control gondola, fuel and ballast systems, control cables and electrical services.

As originally designed *R38* was to be equipped with two types of engines: four direct drive 350hp Sunbeam Cossack engines (weight 1,200lb) mounted in the fore and midships gondolas; and two lighter 275hp Sunbeam Maoris (weight 920lb each), with reduction gearing installed only on the two aft cars situated at frame 9.

During March 1920, for some unexplained reason, this layout was modified to six of the heavier Cossack engines, all equipped with reduction gearing. This change resulted in an increase in weight of 1,300lb and whilst some extra supporting wiring was added at frame 9, no additional strengthening was made in the area of the transverse frame that supported engine cars Nos 5 and 6. (It should be noted that when the airship was lost over the River Humber, the framework failed immediately abaft of frame 9.)

The corresponding increase in weight caused by these alterations resulted in a change of the centre of gravity from the original design, moving it further aft from the centre of lift, which would be a cause of lateral instability in flight and as was indeed experienced in all her flights. At the same time, the installation of bow mooring gear together with the strengthening of the bow girders added an additional 1 ton of weight forward. To a degree, this compensated for the extra weight of the engines aft and again brought the centre of gravity closer to the centre of lift. Ultimately, these changes caused the ship to be bow heavy, a situation that was partially cured by rearranging the positioning of the water ballast bags and petrol tanks.

In late May 1920, with a major portion of the airships frame now erected, the airship assistant constructor, Mr May, conducted a 'deflated gas bag' test designed to prove the strength of the hull. During this test cell No. 8 was fully inflated and various strain gauges were attached to the wiring of that bay and the two adjacent bays.

No. 8 gas cell was kept in an inflated condition over a period of several days, during which any deflection in the girder work or failure of the wiring were recorded. In this condition the No. 8 gas cell bulged into the adjacent bays, putting an additional strain on the transverse wiring and thus causing increased compression loads on the main transverse frames, as required for the test.

A section of the axial cable was installed running through the centre of the cell, to which the radial wiring of both the mainframe and the gas cell mesh wiring were attached. The cable extended into, and was connected to, the radial and chord wiring of the two adjacent empty bays. The installation of the axial cable under these circumstances allowed the compressive loading on the frames to be halved. The compressive loading applied to the transverse frames through the distorted wiring (as the end faces of the cell pressed into the vacant bays) increased as the cube of the frame's diameter, and were of a higher order than the normal loads resulting from those imposed by the lifting force of the gas, or the moveable weights such as fuel, ballast etc.

During these tests the deflection of the upper girders under gas pressure was noted, and also of the keel structure to which weights were added (3,000lb to each mainframe and an additional 4,000lb distributed along the keel between frames 8 and 9), the results being generally accepted as satisfactory to Campbell.

On 1 June one of the triangular trusses of frame 8 at the top of the ship collapsed, together with the B longitudinal failing at its juncture with frame 8. In response to these alarming failures the damaged truss section was replaced and reinforcing plates were fixed to all the joints of frame 8.

On 2 June during further tests the same truss section at frame 8 again collapsed when a rigger in a boatswains chair applied his full weight to the truss. This in turn caused a further fifteen longitudinal and transverse girders to buckle in bay 8 and adjacent bays. Amazingly, these multiple girder failures in static conditions failed to alert the builders or the Admiralty to the dangerous weakness that was obviously inherent in the structure of the ship.

Indeed, following the tests Mr May stated that 'The results of the tests show that the structure is amply strong and a sufficient margin of strength is obtained for any condition that may occur in practice'. Further reinforcing plates were fitted to all the frame joints at main longitudinals between frame 3 and frame 11, and for the time being this was adjudged to be all that was necessary to remedy the problem.

Independent investigation of the strength of *R38*'s structure by the American Bureau of Construction concluded that the 'transverse frames of the *R38* are only just strong enough under normal circumstances, and have NO factor of safety'. Put bluntly, *R38* could barely withstand the estimated static loading let alone the dynamic loads that would be imposed by flight.

By early 1921 the airship was to all intents complete, although much work of a minor nature still required attention. Despite the best efforts of Campbell to encourage completion by offering double pay there was a general slowing down of effort by a workforce who would soon be unemployed, as there was no further contract in sight.

The Air Ministry had chosen Flight Lieutenant A.E. Pritchard, then aged 32, to be the officer in charge of flying trials. This most experienced officer had served in the RNAS on airships during the war, and had been appointed to the airship department of the Admiralty in 1917 as the acceptance officer for the new rigids joining the fleet.

As a member of the Inter Allied Armistice Commission he visited Germany in December 1918, where he had acquired a considerable amount of technical information relating to rigid airship operation. Then in July 1918 he was a crew member aboard the *R34* for the Atlantic crossing. Pritchard was unique among British airshipmen as having the most complete understanding of German operating practice, and was fully cognisant of the dangers inherent with low-level operation on a lightly built craft; vital information that was not fully appreciated by many of his superiors.

In March 1921 Pritchard published, at the request of General Maitland, a memorandum which called for an extended period of flight trials lasting 150 hours and which included a series of flights under rough weather conditions. This protracted programme, although supported in principle by General Maitland and other senior officers, was curtailed on instruction from the Air Ministry acting under government pressure to hand the ship over to the Americans at the earliest possible moment.

This reduction in the length of this test programme was also urged by the nominated commander of the *R38* (or *ZR2* as she was to be known in the United States), Commander Maxfield, who was anxious to make the Atlantic flight before the end of August and the onset of bad weather in the Atlantic. Eventually, on 7 June, the airship was adjudged to be complete and ready for flying trials.

The first flight took place at 9.50 p.m. on 23 June 1921, under ideal and calm conditions, with Flight Lieutenant A.H. Wann in command and a crew of forty on board. The airship cruised in the vicinity of Bedford, rising to a pressure height of 2,500ft testing various combinations of engine settings and reaching speeds of between 35 to 40 knots.

Difficulty was experienced with both the rudder and elevator controls. The control cables became slack, causing steering chains to jump off their sprockets and rendering them ineffective. This was caused by the overbalancing of the Cruiser-type control surfaces, which, although in balance at low speeds, overbalanced at higher speeds. The flight lasted six and half hours with the ship returning at lower speed safely to Cardington in the early morning.

Following a conference on 24 June the leading edges of the fins and elevators were reduced by 14in, in an effort to counter the overbalancing effect. A second trial flight was then undertaken on the evening of 28 June, with Flight Lieutenant Wann again in command, accompanied this time by Commander Maxfield and three other American crewmen.

For this flight the lower rudder was disconnected and fixed in the midships position, then, with the four after engines running at 1,800rpm, speed was increased to 48 knots. Once again control of the airship at this speed became difficult, with both rudder and elevator still being overbalanced.

The next morning, after landing, an inspection revealed structural damage and distortion to the cruciform girders supporting the cantilever fins. Campbell ordered strengthening plates to be fitted to joints and additional external wire bracing installed externally to the fins. To remedy the overbalanced control surfaces a further 24in was removed from both elevators and rudders.

Other changes deemed necessary before the Atlantic flight were made included replacing the air pressure fuel-feed system by gravity tanks over each car. By this measure it was calculated that fuel consumption could be reduced from 13.5 gallons per hour at cruising speed to 11.6 gallons per hour; an essential saving for the proposed crossing.

The third flight, on 17 July, was planned as a forty-eight-hour trial designed to test all aspects of the airship's performance. It was to end with the airship berthing at Howden. The R38 duly departed Cardington at 7.30 p.m., again under British command with Wann in control. It included three representatives of the National Physical Laboratory on board to monitor pressure gradients over the fins, plus four American officers including Maxfield.

The R38 flew north into the night to Howden trying combinations of engine settings and speeds, then turned out over the North Sea to undertake high speed tests. Up to speeds of 43 knots the R38 demonstrated good flight characteristics, being easy to control with no repeat of the difficulties experienced on the earlier flights.

Just after midnight, with the ship flying at 2,000ft, Mr Pannel leading the NPL team requested that Flight Lieutenant Little, then in command of the control car, should increase the engine speed on all six engines to full power at 2,000rpm.

At this point Flight Lieutenant Pritchard raised an objection, fearing failure of the fins under these conditions and citing the earlier, near disastrous experience with the R36 in April when her fins had in fact collapsed. This had followed the application of full power whilst flying at a height of 3,000ft, causing her to fall nose first to 1,500ft. Indeed R36 was only saved from crashing by the immediate dropping of water ballast and stopping of the engines.

With this in mind, Pritchard also recommended that the ZR2 should be taken to a height of at least 2,500ft before a full power trial was undertaken. This advice was refused by Little, who indicated that he was under orders from Wann (then asleep in his bunk) that he was not to take the ship over its present pressure height of 2,300ft.

The speed trial commenced at 12.25 a.m. with the four rearmost engines developing full power, which allowed the air speed to rapidly build up to 50 knots. Almost immediately the American elevator man Collier experienced difficulties with control, and the airship began pitching up and down in a violent manner over a vertical range of 500ft as the elevator problem once more returned.

Flight Lieutenant Pritchard, sensing disaster, took the wheel from the less experienced man and reduced the vertical component of movement to within a 100ft range, at the same time reducing engine revolutions to regain control.

Simultaneously, a report came through to the control car from American crew member Chief Boatswains Mate Charles Aller, who was in the crew space in the lower hull, that the intermediate transverse frames at station 7a and 7b on the port side were failing together with the lower F longitudinal, which was damaged. This information caused Pritchard to further reduce engine rpm whilst Campbell and Wann were summoned to inspect the damage. Course was set for Howden at reduced speed where the ship arrived at 5 a.m. next morning.

Following his inspection Commander Campbell stated that, in his opinion, the damage to the intermediate frames was the result of propeller wash from the forward engine cars Nos 1 and 2 beating on the outer cover, causing the girders to collapse. This was a wholly illogical explanation, as the lower envelope was under all conditions constantly in the slipstream of the propellers.

Even more damning to this explanation was the fact that Nos 1 and 2 engines were over 100ft ahead of the damaged area and were in fact running at idling speed at the time. Despite this, Campbell chose to ignore what must have been overwhelming evidence that the damage had been caused by the girders failing under severe bending loads. Loads imposed by the violent control movements instigated by the overbalanced elevators.

In order that this setback could be quickly overcome, so as not to delay further the handover to the Americans, Campbell recommended that additional strengthening members be applied to the lower longitudinals as a quick fix. This was a wholly inadequate response to the exposure of a serious structural fault in the design, a fault that should have alerted everyone concerned to the seriousness of the situation. Such concerns were instead put aside in the interests of expediency; an attitude that bordered on criminal negligence.

At Howden the *R38* was taken into the shed where the outer cover between frames 5 and 10, and up to the F and G longitudinals nearest the keel, was opened up. The twelve intermediate transverse frames between 5–10 were subsequently all strengthened by adding additional channel sections, whilst the F and G longitudinals were doubled in the same area. In an effort to solve the problem of elevator instability an additional 12in of area was removed forward of the point of balance.

Flight Lieutenant Pritchard again brought to the attention of General Maitland his concerns over the structural integrity of the airship. He requested, for reasons of safety, that further speed trials should be conducted at a greater altitude and that the period of trials be extended to at least seventy hours. However, Maitland chose to ignore these recommendations – he was anxious to have the *R38* off his hands as soon as possible.

Commander Maxfield had already forwarded a report that gave his opinion that the girder breakages experienced on the three flights were of a minor nature, and could be effectively remedied in time for an early departure for the Atlantic flight.

Admiral Griffin of the US Bureau of Steam Engineering, responsible for overall control of the project, took a different view following discussions with Lieutenant Commander Bieg, who had been on board during the third flight and who had witnessed the girder failures. From Bieg's testimony Griffin concluded that the girders had failed 'due to compressive stresses produced under high aerodynamic loading'. Consequently, with the possibility of complete structural failure, the low purity of hydrogen in her cells and the general dissatisfaction with the gas cell wiring, Griffin recommended postponement of the Atlantic flight until the airship had been fully tested and the faults remedied.

In Washington, however, plans were already in train to receive the *ZR2* in late August, with four battleships and other vessels already detailed to take up their stations along the proposed route at the end of the month. By 3 August repairs were complete, with the US colours and her number, *ZR2*, outlined on the bow. Maxfield now indicated that he was ready to leave for America as early as 24 August, provided Washington could persuade the Air Ministry to hand the *ZR2* to the US navy.

Maxfield was insistent that further trials should now be cancelled. This was a view that found some support from those in the British government and at the Air Ministry who were anxious to close down the Cardington works and the airship service at the earliest possible date. An early departure was out of the question, however, as the weather over the next three weeks became unsuitable for flying. In addition, on instructions from Washington Maxfield agreed to one last trial before the Atlantic flight.

Finally, on 22 August, the weather cleared sufficiently for flying, and at 7.10 a.m. on 23 August the *R38* rose into the still air above Howden with a total crew of forty-nine, which included Commander Maxfield and sixteen Americans. Also on board were the chief designer, C.E.I. Campbell, and General Maitland, together with, once more, three representatives of the NPL who were to continue recording their scientific observations during flight.

At the time of this flight the airship, although bearing US markings, was still under the control of Flight Lieutenant Wann and the Air Ministry until she was officially handed over to the US acceptance officer once the trial programme had been deemed to be complete.

The airship flew out to sea where fuel consumption tests were carried out during the morning, followed by turning trials at differing speeds and rudder angle, which during these tests only varied between 5° and 10° of rudder. These manoeuvres were being monitored by the NPL team, who were at that time conducting manometer tests in the vicinity of the fins.

As the airship flew down the coast, thickening low cloud precluded a planned mooring of the *R38* to the high mast at Pulham air station; Flight Lieutenant Pritchard additionally advised against attempting the proposed high-speed trials in such conditions. Accordingly, the *R38* headed north again into better weather, passing over her home base of Howden at around 4 p.m. Here, one of the British officers informed the base via an Aldis signal lamp that they now intended to undertake the high-speed trial before landing at Howden later in the evening.

Under advice from Pritchard the *R38* now climbed to 3,700ft (the greatest altitude she was to achieve) to satisfy the concerns previously expressed in his memorandum to Maitland. With the airship heading towards the mouth of the Humber all six engines were opened up to their full power running at 2,000rpm for a period of ten minutes, where a speed of 62 knots was achieved. The airship behaved well, flying steadily under perfect control with no repetition of the earlier elevator and rudder problems.

With power reduced Campbell was heard to express his satisfaction with the airship's performance to General Maitland, and ordered that a further series of rudder and elevator tests of a more vigorous nature should be undertaken as final proof of the airship's capability prior to acceptance. For this test Mr Bateman from the NPL was asked to take position in the gun arrangement at the extreme tail of the airship, in order to record the movement of rudders and elevators on film.

Accordingly, at 5 p.m. flying at 2,500ft through light cloud at a speed of 54 knots, the *R38* began the additional series of elevator and turning trials.

Flight Lieutenat Wann (one of only five survivors of the subsequent wreck) was as usual on duty in the control car throughout the ten-minute test. He later asserted that the helmsmen were instructed to limit rudder and elevator movement to 15° each way during turns under ¾ power. However Mr Bateman, who was in the rear gun pit and so possibly in the ideal position to observe the elevator and rudder movements, later described this as a 'severe test'. According to his account at the court of inquiry he witnessed the rudders being thrown 'hard over to hard over' in rapid succession, whilst the elevators were also subjected to similar violent movements.

As the tests were coming to an end the *R38* was on a south-easterly course, which took her over the centre of Hull where large crowds had gathered along the river bank to watch the giant silvery airship, glinting in the sunlight as she cruised overhead.

Suddenly, the appreciative murmur of the crowd turned to cries of horror as tragedy unfolded before their eyes. They watched as the airship abruptly altered course through 90° towards the south-west. The first sign of trouble was a deepening wrinkle that developed in the taut outer cover on the starboard side, abaft rear engine cars Nos 5 and 6.

Inside the airship catastrophic structural failure was taking place. The lower, previously strengthened F and G longitudinals had collapsed, together with the main and intermediate transverse frames between 9 and 10, under excessive compressive stresses. From the ground the crowd saw the bow and stern drop as the airship split open at the top, the then ship broke into two parts and a few seconds later the larger forward section caught fire with a huge explosion that was heard over a distance of 20 miles.

The blazing forward portion fortuitously fell into the River Humber, where a second explosion shook the city and left the river covered in blazing petrol. From this forward section only Flight Lieutenant Wann was to be rescued alive, jumping from the control car just before the crash. Although severely injured and unable to attend the court of inquiry he recovered to continue his career in the RAF, later serving as a group captain in the Battle of France in 1940.

The tail section did not take fire and fell comparatively slowly into the river, landing on a sand bar in 5ft of water. In the tail section Bateman, having realised the airship had broken up, attempted to use the fixed parachute in the cockpit, but on attaching his harness and jumping over the side he found his lines had fouled the structure and was left dangling several feet below the falling airship.

Bateman's assistant, Corporal Potter, had already run forward to another parachute position but found his path blocked by wreckage, so he made his way back to the gun pit where he discovered Bateman's predicament and attempted, unsuccessfully, to pull him back aboard. By this time, however, the stern had hit the water, only Potter was quickly picked up by one of the many boats that rushed to the scene.

Lance Corporal Davies was at frame 10 when the girders started to collapse around him. Fortunately, he also ran aft and was in the lower catwalk as the airship struck the water. Here, after a short swim in the fast-flowing, petrol-covered river he was also rescued, along with the sole American survivor A.C.R. Norman Walker.

Two or three parachutes were seen to fall from the airship, but their occupants failed to survive: two being seen to fall into blazing oil and, despite the best efforts of the rescue boats who managed to recover several bodies, no further survivors were found in the rapidly rising tide.

The forty-four lives lost included sixteen American aircrew including Commander Maxfield, Lieutenant Commander Bieg and the senior US engineering officer. Whilst on the British side Commander Campbell, General Maitland, Bateman of the NPL and Flight Lieutenant Pritchard were among the dead.

Over the following week portions of the wreckage were recovered from the river with the bodies of the American crew members, however six of the British crew, including Pritchard and the captain of the *R 33* Flight Lieutenant Thomas, were not found.

On 7 September a memorial service was held at Westminster Abbey. The American dead were meanwhile transported to the United States aboard the British cruiser HMS *Dauntless*, being met by ships of the US navy off New York harbour and escorted into Brooklyn Navy Yard, where the bodies were turned over to relatives for individual burial in their home towns.

On 27 August a court of inquiry was convened at Howden to inquire into the loss of the *R 38*, with Air Vice-Marshal Sir John Salmond as president and six other RAF officers experienced in airship operations in attendance. Five of the survivors gave testimony, along with those who had witnessed the disaster on the ground and many other expert witnesses. The purpose of the official inquiry was only to establish the order of events that led to the tragedy and its outcome, without arriving at any conclusions as to the cause of the accident – this aspect being left to an independent committee of experts.

In order to establish the technical reasons that led to the disaster, the Air Ministry asked the Aeronautical Research Committee to implement an independent investigation of eleven members who comprised the leading experts in the fields of aerodynamics and structural elements. This committee was led by Mervyn O'Gorman, former director of Farnborough, and included Major G. Scott, the commander of the *R 34* on her Atlantic flight, and Professor L. Bairstow, whilst the US Bureau of Aeronautics detailed Commander Dyer and C.P. Burgess to attend.

The committee studied the history of the *R 38*'s construction, noting the lack of aeronautical data available to the British relating to the calculations of aerodynamic loads that would be applied in flight. In addition they looked at the absence of a continuous narrative of technical log books in which the definitive stress calculations of the structure would be contained. They also commented on the fact that, following the Armistice and her sale to the US navy, no measures had been taken to strengthen the ship for low-altitude operation, which in Professor Bairstow's opinion would have only required some 4 tons of additional material to double the strength of the structure.

The professor's calculations showed that the factor of safety within the total structure was not 'FOUR', as Campbell had asserted, but only 'ONE', and that due to her greater size and speed in relation to *R 34* she should have been 65 per cent stronger than the earlier ship. In fact the *R 38* was, in the committee's opinion, by comparison only half as strong as the *R 34*.

Further, in relation to the airship's factor of safety Professor Bairstow calculated that the *R 38* flying at an angle of 10° nose-up at a speed of 54 knots, with her elevators set at a corresponding down angle (that is, flying and heavy, developing dynamic lift), would produce a powerful bending moment on the hull structure that would be sufficient to cause structural failure. In addition, Professor Bairstow, after having examined the structure of the surrendered *L71*, had surprisingly adjudged her hull to be structurally weaker even than that of the *R38*.

In the German ship all the rudder area was behind the hinge line, whereas the *R38*'s rudders were of the balanced Cruiser type, with a large balancing area of surface projecting ahead of the hinge line. It would appear in the case of the *R38* that no scientific calculations were made to determine the optimum proportion of area required to correctly balance the rudders or elevators, this being left to a trial-and-error basis during test flying.

In all, her first three flights' problems were experienced with the control surfaces overbalancing, and although the surfaces were balanced at low speeds, when speed was increased beyond 35 knots control became harder to maintain. Flight Lieutenant Pritchard had earlier asserted that the centre of pressure on the rudders moved forward with increasing speed.

Constructor-Commander Campbell was severely criticised for his management of the whole project: his lack of supervision at key stages in her construction, his failure to recognise and act effectively upon the damage sustained during the first three flights, and his inability to keep adequate records during construction. The inquiry also referred to his wholly inadequate response to the serious structural faults in the design, faults that should have alerted everyone concerned to the seriousness of the situation.

The committee summarised its conclusions in its report of March 1922 by stating that:

The weather did not contribute to the accident.
The accident was due to structural weakness in the design of the airship.
That the process of failure was gradual, the final collapse of the structure being due to a number of reversals of stress of a magnitude close to that of the factor of safety.
Owing to her instability any movements of her controls necessary to keep her on course were large and rapid.
That the *R38* would have almost certainly crashed in the first bad weather she would meet due to her control instability and general structural weakness.

At the time of the accident three quarters of the total purchase price had been paid to the British government, and as the airship had not been officially accepted by the US navy – she had been lost before delivery – the question of compensation was raised. The US naval attaché in London presented a claim for $500,000, which was promptly paid by the Treasury in April 1922, bringing to an end this unhappy chain of events.

18

The Airship vs the Flying Boat

The Royal Navy had shown a great deal of interest in hydro-aviation before the outbreak of the war, indeed they actively fostered the development of seaplanes and flying boats, seeing this form of flying machine as principally suited to naval purposes. Private firms, Short Brothers in particular, were encouraged in the development of seaplanes for naval use, with the navy trying out a wide variety of aircraft from the various manufacturers to determine those types most suitable for the particular demands of service at sea.

Importantly, the navy's approach to the acquisition of aircraft from diverse sources was in contrast to the army's almost sole reliance on the products of the Royal Aircraft Factory.

The first occasion an aircraft was to fly from the deck of a warship took place as early as November 1910, when a US navy Curtiss biplane was launched from a platform built over the forward 5in gun of the cruiser USS *Birmingham* in Hampton Roads. Later, this feat was emulated by the Royal Navy when the dashing Lieutenant C.R. Samson flew a Short S27 off the foredeck of the battleship HMS *Africa*, moored in the Medway in December 1911. In May of the following year he repeated the feat from a staging fixed over the bows of the battleship HMS *Hibernia*, as the ship steamed at 15 knots in the English Channel.

The Royal Navy took the lead in the development of specialised ships for the carrying and launching of aircraft, when in early 1913 the old cruiser *Hermes* was converted to carry two seaplanes from a platform fitted forward. A rather precarious method of launching was accomplished by mounting the seaplane on a wheeled trolley, which was guided by rails during take-off to be jettisoned into the sea, with the aircraft being recovered from the sea by crane upon its return.

It is difficult for us today to realise just how revolutionary and extraordinary these developments appeared to be in the 1910s, when the aeroplane had literally only just emerged as a practical proposition capable of being effectively controlled in the air. It says much for what was often considered to be a hidebound and inflexible organisation that the British Admiralty demonstrated considerable foresight in condoning these experiments at such an early date.

Encouraged by these experiments, a collier was converted whilst building in 1914 to become the first purpose-built seaplane carrier. This ship, bearing the illustrious name HMS *Ark Royal*, carried its aircraft in her holds, from where they were craned out for launching either from the short flying-off deck at the bows, or by being lowered into the water when the ship was stationary for take-off and recovery.

Similar arrangements of accommodation and launching methods were employed in the second generation of carriers, which were the faster Isle of Man (IOM) steam packet, and railway cross-Channel steamers impressed into service in 1914. These handy ships gave excellent service; their relatively high speed enabling them to keep station with the fleet. The *Engadine* even took part in the Cuxhaven raid of December 1914.

Later at Jutland the same ship, whilst attached to the Battle Cruiser Squadron, sent up an aeroplane that spotted the German High Seas Fleet, this being the aircraft reconnaissance made during the engagement.

Other actions involving these plucky ships included the IOM steam packet *Ben-My-Chree*, which, while in service during the Dardanelles campaign, launched the first successful air torpedo attack on a ship, before being sunk by Turkish shore batteries.

The old Cunard record-breaker *Campania* of 1893 was also converted in 1914–15 to carry ten aeroplanes. She was stationed with the Grand Fleet at Scapa and often accompanied the fleet on sweeps of the North Sea, launching aircraft in favourable conditions on reconnaissance duties. Unfortunately *Campania* was absent at Jutland – due to an oversight when she did not receive the order to put to sea at her remote location in the anchorage. Thus her aircraft were unavailable to Jellicoe at a critical time when their observations could have had a positive tactical influence on the outcome of the battle.

Although the Royal Navy pioneered the development of the aircraft carrier, in these early days the main function of their aircraft was seen, primarily, to be in the role of a scouting adjunct to the battle fleet. Their offensive capability was regarded as a secondary consideration. During the interwar years this philosophy was maintained and, while torpedo-carrying aircraft were a part of the carrier's complement, the primary duty of the fleet's aircraft continued to be in the area of reconnaissance.

The United States and Japan, on the other hand, recognised and developed the offensive potential of carrier-borne aircraft that could strike at a distance far beyond the range of the guns of battleships. Extending this philosophy, some American and Japanese naval planners even foresaw the day when carrier-borne aircraft would launch massive attacks on an enemy fleet whilst the opposing fleets were hundreds of miles apart, and might even deal a decisive blow without the fleets ever meeting.

In pursuit of these revolutionary ideas both the American and Japanese began building large fleet carriers, equipped with large squadrons of dive bombers and torpedo-carrying planes, to be used as long-range artillery. In both navies this policy was pursued with determination and employed with devastating effect during the Second World War.

By 1918 the earlier British ships had been joined by HMS *Argus*, the first flush-decked carrier, formerly an Italian liner, that had been appropriated by the Admiralty and equipped to carry twenty aircraft. In design it foreshadowed the Woolworth cargo ship conversions of the Second World War. Smoke from the boilers was ducted through vents on the sides aft in an effort to reduce the air disturbance from normal funnels to aircraft landing on.

Argus was also joined by the massive HMS *Furious*, a converted large, light battlecruiser capable of 35 knots and equipped with a complement of twenty planes. Two sister cruisers, the *Glorious* and *Courageous* were similarly converted after the war. These three ships had originally been designed to support a landing on the coast of Germany, being heavily armed with 15in guns and of shallow draught in order to operate in the waters off the Friesian Islands. Fortunately for all concerned, the plan was not proceeded with.

The first ship to be designed as an aircraft carrier from the keel up was HMS *Hermes*, completed in 1923. She was the first to establish the now accepted design feature of setting the funnels and superstructure out on the starboard side to allow for continuous, unobstructed flying off deck.

The provision of aircraft for use by the fleet grew rapidly as the war progressed, so that by November 1918 the navy possessed over 130 aircraft operating at sea and further supported by 104 airships. In addition, many of the battleships were by that date fitted to carry a two-seat reconnaissance biplane over the forward gun turret, while twenty-two of the light cruisers were similarly equipped.

Excellent though the seaplanes were, a larger and more robust type of aircraft capable of flying extended patrols deep into enemy controlled waters was required to counter the menace of the U-boat, and to disrupt the reconnaissance activities of the scouting Zeppelins. Following the sinking of the Cunard liner *Lusitania* in May 1915, and that of the White Star *Arabic* in August, the German Foreign Office were conscious that these acts had brought Germany and America to the verge of war.

Beckman Holveg, the Foreign Minister, prevailed on the high command to suspend the policy of unrestricted submarine warfare, which had been introduced as a counter to the British economic blockade. With the danger of war with the United States temporarily averted, a number of U-boats were transferred from their North Sea bases to the Mediterranean. Here, for a period, they caused havoc amongst Allied merchant shipping and inflicted disastrous losses on the British and French warships in the Dardanelles.

By March 1916, overriding the misgivings of their own Foreign Office, the policy of unrestricted submarine warfare was again resumed by the Germans, only to be suspended once more in April following further strong protests from the United States. This situation lasted until February 1917, when throwing caution to the wind the ultimate phase of the policy of unrestricted submarine warfare was implemented regardless of the protests of both neutrals and the United States.

During this period the German navy had some 112 U-boats operational, with 80 stationed at the North Sea ports of Ostend and Zeebrugge and a further 22 operational in the Mediterranean and Adriatic in the Austrian port of Pola. In a single week in February 1917 thirty-five ships were sunk in the English Channel and the Western Approaches; in this desperate situation extreme measures were required to turn the tide in the entente favour.

As a result of these mounting losses severe food shortages were experienced throughout the British Isles, with actual starvation being a real possibility in some areas. The situation led to civil disorder and food riots in several major cities, requiring the authorities to suppress these outbreaks with force.

Remarkable as it may seem, despite the seriousness of the situation the government still allowed market forces and private suppliers to control the supply and distribution of food, the effect of which ensured that the poorer section of the community were to be those who suffered most. Food shortages continued to be of concern throughout the war, until, belatedly, the government introduced food rationing in mid-1918; a long-overdue measure to finally ensure a fairer distribution of the nations' food resources.

At sea, intensive patrols were mounted by aeroplanes and airships, together with the provision of more surface patrol craft carrying depth charges and equipped with hydrophone detection equipment. These measures, alongside the introduction of the convoy system, were ultimately to control the submarine menace.

The RNAS had early in the war acquired several examples of the Curtiss H4 Small America flying boat, which although of limited range and performance demonstrated the promise of further development. These aircraft had been obtained through the efforts of a remarkable and dedicated reserve office, Commander J.C. Porte, who in October 1915 was commanding the RNAS air station at Felixstowe.

Porte was well suited to the task in hand as he possessed considerable experience with flying boats, having earlier joined Glenn Curtiss, the aviation pioneer, at Hammondsport in the US in 1913. He had assisted Curtiss in the development of a twin-engine (later three-engine) flying boat that was being designed for the wealthy storeowner Rodman Wanamaker, specifically for a transatlantic flight attempt.

The attempt was scheduled for the autumn of 1914 with Porte as the pilot, the machine being delivered and tested before the outbreak of war, at which point the undertaking was abandoned. Porte had previously held a commission in the Royal Navy from which he had been invalided out in 1909 with tuberculosis. He rejoined in 1914, where he argued the case forcefully for flying boats to be employed for anti-submarine and long-range reconnaissance work. His proposals so impressed their lordships that not only was he appointed to command the RNAS base at Felixstowe in September 1915, but was also given a brief to develop the Curtiss boats along the lines he had propounded.

The Admiralty had bought the original *America* along with several examples of the smaller Curtiss H4 boats of 2 tons displacement. Powered by two 180hp Curtiss engines, they were generally adjudged to be inferior in performance to the contemporary British seaplanes then in use.

Porte's first design was an enlarged and more powerful version of the Small America, the Large America H12. This was, as the name suggests, a much larger boat, with a wingspan of 90ft and powered by two Rolls-Royce Eagle1 engines of 275hp. It gave a speed of 85mph and had the ability to climb to 11,000ft in thirty minutes. It was armed with four machine guns and four 100lb bombs. These boats came into service in mid-1916, and although the design of the planing hull displayed a structural weakness, which required great caution during take-off and landing, the type proved to be a successful reconnaissance and anti-Zeppelin fighter.

The first success against the latter by one of these machines took place on 14 April 1917, when Large America H12 *No. 8666* left from Great Yarmouth air station to patrol the waters around the Terschelling lightship, observing radio silence to avoid detection by the enemy. She was under the command of Flight Lieutenant J.C. Galpin with Flight Sub-Lieutenant R. Leckie as pilot, Chief Petty-Officer V. Whatling as wireless operator and Air Mechanic O. Laycock as engineer. After one hour and thirty minutes Galpin and his crew spotted a Zeppelin dead ahead and some 10 miles distant. *No. 8666* was at that time flying at 6,000ft, some 3,000ft higher than the enemy airship. Their prey was the *L22* commanded by Kapitän Dietrich-Bielefeld, and was just turning to the north-east having reached the southern limit of her patrol line.

Dropping their bombs to lighten ship, Leckie opened the throttles and, using broken cloud as cover, put her nose down. Diving towards their quarry at over 100 knots she levelled out at her altitude at 75 knots and overhauled her on the starboard quarter. From a range of 150ft, *No. 8666* opened fire with the twin bow and midships guns, firing a complete tray of ZPT tracer from the forward gun and half a tray from the midships position before it jammed.

As the flying boat banked away to clear the gun they saw a glow inside the envelope and within seconds the rear portion was in flames, quickly engulfing the entire framework, which fell stern-first into the sea. The cause of the loss of *L22* was unknown to Strasser as the Zeppelin had no time to send out a wireless message, so complete was the surprise of the attack. This left Strasser to assume that the *L22* had been lost to surface fire from British warships.

In his report Galpin stated that 'the *L22* had been set alight before the crew had realised the nature of the attack' and the element of surprise, together with their greater speed, gave the flying boat the advantage. Galpin went on to say 'that even under normal conditions this type of flying boat should prove superior in every way to a Zeppelin if one can judge from the amount of power left untouched, and she proved an exceptionally steady gun platform'.

After several unsuccessful attacks on Zeppelins off the Dutch coast, which alerted Strasser to the danger posed by the flying boat, further success came on 14 June when Large America H12 *No. 8677*, commanded by Flight Sub-Lieutenant Hobbs with Flight Sub-Lieutenant Dickey as navigator, left Felixstowe at 5.15 a.m. to patrol off the West Friesian Islands.

At 8.40 a.m., while cruising off Vlieland at 500ft, a Zeppelin was sighted 5 miles on the starboard bow heading north at 1,500ft. All three gun positions were manned and Hobbs climbed to 2,000ft, endeavouring to get into an up-sun position of the enemy. Levelling out on her port quarter, Hobbs made a diving pass at over 100 knots over the stern of the airship, opening fire with Brock & Pomeroy tracer ammunition from all guns. He noted the airship's number, *L43*, as they passed over her.

Turning to starboard for a second attack, the crew of the flying boat saw that the Zeppelin was already completely enveloped in flames; the wreckage falling into the sea where flames and smoke continued to rise from the spot for some time. Kapitänleutnant Kraushaar and his crew were taken by surprise, not seeing their attacker until the last moment, and were unable to transmit a warning message. There were no survivors from the doomed airship. After circling the spot the H12 *No. 8677* turned for home, landing safely at her base at 11.15 a.m.

On the same day another H12, *No. 8660* with Galpin and Leckie on board, was also on patrol from Great Yarmouth air station. At 9.08 a.m. off the Nord Hinder lightship they also sighted a Zeppelin, 15 miles distant steering west at 10,000ft. Leckie turned in pursuit climbing hard, but after ten minutes Kapitän Hollender of the *L46* had sighted the British plane and, dropping water ballast, rose quickly to 15,000ft. Leckie continued the chase, reaching 12,500ft and firing four trays of explosive tracer at the airship. Although hitting her, the incendiary mixture had burnt out due to the extreme range.

The airship returned their fire, whilst the executive officer Oberleutnant Frey closely examined the flying boat through binoculars. As there was no possibility of overhauling the airship at that altitude after a chase of forty minutes, and having expended the majority of his ammunition, Leckie broke off the action and disappointedly turned his flying boat for home.

Following this engagement and their fortunate escape, Kapitän Hollender was able to report the exact nature of the threat posed to patrolling Zeppelins in the German Bight. Over the coming weeks Strasser issued orders to his commanders to exercise great caution when operating in, what up to then, had been considered friendly waters.

The superior performance of these new flying boats forced Strasser, after consideration of the implications, to abandon low patrol patterns in the German Bight and along the Dutch islands. From now on reconnaissance was to be conducted above 10,000ft for safety, providing a margin of greater protection to the vulnerable airships but lessening their ability to observe surface details effectively.

Further development of Porte's original concept followed, with the introduction of the larger and more seaworthy F2a boats built, as they were, with a much stronger double-stepped hull. The production of these craft was entrusted to the Aircraft Manufacturing Co. Ltd and its energetic owner. George Holt Thomas. But as the company were fully committed to the building of DH4 bomber aircraft, the order was sub-contracted to May. Harden & May were boat builders of Hythe on Southampton Water. Here, some sixty or so machines were built. Later, several other contractors assisted in building a further forty aircraft.

The F2a spanned 98ft with a loaded weight of 5 tons and two 350hp Rolls-Royce Eagle III engines. Producing a top speed of 90 knots, these formidable craft mounted no less than six .303 Lewis machine guns plus bombs, and had an endurance of eight hours. Carrying a crew of four they were the first true long-distance over-water reconnaissance aircraft. A range of 600 miles enabled them to scout large areas of the North Sea with rapidity.

In the offensive role the F2as could more than hold their own against the nimble and equally effective Brandenburg sea monoplanes and biplanes they encountered on their incursions into the German Bight and along the Friesian Islands. The F2as would sometimes form an attacking squadron of four or five boats against the German seaplane bases, during the course of these forays long-running aerial duels were often fought.

When attacked by the German aircraft the F2as would form into line astern, flying straight and level to engage the enemy machines with intense, combined broadsides of machine-gun fire from their superior armament of up to six machine guns – in the same manner as that employed by Nelson 'Wooden Walls' a century earlier. As a Zeppelin destroyer the F2a proved to be an efficient and steady gun platform, with a fair turn of speed and a respectable rate of climb, able to bring its powerful armament to bear on a Zeppelin with every chance of success.

F2as were responsible for the destruction of two Zeppelins and made several other attacks, in which, although the airships managed to avoid their fiery fate, they were forced to abandon their patrol duties early, or had to climb to altitudes where their observations were rendered ineffectual. The F2a's greatest contribution possibly lay in its anti-submarine warfare role, together with its valuable contribution to mine barrage spotting duties, and the part it played in enforcing the blockade by reporting ships to the surface patrols on the lookout for blockade runners or contraband cargoes.

The U-boat posed a continuous and increasing threat to these islands throughout the war period, with on more than one occasion causing the losses of merchant shipping with their vital cargoes of war supplies, raw materials and, most importantly, food. This reached crisis proportions, requiring an unprecedented concentration of naval and air forces to remedy the critical situation the country found itself in.

Alongside the numerous destroyers, mine hunters, coastal motor boats, other patrol craft, airships and seaplanes, the F2as operating from Felixstowe, Great Yarmouth and other stations along the east coast were selected to fly the 'Spider Web Patrols'. Starting in early 1917, these patrols were designed to counter the U-boat menace. The central point of the patrolled area was based on the Nord Hinder light-ship, from where eight patrol lines radiated out for a distance of 30 miles with concentric lines joining them at distances of 10, 20 and 30 miles from the centre. This allowed 1,000 square miles of ocean to be systematically scoured with rapidity.

As the war progressed the F2as and the earlier H12s became ever more active against the German seaplane bases along the Friesian Islands, and also in flying long reconnaissance missions over the North Sea attacking submarines, enemy merchant shipping and protecting convoys. The flying boat by mid-1917 was able to perform all the duties with greater efficiency and reliability than those that had been attributed as the main role of the airship a few short years before.

Although at this stage of the war naval airships still made a valuable contribution in the struggle for command of the sea routes, it was becoming obvious to far-sighted naval planners that the flying boat could fulfil all of the airships' functions and duties more quickly and more effectively. The flying boat and its base required less supportive infrastructure to operate in terms of massive sheds, potentially dangerous gas plants and large ground crews, and importantly it could perform the required tasks with considerably less expenditure.

During 1917 the ability of the Royal Navy to launch attacks on the North Sea Zeppelin bases from their growing number of aircraft carriers became an increasing threat to Strasser's ability to maintain reconnaissance flights over the German Bight. This threat was realised when, in June 1918, aircraft from the *Furious* successfully raided the Tondern airship base, destroying two Zeppelins with bombs without loss to the attacking craft.

The two airships destroyed were the *L54* and *L60*, both of the advanced Height Climber class, and a severe blow to the naval airship division. Their destruction rendered the Schleswig-Holstein base and coastline untenable to further use.

Other devices designed to disrupt or destroy Zeppelins were also employed, including the towing of specially adapted lighters behind destroyers. These carried a single Sopwith Camel scout, which could be rapidly brought to a suitable radius of action within enemy waters. When a Zeppelin was sighted the destroyer turned into the wind, working up to 30 knots. Here the Camel, after a run of 10ft, was airborne.

By this method on 11 August 1918 Lieutenant S.D. Cully RN attacked and destroyed a patrolling Zeppelin, the *L53*, at the great height of 18,000ft off Terschelling, this being the last such airship destroyed in the war. Only six days prior to this incident the leader of naval airships, the redoubtable Korvettenkapitän Peter Strasser, had been killed when the *L70*, the most modern airship of the fleet, had fallen in flames off the Norfolk coast with the loss of all her crew.

These losses marked the end of the Imperial German Navy's Airship Service as a fighting force, and by these actions in the last year of the war the aeroplane had proved its undoubted ascendancy over the airship in all aspects of operation, apart from that of endurance.

The increasing use of carrier-borne bombers and torpedo-carrying aircraft with the fleet faintly warned of the coming change in the balance of power from sea to air; to a time when air power would determine the outcome of great sea battles in which the opposing fleets would launch their attacks hundreds of miles apart.

With the ending of the war the Admiralty lost control of long-range reconnaissance duties, these being taken over by the Royal Air Force. Throughout the 1920s and 30s, despite severe financial stringency in the defence estimates, the RAF continued to develop the flying boat to a high state of technical efficiency, establishing bases around the world in pursuance of the protection of maritime trade.

During this period the RAF continued the pursuit of excellence by providing superb training and honing crew skills by a series of long-distance flights across the empire. This established the sound base on which Coastal Command was able to build with its modern Sunderland and Catalina flying boats, which gave such sterling service in the protection of trade routes throughout the six-year struggle that was to come.

Following the end of the Great War flying boats were also being developed concurrently for civil transport purposes. In Germany in particular the Dornier Company produced the highly successful Wal twin-engine all-metal flying boat. This sold worldwide in large numbers and did much to create modern airline services during the 1920s and '30s.

Later the Sikosky, Martin and Boeing companies in the United States developed a series of large, powerful flying boats with huge carrying capacity and range. For a while these seemed to presage the superior qualities of the flying boat over the land plane as the main carrier of the future. In Great Britain this view also prevailed with Imperial Airways largely re-equipping with the modern four-engine Short C class flying boat.

Ultimately, in both the military and civil field, the flying boat, which for a short period was seen as the successor to the airship, was also bypassed by technological developments in the landplane.

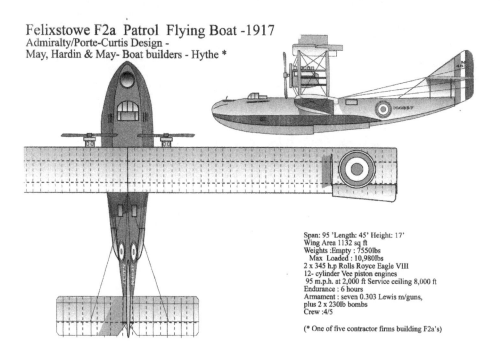

Felixstowe F2a Patrol Flying Boat -1917
Admiralty/Porte-Curtis Design -
May, Hardin & May- Boat builders - Hythe *

Span: 95 'Length: 45' Height: 17'
Wing Area 1132 sq ft
Weights :Empty : 7550lbs
   Max Loaded : 10,980lbs
2 x 345 h.p Rolls Royce Eagle VIII
12- cylinder Vee piston engines
 95 m.p.h. at 2,000 ft Service ceiling 8,000 ft
Endurance : 6 hours
Armament : seven 0.303 Lewis m/guns,
plus 2 x 230lb bombs
Crew :4/5

(* One of five contractor firms building F2a's)

19

Airship Development in the
Interwar Years

Most of the airships in this section have already been described in great detail in numerous books by other authors over the years, consequently their stories are well known. Accordingly, I have included only a brief review of the more famous craft such as *R100*, *R101*, *Graf Zeppelin*, *Akron*, *Macon*, *Hindenburg* etc.

At the end of the First World War the remaining German airships that had not been scuttled by their crews were distributed between the victorious Allies under the terms of the peace treaty; the ships being safely delivered by their German naval crews to the entente cordiale bases in England, France and Italy. These delivery flights must have been a humiliating experience for the German crews, but were carried out with a professionalism that was a credit to their former service.

In June 1920, Great Britain received one of the most technically advanced German naval airships, the giant *L71*, together with the earlier Height Climber *L64*. Both airships were delivered by German crews to Pulham Air Station in Suffolk in July 1920, where they were emptied of their lifting gas and hung up in the No. 2 shed.

Whether the Air Ministry ever intended flying the two ships with British crews is doubtful, rather they remained as instructional airframes with no attempt ever being made to make them airworthy again. The British made detailed examination of both airships, from which the exceptionally specialised nature of their construction was revealed, and, whilst confirming their wise decision not to fly them, contributed little to British hopes implementing a commercial airship programme.

The two German airships were left to moulder in the No. 2 shed, where they were occasionally viewed by British engineers and the America aircrew of the Howden contingent waiting to take over the *R38*. Throughout this time they were systematically pillaged for spare parts and engines (two of *L71*'s engines being installed in the civil *R36*) until they were eventually scrapped, with *L64*'s denouement being of a particularly violent nature.

France, in its turn, was awarded the naval airship *L72* and two army airships, the *LZ-113* and *LZ-83*. They also received the small, streamlined modern commercial Zeppelin, *LZ-121 Nordstern*, which, although it had been completed after the war, was seized by the Allied Control Commission. The French took a more active role in airship operations than the British and had plans for the development of an airship service between metropolitan France and their possessions in North Africa.

In accordance with this plan, the French navy operated the *L72* as the *Dixmude* under the command of the capable Lieutenant Jean du Plessis de Grenédan, making a number of successful and extended flights from their base at Toulon over the Mediterranean and French

40. Commercial Zeppelin *Nordstern*'s shed at Friedrichshafen, 1919.

North Africa. The French were aware of the specialised nature of the *L72* and its unsuitability for commercial operation, but were prepared to take the risk in order to advance their grand plan to develop an aerial communications network linking their possessions around the world.

After a long over-water flight the *Dixmude*, whilst cruising off the coast of Sicily on the night of 21 December 1923 in stormy conditions, appears to have suffered a catastrophic structural failure. The giant airship was seen by the crew of a fishing boat to fall in flames into the sea with the loss of her commander and her entire crew of fifty.

The commercial airship *Nordstern*, now renamed *Mediterranee* by the French, was also operated from Toulon by the navy. Fortunately, due to being of a more structurally robust design than the *L72*, she enjoyed a trouble-free existence under her new owners. The *Mediterranee* was in fact considered to be so reliable that during 1923 she was chosen to carry the French Air Minister M. Eynac and his staff in some luxury to a governmental meeting in Rome. Here, she was moored overnight at Ciampino before returning to Toulon the next day.

After long service and due to the effects of both economy and the aftermath of the loss of the *Dixmude*, she was eventually scrapped in September 1926.

Italy received the naval Zeppelin *L61*, an advanced 15m bay Height Climber, together with the older army ship *LZ-90*, both of which were delivered to Ciampino near Rome in

French Naval Airship "Dixmude" - 1920
ex Zeppelin"X" Type LZ114 - "L72"
Luftschiffbau Zeppelin GmbH, Friedrichshafen

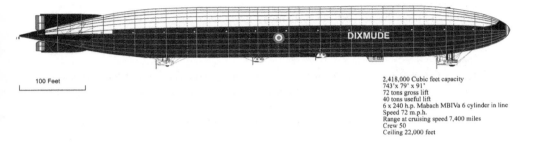

100 Feet

2,418,000 Cubic feet capacity
743'x 79' x 91'
72 tons gross lift
40 tons useful lift
6 x 240 h.p. Mabach MBIVa 6 cylinder in line
Speed 72 m.p.h.
Range at cruising speed 7,400 miles
Crew 50
Ceiling 22,000 feet

German Civil Airship LZ 120 "Bodensee" - 1919
Zeppelin "y" type
Luftschiffbau Zeppelin GmbH, Freidrichshafen

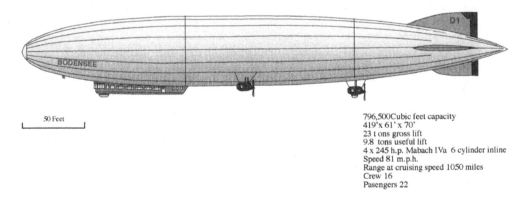

50 Feet

796,500Cubic feet capacity
419'x 61' x 70'
23 t ons gross lift
9.8 tons useful lift
4 x 245 h.p. Mabach IVa 6 cylinder inline
Speed 81 m.p.h.
Range at cruising speed 1050 miles
Crew 16
Pasengers 22

December 1920. The *L61* was initially renamed *Italia* (not to be confused with the later polar airship of the same name), which would appear to indicate that great things were expected of her. In fact she made only two flights under Italian naval command, proving to be difficult to handle and totally unsuitable for their purposes. After only five months under their control the Italian navy decided better use could be made of the hangar space she occupied and she was demolished during 1921.

Meanwhile, the older *LZ-90*, renamed *Ausonia*, entered service with the army, but her flying career came to an abrupt end after six months when she was severely damaged during manoeuvres by flying into the ground due to jammed elevator controls. Despite the damage sustained she managed to return to base safely and was taken into her shed, but following inspection of her structure she did not fly again and was dismantled during June 1921.

The Italians also gained the smaller of the two modern civilian airships, the *Bodensee*, which the Zeppelin Company had operated through DELAG between Friedrichshafen and Berlin from August to December 1919, before the inter-Allied commission interceded and confiscated the airship.

In Italian hands the *LZ-120*, now renamed *Esperia*, also entered service with the Italian navy. Here she enjoyed a successful career and was in continuous use until the loss of the semi-rigid polar airship *Italia* in July 1928, when she was withdrawn from service and broken up.

Three other earlier L30 class Zeppelins were apportioned respectively to Belgium (*L30*) and Japan (*L37* and *L41*), but as neither country had suitable sheds available to house them they were delivered in the form of disassembled parts.

Japan, however, later negotiated the delivery of an appropriated large double shed, which was erected at Kasumigaura 40 miles north of Tokyo, and which in 1929 was fortuitously placed to house the *Graf Zeppelin* on her round-the-world flight.

Out of this group of surrendered wartime Zeppelins that had escaped being wrecked by their crews, only the post-war ships *Bodensee* and the *Nordstern*, being structurally sound and having been built solely for commercial purposes, were to be employed effectively by the victors. Of the wooden-framed Schütte-Lanz airships employed by the German navy only the *SL22* survived the war, to be dismantled at her Baltic base in 1920 with examples of her structure distributed between the Allied nations.

The older naval and army airships had already been subjected to hard use and in most cases their gas cells were nearing the end of their life, so to a large extent the whole process of seizing the ships was a wasted and expensive exercise for the entente cordiale.

The US was determined to receive her due share of the spoils of war, but was faced with the problem that the two most suitable airships of the L70 class ships had already been apportioned to France and Britain. Furthermore, the marginal performance of the remaining earlier German wartime ships, if awarded to the US, would make an Atlantic crossing an extremely hazardous venture.

At an early stage the US government proposed that the Zeppelin Company should be allowed to build a large 'repatriations ship' suitable for an Atlantic crossing, despite the articles in the peace treaty that forbade such activity. Needless to say, this suggestion found little favour with the other Allies who were anxious to see the end of the Zeppelin companies' domination of worldwide airship construction.

With this avenue temporarily closed to them, the response of the American navy was to seek to obtain airships from other Allied sources. Initially they approached the British, where they were to secure the purchase of the *R38*, then the largest airship in the world at 2,740,500cu ft capacity with a gross lift of 81 tons (whose story and denouement are described in Chapter 17).

Under the stringent terms of the peace treaty the Germans were forbidden to build any further airships at this time and the Zeppelin Company, in order to keep the Friedrichshafen works open and maintain a workforce, turned to the manufacture of aluminium kitchenware as a temporary expedient. At the same time the US Army Air Service, led by the energetic General 'Billy' Mitchell, made its own separate attempt to acquire a rigid airship from the Zeppelin Company.

Colonel William Hensley, who had been the US army representative aboard the *R34* on her return transatlantic voyage, entered into negotiation with the Zeppelin Company for a giant

41. US airship *ZR 3 Los Angeles*, 1924.

3,532,000cu ft airship then being designed, the *LZ-125*. With a length of 774ft and powered by twelve 240hp engines, the contract for the airship was signed in November 1919 for a price of $360,000.

At this stage no facilities existed in the United States to accept or house such a huge ship and this, together with the extreme annoyance of their allies at this flaunting Article 201 of the treaty, caused the contract to be cancelled in January 1920. Frustrated by this turn of events the US army next decided to make an approach to the Italian airship industry to supply a smaller airship suitable for army use.

The Italians had during the war produced a small number of semi-rigid craft of a very successful design, built by the government-run *Stabilimento di Costruzioni Aeronautiche* near Rome. This group continued the development of the large and successful M class of 441,000cu ft capacity using the same streamlined metal keel structure and suspension system in the later N class, which culminated in the building of the polar airships *Norge* and *Italia* in 1926 and 1928 respectively.

Under the direction of chief engineer constructors Crocco and Usueli at the wartime government facility at Ciampino, a large semi-rigid, the *Roma*, was built during 1921. The 410ft-long airship was of 1 million cu ft capacity; this was one of the largest semi-rigid airships ever to be constructed.

After seeing brief service with the Italian army the *Roma* was sold to the American government. She was transported by sea to America where, following reassembly and substitution of two of the six 200hp Ansaldo engines with 240hp Liberty engines, the airship entered service with the US army.

In February 1922 after take-off from its base at Langley Field, Virginia the airship suffered a failure of the horizontal steering planes causing her to nose slowly down to the ground, where, unfortunately, she ran on to high-tension power lines at the end of the field.

The *Roma*, inflated with hydrogen, immediately took fire. This resulted in the deaths of thirty-four crew members, while many of the survivors were badly injured in the conflagration. As a result of this tragedy following so closely on from the loss of the *ZR2*, the use of hydrogen in American airships was henceforth banned, with helium being substituted for all future US airships.

42. Polar airship *N1 Norge* at Kings Bay, Spitsbergen, 1926.

In Italy an even larger semi-rigid was designed in 1921 under the direction of engineers Crocco and Usueli at Ciampino, being of a massive 1,942,600cu ft capacity. Known as the T type, or Trans-Atlantic, the giant airship was intended for a proposed South American passenger service carrying fifty passengers with a range of 12,000 miles.

In retrospect, the design of this huge airship was an impractical application of the semi-rigid principle. Indeed it had more to do with political manoeuvring and a demonstration of nationalistic fervour to the new Fascist government, than a serious attempt to build a real transatlantic passenger craft.

ZR1 Shenandoah, 1923

See Chapter 15 for a more detailed account.

The United States first home-produced dirigible was the *ZR1 Shenandoah*, which was based on the Zeppelin *L49* that had been brought down intact at Bourbonne-les-Bains during the Silent Raid of October 1917.

From the plans provided by the French, the British had begun construction of the *R36*, which was almost a direct copy of the captured Zeppelin, merely lengthened by one bay.

Whereas the *Shenandoah* was a better engineered extrapolation of the original design, being stronger frame for frame than the *L49*.

The *ZR1* was launched in 1923 and completed a successful transcontinental flight of the United States during 1924. She later operated with the fleet on exercises in both the Atlantic and the Pacific, flying for a total of 730 hours in her two-year life. With the arrival of the *Los Angeles* from Germany in October 1924 there was insufficient helium to fill both airships simultaneously. Consequently, the *Shenandoah* was hung up in the Lakehurst shed whilst her helium was transferred to the *Los Angeles*, and vice versa, whilst both rigids were in commission.

The *Shenandoah* was also employed on mooring experiments to both High and Stub Masts, and at sea to the converted fleet oiler *Patoka*, which was fitted with a mooring mast aft.

During 1924 a proposal was made that the *Shenandoah* be made ready for polar flight, starting from Lakehurst, flying over the North Pole and on to the US Pacific coast. Although preliminary plans were made for the flight and forward bases established, the plan was finally rejected by the navy department, being considered too dangerous.

In September 1925, whilst under the command of Lieutenant Commander Zachary Lansdowne, the *Shenandoah* was lost in a severe storm on an ill-advised publicity flight to the Midwest. Over Ava, Ohio the airship encountered a fierce weather front during a storm, here, despite the great strength, the hull tore the airship into three sections resulting in the loss of Lieutenant Commander Lansdowne and thirteen of the crew.

U.S Naval Airship ZR-3 "Los Angeles" - 1924
Zeppelin LZ 126
Luftschiffbau Zeppelin GmbH, Freidrichshafen

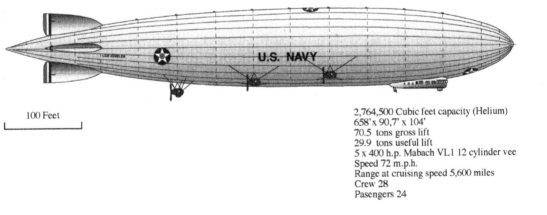

100 Feet

2,764,500 Cubic feet capacity (Helium)
658' x 90,7' x 104'
70.5 tons gross lift
29.9 tons useful lift
5 x 400 h.p. Mabach VL1 12 cylinder vee
Speed 72 m.p.h.
Range at cruising speed 5,600 miles
Crew 28
Pasengers 24

ZR3 Los Angeles, 1924

During 1922 the former Allies waived the restrictions imposed on the Zeppelin Company to allow the building of an airship for the United States navy, in lieu of those that would have been awarded under the terms of the peace treaty.

Construction of the *LZ-126* subsequently began in 1922, on the understanding that although it would be a unit of the US navy it would not be fitted out for war purposes. The 658ft-long airship was of 2,764,500cu ft capacity with a gross lift of 70.5 tons (helium), it could accommodate twenty-four passengers with a crew of twenty-eight and a range at cruising speed of 5,600 miles.

Upon completion the *LZ-126* was delivered to Lakehurst, New Jersey by Dr Eckener after an uneventful transatlantic flight of eighty-one hours on 15 October 1924. Here the airship received its American designation ZR3 and was named the *Los Angeles*.

The airship was used for training and fleet exercises in both the Atlantic and Pacific, making extended voyages across the United States, the Caribbean and to South America. She was additionally used in the experimental dropping and recovery of aircraft.

Whilst in service the *Los Angeles* flew for some 4,300 hours covering 230,000 miles. She remained in commission for eight years before being retired in 1932 on the grounds of economy. Following her retirement, although she no longer flew, the *Los Angeles* was used as an instructional airframe and was utilised in mooring-out experiments, before being finally dismantled during 1941.

Norge, 1926

During 1921 Colonel Umberto Nobile, whilst employed at the *Stabilimento di Costruzioni Aeronautiche* near Rome, produced a highly advanced and practical semi-rigid design, the *N1*. This airship was of 654,000cu ft capacity and powered by three 245hp Maybachs, which gave a speed of over 70mph. In its original form it included a cabin for twenty-five passengers. The envelope was divided into ten separate gas compartments broken up by vertical fabric partitions, with a rigid nose cap and a similar structure aft to carry the stabilising fins.

Due to political events in Italy at that time the construction of *N1* was delayed and did not commence until 1923, with the ship first flying the following year and being assigned to the army.

In 1925 the *N1* was purchased by Lincoln Ellsworth for $100,000 on behalf of the Amundsen-Ellsworth expedition, who planned to fly an airship over the North Pole the following year.

Under the terms of the contract, endorsed by Mussolini who believed that a successful flight would bring credit to his Fascist regime, Nobile would pilot the craft with a largely Italian crew. He would effectively be in command of the expedition due to his experience of airship handling. However, his co-expedition leaders uncharitably regarded him merely as a 'hired pilot', with the expedition itself being referred to as the 'Amundsen-Ellsworth Polar Expedition', with no mention being made of Nobile himself.

This demeaning snub by the two principals, who were in effect no more than passengers to the man on whose skill the safety and successful outcome of the expedition depended, was rightly much resented by Nobile.

The plan was to fly from Italy to Spitsbergen and then from there over the Pole; flying over thousands of square miles of unexplored territory to land in Alaska or to fly further down the Pacific coast if sufficient fuel remained. In order to make the *N1* suitable for

Polar Airshp "NORGE" - 1926
Italian Semi rigid,"N" Class
Stablimento di Construzoni Aeronautiche
Corps of Engineers - Ciampino, Rome

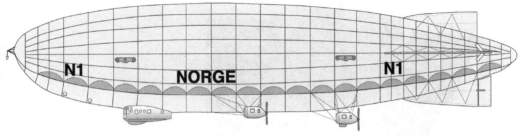

654,000 Cubic feet capacity
348' x 72' x 87'
19.1 tons gross lift
9.2 tons useful lift
3 x 240 h.p. Ansaldo
Speed 70 m.p.h.
Range at cruising speed 3000 miles
Duration 70 hrs
Crew 16 (in two watches)

In May 1926 the "Norge" with a crew of sixteen flew from
Kings Bay Spitzbergen to Teller Alaska, via the North Pole
covering 3,180 miles non stop in 70 hours 40 minutes

such a long and potentially dangerous trip, the ship required considerable modification
and lightening of the structure: the long passenger cabin was removed and replaced with
a small car of sufficient size to contain the steering positions, navigation and wireless
room.

Finally, Nobile was satisfied with the changes that endowed the *N1* with an extra 2 tons of
useful lift. This would be essential to the success of the proposed flight due to the marginal
performance of the little airship.

N1 was handed over to its new owners on 29 March 1926 at Ciampino outside Rome at
a lavish ceremony attended by Mussolini, where the ship was formally named the *Norge*. At
9.30 a.m. on 10 April 1925 the *Norge*, with Nobile in command and Ellworth, Amundsen
and a crew of sixteen plus Titina (Nobile's dog), lifted off. After circling Rome they headed out
over the Tyrrhenian Sea on course for France.

Flying via the Rhone valley and across France in adverse and squally weather conditions,
the *Norge* eventually dropped her landing lines over Pulham air station in Suffolk, where
she was berthed alongside the ageing *R 33* after a flight of thirty-two hours.

The next stage of the flight was from Pulham to Oslo, where a short stop was made,
then on to Gatchina Airfield near Leningrad. From here, after a two-week layover, the
Norge lifted off on 5 May, passed over Leningrad then flew on to Vadose in northern
Norway. In Vadose, following a brief stop and despite having one engine out of order,
Nobile decided to make the 780-mile crossing to Spitsbergen without waiting to repair
the engine. Arriving at King's Bay on 7 May *Norge* was taken into the roofless hangar
having covered a total distance of 3,900 miles since leaving Rome; in itself a remarkable
achievement.

43. Polar airship *N4 Italia* at Stolp, Germany en route to the North Pole, 1928.

Four days later on 11 May the *Norge*, with a crew of sixteen including Ellsworth and Amundsen, rose from the landing ground and headed north. Initially they met clear weather, but this later turned to freezing fog. At 1.20 a.m. the next morning, navigating with a special sun compass required in such high latitudes, the *Norge* reached the North Pole where the flags of Italy, Norway and the United States were dropped on the ice amid celebrations in the tiny gondola of the airship.

Course was now set for Alaska flying over previously uncharted territory. No new lands were sighted as the dirigible flew on, only monotonous fields of broken pack ice lay gleaming white below them. The airship flew on at 3,500ft above fog that came and went, and which formed a coat of ice on the outer cover that in turn threw splinters of ice from the propellers into the gas cells.

At 1.30 a.m. the following day the coast of Alaska came into sight and wireless contact was made with Nome, then, following the coast, the airship made its way slowly south against a rising wind.

Eventually, after seventy-two hours in the air, Nobile ordered the ship to land at the village of Teller, where it made a perfect landing aided by local Inuit trappers, just 90 miles short of their intended destination of Nome. The *Norge* was subsequently dismantled where she lay and shipped back to Italy. The epoch-making flight was justly applauded around the world, with Nobile receiving honours from Mussolini and a hero's welcome for him and his crew on their return to Italy.

Polar Airship "ITALIA" (N4) - 1927
Italian Semi rigid, "N" Class
Stablimento di Construzoni Aeronautiche
Corps of Engineers - Ciampino, Rome

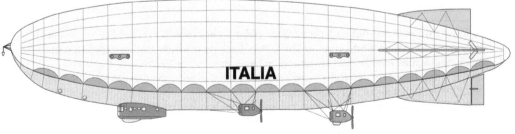

ITALIA

In May 1928 the "ITALIA" with a crew of sixteen successfully flew from Kings Bay Spitzbergen to the North Pole, but on the return flight crashed on the ice 20 miles off the N.E. coast of Spitzbergen.
An international effort finally rescued eight of the crew after an ordeal of 49 days.

654,000 Cubic feet capacity
348' x 72' x 87'
19.1 tons gross lift
10.5 tons useful lift
3 x 240 h.p. Ansaldo
Speed 71 m.p.h.
Range at cruising speed 3500 miles
Duration 70 hrs
Crew 16 (in two watches)

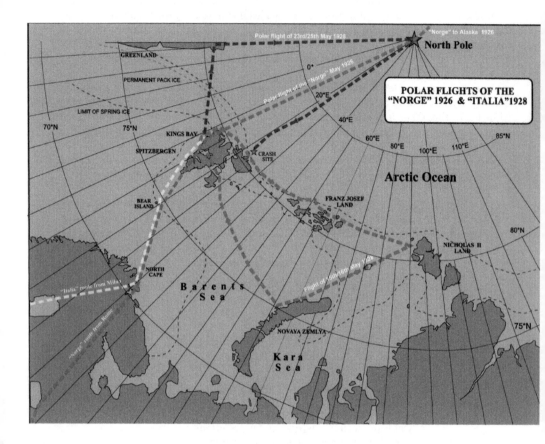

N4 *Italia*, 1928

Following the success that had attended the polar flight of the *Norge*, the Fascist regime in Italy resolved to repeat and emulate the feat by backing a wholly Italian expedition to the polar regions that would reflect favourably on Mussolini and the Italian people in general. The airship chosen was the N4 *Italia*, which was an improved version of the successful *Norge* – being lighter and with an increased useful lift.

The expedition was carefully planned and was to involve five flights of a scientific and exploratory nature in the polar regions, which, if they had all been successful, would have made a valuable contribution to Arctic exploration, meteorology and other areas of scientific study.

The *Italia* departed Milan on 19 April 1928 with a complement of twenty persons and Nobile's dog Titania, heading north over the Alps towards Germany via Vienna. After a stormy trip across the mountains of Czechoslovakia the *Italia* arrived at Stolp to the north of Berlin. Here, after a ten-day layover whilst the fins were strengthened, the airship departed for Stockholm on 2 May, then on over Finland to the mooring mast at Vadose in northern Norway that had been used by the *Norge* two years before.

Leaving Vadose at 8.45 in the evening on 5 May, the airship passed over Bear Island to drop their lines at the King's Bay Ny Alesund mast at 12.45 next morning.

44. Airship USS *Akron* approaching mast, 1931.

Since departing Milan fifteen days before the *Italia* had traversed 3,200 miles in far from ideal conditions in a demonstration of skilful airmanship on behalf of the Italian crew. The first flight of the five-part exploration programme began on 11 May, but had to be abandoned after a few hours due to frayed rudder cables.

With a crew of fourteen, and two scientists plus their instruments, the *Italia* rose into a clear blue sky on 15 May, heading first north then north-east past Spitsbergen's North Cape along the 81st parallel towards the little-known archipelago of Franz Joseph Land. The airship was flying over the area where the supposed 'Gillis Land' had been reported by a Dutch sea captain some 200 years earlier, but no such land was seen by the *Italia*'s crew as the flew on in clear weather.

Passing the northerly point of Franz Joseph Land the airship continued eastward over an unknown waste of jumbled ice and open leads that glistened in the sunlight. At midnight on 16 May, with Nicholas II Land only two hours' distant (which had only been discovered fifteen years previously), Nobile reluctantly turned the airship towards the south due to an increasing headwind that would have threatened their return flight to Spitsbergen.

It would have been a major coup for Nobile if he had continued to explore the unknown island from the air, but he demonstrated his professionalism by putting the safety of his crew before the prestige and the plaudits that new discoveries would have brought him. Taking advantage of a strong tailwind the *Italia* flew down the west coast of the great island of Novaya Zemlya in perfect visibility as far as 76°N, then turning to the north-west crossed the unexplored portions of North-East Land and the almost unknown area of northern Spitsbergen, before completing a highly successful exploratory flight of sixty-three hours covering over 2,000 miles.

Prior to these record-breaking flights, Nobile and the Italians in general were not held in any great esteem as Arctic explorers by the more traditional nations involved in polar exploration. The next trip, however, was also to be a high-profile affair. Nobile planned to fly from King's Bay to Greenland, then follow the 27th meridian directly for the North Pole. From here the plan was to fly on to North America.

Loaded with scientific instruments, three scientists, survival gear, a crew of sixteen and the faithful Titania, the *Italia* lifted off in the early morning of 23 May 1928, heading directly for Cape Bridgman on the north-east cost of Greenland. At 5.30 p.m., in bright sunshine, they accurately reached the Greenland coast at Cape Bridgman, then turned north to the pole with a strong following wind.

Just after midnight on 24 May the *Italia* circled the North Pole, where they dropped the Italian and Milanese flags – the city that had put up money to fund the expedition together with a heavy wooden cross presented to them by the pope.

The expeditions' Swedish meteorologist, Malgren, had recommended returning to King's Bay, which would allow them to continue with their programme of exploratory flights, whereas Nobile wanted to press on to the United States. Eventually Malgren's argument won the day and the airship turned back towards Spitsbergen. Leaving the Pole at 2.30 a.m. she began the return flight, but as they flew on the sky gradually clouded over and ice began to form on the envelope as the increasing strength of the wind began to oppose their progress southwards.

After flying in cloud and fog for eight hours, all the while struggling against the rising wind, the elevator controls suddenly stuck fast in the up position. Nobile ordered the engines stopped while repairs were carried out, allowing the airship to rise above the cloud. Here, they took the opportunity to take a sun sight to check their position. Emerging from the fog bank at 1,100m the airship was in bright sunshine that allowed the gas to expand and to be expelled through the automatic valves.

Nobile ordered the airship down in the fog bank to limit the loss, an action that made the remaining gas contract, causing the airship to become heavy and fall towards the ice below. Finally, despite strenuous efforts to maintain the airship in the air, at 11 a.m. on 25 May the

Italia crashed on to the ice, smashing off the control car and the rear engine gondola.

On the pack ice one man was dead with nine left alive and injured, while six others were missing, carried away in the hull of the airship that rose into the air drifting rapidly to the east never to be seen again.

The *Italia* had crashed in a position 140 miles north-east of Spitsbergen and when news of the disaster reached the outside world it prompted a massive international rescue effort, during which the famed explorer Roald Amundsen was lost during an air search.

The ordeal of the survivors lasted for forty-five days, before they were rescued by the Russian icebreaker *Krassin*. The injured Nobile had been earlier rescued by air in a two-seat Fokker, on the understanding that his crew would be taken off immediately after him; a promise Lundberg the Swedish airman was unable to keep as he crashed beside the survivors tent on the next return flight.

The disaster was a major embarrassment to the Italian Fascist regime, and at a hastily convened court of inquiry headed by the Air Minister Balbo, all the blame was unfairly laid at Nobile's door – without summoning him to court to defend himself.

Disillusioned, Nobile resigned his army commission and left the country to work in Russia on airship projects until the outbreak of the Second World War.

LZ-127 Graf Zeppelin, 1928

Following the successful delivery to the United States of the *Los Angeles*, Dr Eckener began the construction of the *LZ-127 Graf Zeppelin* during 1927. The size and shape of the new vessel was determined by the dimensions of the existing shed at Friedrichshafen and, consequently, did not represent the optimum design envisaged by Eckener.

The airship was of streamlined form, being 776ft in length with a diameter of 100ft and a capacity of 3,036,660cu ft (hydrogen), giving a gross lift of 105 tons. A unique feature of the design was that twelve gas cells, separated from the hydrogen lifting cells and positioned below

German Civil Airship "Graf Zeppelin" - 1928
Zeppelin LZ 127
Luftschiffbau Zeppelin GmbH, Freidrichshafen

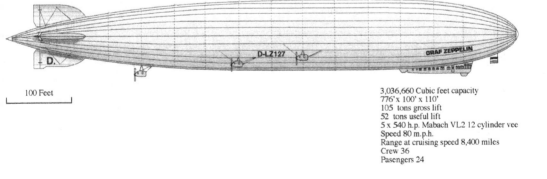

100 Feet

3,036,660 Cubic feet capacity
776' x 100' x 110'
105 tons gross lift
52 tons useful lift
5 x 540 h.p. Mabach VL2 12 cylinder vee
Speed 80 m.p.h.
Range at cruising speed 8,400 miles
Crew 36
Pasengers 24

the axial girder, were filled with a gaseous fuel, *Blaugas*, which weighed only slightly more than air. This saved the carrying of around 30 tons of petrol. A small quantity of petrol was carried for emergency purposes, however, and the five 530hp Maybach engines were capable of running on either type of fuel.

Accommodation was provided by a long integral gondola at the bow containing control, wireless and navigation rooms, with a passenger saloon, galley, dining room and ten cabins for twenty-four passengers. The crew of thirty-six were accommodated in the keel area when off-duty. The *Graf Zeppelin* was capable of a top speed of 80mph and had a still air range of 8,400 miles.

In a life of over ten years the *Graf Zeppelin* made over 590 flights, covering over 1,053,000 miles: she made numerous flights to both North and South America, a round-the-world flight in 1929, and two exploratory flights into high Arctic regions of a scientific nature.

Following the loss of the *Hindenburg* in May 1937 the *Graf Zeppelin* was withdrawn from service and hung up in the Frankfurt am Main shed until being broken up in 1940 on the orders of Reichsmarschall Herman Goering.

The Airship Guarantee Company *R100*, 1929

The *R100*, as previously mentioned, was constructed by the Airship Guarantee Company between 1925–29. Generally, it was of similar appearance and size to the *R101*, being 709ft in length by 133ft at maximum diameter, with a capacity of 5,150,000cu ft imparting a gross lift of 156 tons and a creditable useful lift of 60 tons. The six Rolls-Royce Condor IIIB twelve-cylinder petrol engines were positioned in pairs in three engine cars, each driving a

Civil Airship H.M.A. R 100 -1929
Airship Guarantee Company, Howden, Yorkshire

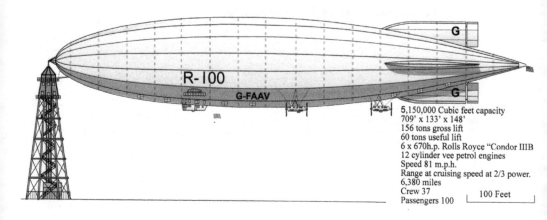

5,150,000 Cubic feet capacity
709' x 133' x 148'
156 tons gross lift
60 tons useful lift
6 x 670h.p. Rolls Royce "Condor IIIB
12 cylinder vee petrol engines
Speed 81 m.p.h.
Range at cruising speed at 2/3 power.
6,380 miles
Crew 37
Passengers 100   100 Feet

pusher propeller and a tractor airscrew to give a top speed of 80mph and a still air range of 6,380 miles.

The structure, however, more closely followed Zeppelin practice than did the *R101*, with the hull framing consisting of fifteen main transverse frames and sixteen triangular longitudinals, without intermediate transverse or intermediate longitudinals being fitted. A central, deep triangulated axial girder ran from bow to stern through the centre of the gas cells in place of the axial wire employed in Zeppelins, and to which the radial wiring of the transverse frames met forming a retaining bulkhead between adjacent gas cells. Duralumin was the main structural component employed, with the components made up from eleven standard sections that were formed from helically wound tubes, constructed on a special machine from riveted flat strip.

The fifteen gas cells were composed of a layer of cotton fabric with two layers of goldbeater's skin, which were fabricated by Zeppelin GmbH and equipped with automatic valves in the bottom of the cells that emptied into fabric trunks, discharging to the atmosphere on the upper portion of the hull. Hand-operated manoeuvring valves were arranged on the top of the ship for the quick release of gas in an emergency

Due to the reduced number of longitudinal girders forming a sixteen-sided polygon in *R100*, as opposed to the thirty-sided cross-section of *R101* (where the intermediate 'reefing booms' could be adjusted to tighten the outer cover), resulted in a larger area of the outer canvas cover requiring support. A system of cross-wiring and tapes were employed to keep the cover tight, but this proved to be an unsatisfactory solution as harmonic ripples could form in the fabric at high air speeds, and on occasion long and dangerous tears developed in the outer cover during flight.

The passenger accommodation was situated between frames No. 5 and No. 6 on three decks; the airship being designed to carry one hundred passengers and a crew of thirty-seven.

The *R100*'s first flight took place in December 1929, and early trials proved highly satisfactory with the airship making a cruise of fifty-three hours over the British Isles. It had been intended to fly as far south as the Azores to demonstrate her efficiency, but the Air Ministry forbade the *R100* to fly beyond a line from the Lizard to Finistère, as the government did not wish to be upstaged with such a successful demonstration by their rivals.

On 29 July 1930, the *R100* departed the Cardington mast with Captain Ralph Booth in command – together with Major Scott, thirteen passengers and a crew of forty-two – bound for Canada. The *R100* moored safely at St Hubert airport, Montreal after a flight of seventy-eight hours covering 3,300 nautical miles. It was a voyage that was not without incident: the fabric of the port horizontal fin was damaged in a squall over the St Lawrence River.

The *R100* stayed in Canada for twelve days making several local flights, carrying Canadian government dignitaries and Canadian and American businessmen, including an extended flight over Toronto, Niagara Falls and into New York state. The visit was adjudged to be a tremendous success for the imperial airship plan. The *R100* departed for England on 13 August, completing the return journey in fifty-six hours.

On arriving at the Cardington mast at 11 a.m. on 16 August, the crew found, to their disappointment, that no official welcome had been laid on to celebrate the epoch-making voyage.

This was to be the last flight of what was possibly the most successful and efficient British airship ever to fly, with a total of 300 hours' flying time to her credit. The next day the *R100* was taken into the No. 2 shed for an overhaul to prepare her for a flight to India in the following year, 1931.

Following the loss of the *R101* she was emptied of her gas and hung up on a care-and-maintenance basis, until the government ordered her to be broken up for scrap during 1931.

ZMC-2, 1929

An interesting development was the construction in 1929 of a metal-clad airship, similar in principle to that built by Schwartz in 1898. The airship, the ZMC-2, was designed by Ralph Upson and built under the direction of Carl Fritsche of the Metalclad Airship Corporation of Detroit, Michigan.

This small experimental craft was of a teardrop shape, 150ft in length with a beam of 52.5ft and a capacity of 202,000cu ft (helium). The metal envelope formed a single rigid cell over a metal framework, was composed of a series of narrow transverse rings of Alclad Duralumin sheets 0.008in thick and joined by rows of rivets to form a gas-tight cell. The airship was powered by two 220hp Wright radials and was capable of a top speed of 60mph.

The ZMC-2 was delivered in September 1929 to the US navy, who found the unusual method of construction to be successful and trouble free in operation. The little airship remained in service with the navy for ten years until struck from charge in 1941, having flown over 4,000 miles in that time

The ZMC-2 was intended to be the prototype for a series of very large metal-clad airships for commercial use, but the economic situation following the Wall Street Crash of 1929 militated against its further development.

U.S. Naval Airship Metalclad ZMC-2 - 1929
Airship Development Corporation, Detroit, Michigan

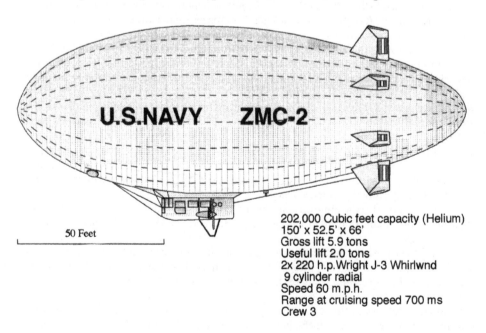

50 Feet

202,000 Cubic feet capacity (Helium)
150' x 52.5' x 66'
Gross lift 5.9 tons
Useful lift 2.0 tons
2x 220 h.p. Wright J-3 Whirlwnd
 9 cylinder radial
Speed 60 m.p.h.
Range at cruising speed 700 ms
Crew 3

The Royal Airship Works *R101*, 1930

In 1924, the newly elected Labour government of Ramsay MacDonald proposed the development of an airship service for the carriage of passengers and mail throughout the empire, authorising the construction of two large experimental airships for the purpose. One, the *R101* was to be a government-funded undertaking, to be built by the Royal Airship Works (RAW) at Cardington. Whilst the other, the *R100*, was a private venture project to be constructed by the Airship Guarantee Company Ltd, a subsidiary of Vickers Ltd, at Howden in Yorkshire.

The *R101* was designed by Colonel Vincent Richmond of the RAW and marked a major departure from current Zeppelin practice, incorporating as it did many novel and previously untried features. Construction commenced in 1926.

As initially conceived and built the *R101* was 714ft in length with a maximum diameter of 131ft and a gas capacity of 5 million cu ft, giving a gross lift of 146 tons and a projected useful lift of 60 tons. The tear-shaped, highly streamlined hull was largely built from a steel framework, consisting of fifteen main and fifteen intermediate longitudinal girders supported by deep-section, triangular main transverse frames. These were sufficiently rigid to obviate the need for the normal wire bracing in the plane of the ring structure.

Originally, both the *R101* and the *R100* were to be powered by four of the revolutionary 1,200hp Ricardo hydrogen/kerosene engines, but after much experiment these engines proved to be a disappointing failure and resulted in the *R101* being fitted with five Beardmore Tornado twelve-cylinder diesel engines instead.

These engines, formerly designed for service on Canadian railway trains, were supposed to develop 700bhp, but in practice they only delivered around 580bhp. Added to this, the five engines weighed over 16 tons, this being almost twice the weight of conventional petrol engines. In contrast, the *R100* designers substituted six proven Rolls-Royce Condor III petrol motors with a gross weight of 9 tons to power their craft.

The passenger accommodation of the *R101* was situated on two decks within the hull, consisting of a spacious lounge, dining room, twenty-six double cabins and promenade areas on the upper deck. The main control room, chartroom, wireless cabin, crews' quarters, galley

Civil Airship H.M.A. R 101 - 1929
Royal Airship Works, Cardington, Bedfordshire

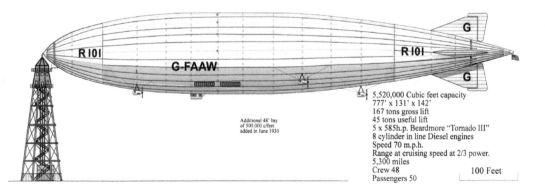

R 101

R 101

R 101

G-FAAW

G

G

Additional 48' bay
of 500,000 c/feet
added in June 1930

5,520,000 Cubic feet capacity
777' x 131' x 142'
167 tons gross lift
45 tons useful lift
5 x 585h.p. Beardmore "Tornado III"
8 cylinder in line Diesel engines
Speed 70 m.p.h.
Range at cruising speed at 2/3 power.
5,300 miles
Crew 48   100 Feet
Passengers 50

45. *R100* at the Cardington mast, 1930.

and even a smoking room were positioned on the lower deck, with a small streamlined control car projecting directly below the main control room.

The *R101* was to carry fifty passengers and a crew of forty-five, but following its first flight in October 1929, and after early test flying, it was apparent that the airship was too heavy and her useful lift was woefully inadequate at only 35 tons against the projected 60 tons.

A thirty-hour endurance flight over England, Scotland and Ireland in November 1929 demonstrated control and stability problems, while the leakage of gas from the automatic valves, together with severe chafing of the cells against the structure, resulted in a further loss of lift. The automatic valves were also a cause of concern as they were considered to be too sensitive and could open, discharging gas when the airship rolled. This contributed to a continuous loss of lift; a problem that was never satisfactorily solved.

Following an appearance at the RAF display at Hendon in June 1930, and in order to remedy these problems and the lack of lift, the airship was taken into her shed at the end of the month to have an extra bay of 500,000cu ft capacity inserted, increasing her useful lift to almost 50 tons. Upon emerging from the shed on 1 October 1930 the airship was 777ft in length with a capacity of 5,500,000cu ft, giving a revised gross lift of 167 tons.

The *R101* undertook only one trial flight with the new bay installed on 1/2 October of sixteen hours' duration (which could hardly be considered a thorough test programme) before being awarded a certificate of airworthiness to allow it to undertake a proving flight to India. Political pressure dictated these ill-judged decisions; they wanted the airship ready to depart for India at the earliest possible date, carrying the Secretary of State for Air Lord Thompson, who confidently expected to be appointed the next Viceroy of India upon his return. Also flying aboard the airship was the charismatic and energetic Director of Civil Aviation Sir Sefton Brancker – who had done much to encourage civil aviation in Britain – together with ten other officials and passengers, and forty-two officers and crew.

The *R101* slipped the Cardington tower at 6.36 p.m. on 4 October heading south in deteriorating weather, bound for the Ismailia mooring mast in Egypt on the first leg of her epic journey. After flying over London at 8.20 p.m., where the lights of her lounge were visible to those below in the rain-soaked streets, she crossed the coast near Hastings against the driving rain and the increasing force of the wind.

At 2.09 a.m., just after sending a last wireless message saying that all was well, the airship nosed gently into rising ground at Alone (a small village south of Beauvais, northern France), where she burst into flame with the loss of all but six of her complement.

The effect of the disaster was to cause the government to abandon any further development of airships.

'Project H', *R102*, 1931

During the mid-1930s, before the disaster to the *R101*, a schedule of airship services to Canada and India had been produced, initially employing the *R100* and the *R101*, with each ship running two services to India and two to Canada in the year 1931.

For the Indian route three mooring towers were sited at Ismailia, Baghdad and Karachi (where a shed of similar dimensions to that at Cardington was also erected). Whilst on the Canadian route, which was naturally a non-stop journey between Cardington and St Hubert, Montreal, the existing tower was to be augmented by the provision of a shed. Similarly, on the South African section towers at Johannesburg and Cape Town were initially planned.

At the same time the Air Ministry had authorised three further larger ships, the *R102*, *R103* and *R104*. Preliminary design work was being carried out on the first of these, the *R102*, or 'Project H', as it was known. The *R102* was to be of 8.3 million cu ft capacity, 822ft in length with the same 134ft diameter as the *R101*. She would carry 150 passengers and mails on the power of seven improved Beardmore Tornado diesel motors for 10,000 miles at cruising speed.

The later *R103* and *R104* were to have been completed by 1935, being of 9.5 million cu ft and designed to carry 200 passengers.

ZRS-4 Akron, 1931 & *ZRS-5 Macon*, 1933

During 1926 the US Congress authorised the construction of two large, rigid airships for the US navy, the ZRS-4 *Akron* and ZRS-5 *Macon*, with the S indicating the scouting function intended. Following a competition the contract was awarded in October 1928 to the Goodyear-Zeppelin Corporation of Akron, Ohio. The chief designer, under the direction of the Bureau of Aeronautics, was Dr Karl Arnstein formerly of Luftschiffbau Zeppelin.

U.S Naval Airship ZRS-4 USS "Akron" - 1931
Goodyear - Zeppelin Corporation, Akron, Ohio

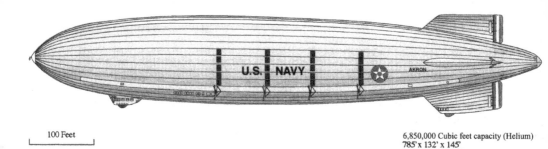

100 Feet

6,850,000 Cubic feet capacity (Helium)
785' x 132' x 145'
179 tons gross lift
68 tons useful lift
8 x 560 h.p. Mabach VL2 12 cylinder vee
Speed 85 m.p.h.
Range at cruising speed 5,900 miles
Crew 60
Armament 8 x .30 cal m/guns plus
3 single seat fighters
(5 Aircraft carried by "Macon")

F9C-2 "Sparrowhawk" Pursuit Airplane - U.S.S. "Akron" - 1931
Airplane launch and recovery proceedure to hanger within airship hull

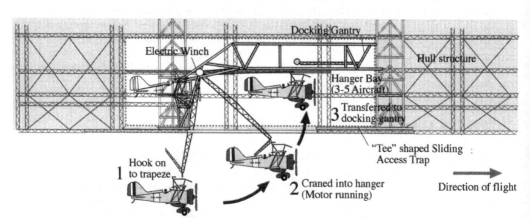

Docking Gantry

Electric Winch

Hull structure

Hanger Bay
(3-5 Aircraft)

3 Transferred to
docking gantry

"Tee" shaped Sliding
Access Trap

1 Hook on
to trapeze

2 Craned into hanger
(Motor running)

Direction of flight

The "Akron" carried three Curtis F9C-2 scout planes, the "Macon" five
in the latter case the undercarriges were removed, and a drop tank installed
to increase the range

The two airships were to be built in the Akron Airdock, a huge building 1,175ft in length, 325ft wide and 197ft in height, which was completed in November 1927. The contract price for the *Akron* was set at $5,375,000 with the later *Macon* being built at the lower price of $2,450,000.

The first airship, the *ZRS-4 Akron*, had a helium capacity of 6.85 million cu ft contained within a streamlined hull 785ft in length by 133ft in diameter. It was constructed from thirty-six longitudinal girders and twelve deep, triangulated mainframes set at a 22.5m spacing – similar to those employed in the *R101* – with three lighter intermediate frames between. In place of the usual single lower keel, the *Akron* had three internal keels, with one being situated along the top of the airship and two side keels set at 120° from the upper keel.

The use of helium allowed the installation of engine rooms, of which there were eight situated within the hull along the lower keels, with each containing a 560hp Maybach VL-12 engine. These drove bracket-mounted propellers through long drive shafts mounted outboard, which could be rotated through 180° to provide thrust in any direction. Above the engine positions banks of very efficient condensers were installed to act as a water ballast recovery system, to compensate for the weight of fuel burned.

Twelve large gas cells were installed, made from cotton fabric coated with a gelatine latex sealant in place of the more usual goldbeater's skin. The three keels allowed easy access to all gas valve and discharge shafts, and for overall inspection of the gas cells. A small control car, integral to the hull, was positioned near the bow, which was augmented by an auxiliary control position situated in the lower fin for use in an emergency.

An internal hangar, measuring 75ft by 60ft, was built into the lower hull that could accommodate five single-seat fighter aeroplanes, whose function was both to protect and widen the area the airship could scout by flying a search pattern ahead and on the beam of the airship. The aeroplanes were lowered from a gantry into the slipstream for take-off and recovered by means of a trapeze, to which the aircraft hooked on upon return to be lifted back into the hangar space.

With a gross lift of 180 tons and a useful lift of 68 tons, the *Akron* had a total horse power of 4,480, giving a still air range of 5,900 miles and a top speed of 84mph.

The *Akron* was completed in 1931 and took part in fleet manoeuvres in both the Atlantic and on the Pacific coast, making seventy-three flights totalling 1,695 hours.

The *Akron* was lost when she flew into the sea in a storm at night off the New England coast on 4 April 1933, with the loss of seventy-four lives. The loss was attributed to the incorrect setting of the ship's barometric altimeter.

The *ZRS-5 Macon* was launched on 21 April 1933 and incorporated several improvements over the *Akron*, being lighter and slightly faster than her predecessor. The *Macon* was stationed at Sunnyvale, California operating with the fleet. She notched up fifty-four flights and 1,796 hours' flying time before she, too, was lost in an accident off the Pacific coast. Her loss resulted from earlier damage that had been caused by turbulent weather to the upper fin, which had not been properly repaired due to the necessity of having the airship available for fleet exercises.

On 12 February 1935, as the giant airship was cruising off Point Sur, California, she was struck by a severe gust of wind that wrenched the damaged fin out of the ship, deflating the rear three cells. Initially the airship rose to 4,300ft where, despite the damage, she could possibly have been saved, but she lost further gas through the automatic valves due to going over pressure height.

With the airship now too heavy and unable to maintain herself in the air, the *Macon* settled gently on to a calm sea where the crew had time to launch life rafts and only two lives were lost out of the crew of eighty-four.

The loss of the *Macon* marked the end of United States' involvement with the development of the rigid airship, although the navy actively pursued the building of pressure airships for maritime patrol purposes over the coming twenty-five years.

LZ-129 Hindenburg, 1936

Work commenced on the *LZ-129 Hindenburg* in late 1934, the new airship being 803ft in length with a capacity of 7,062,000cu ft of hydrogen. She was powered by four 1,030hp Mercedes-Benz twelve-cylinder diesel engines, capable of driving her through the air at 86mph. Originally, the *Hindenburg* was designed to use inflated helium as the lifting agent, but with the rise to power of Adolf Hitler in Germany Congress placed an embargo on the export of helium, fearing that Nazi Germany could use it for war purposes.

The airship was completed in March 1936, entering service on the North Atlantic and South American service carrying fifty passengers in luxurious accommodation on two decks within the hull. She completed ten double North Atlantic crossings from Frankfurt am Main to Lakehurst during the 1936 season, averaging fifty-four hours westbound and seventy-four hours eastbound, plus seven flights to Rio de Janeiro.

During the winter of 1936/37 the *Hindenburg* was overhauled and had her passenger accommodation increased and improved, affording even greater comfort. At the start of the 1937 season the idea of a regular airship service between the United States and Europe seemed an established fact, when the *Hindenburg* departed Frankfurt on 3 May bound for Lakehurst, New Jersey.

After an uneventful crossing the *Hindenburg* arrived over the field in the early afternoon, but was advised to postpone landing until later due to unsettled weather conditions. Finally, after cruising for some hours along the east coast and Long Island, an approach was made to the mast and landing lines dropped.

German Civil Airship "Hindenburg" - 1936
Zeppelin LZ129
Luftschiffbau Zeppelin GmbH, Freidrichshafen

100 Feet

7,062,160 Cubic feet capacity
803' x 135' x 148'
216 tons gross lift
100 tons useful lift
4 x 1030 h.p. Mercedes Diesel 12 cylinder vee
Speed 81 m.p.h.
Range at cruising speed 8,400 miles
Crew 60
Passengers 50

46. *R101* after lengthening at the Cardington masthead, September 1930, prior to departure to India.

At 7.25 p.m., at a height of 200ft, the almost motionless airship hung under a troubled, darkened and threatening sky. Suddenly, a burst of flame erupted from the top of the airship just forward of the upper fin. Within a minute and a half the great airship was consumed by flame and lay a smouldering wreck within a few yards of the mooring mast. Miraculously, sixty-two persons survived the inferno, whilst thirty-three who were less fortunate died in the disaster.

Various causes including sabotage were attributed to the fatal fire, but the most likely cause is an initial leak of gas from a rear cell accumulating under the envelope fabric then being ignited by a brush discharge from the electrically charged atmosphere in the stormy conditions prevailing at the time.

LZ-130 Graf Zeppelin II, 1937

Following the loss of the *Hindenburg* the old *Graf Zeppelin* was withdrawn from service and flown to Frankfurt am Main in June 1937, where it was emptied of its gas and hung in the shed pending a decision on its future.

47. The *Hindenburg LZ-129*'s first flight.

Meanwhile, the *LZ-130* was being built at Friedrichshafen, making its maiden flight on 14 September 1938. The new airship, now named *Graf Zeppelin II*, was generally similar in design to the *Hindenburg* but with revised and improved passenger accommodation, and tractor airscrews in place of the pusher arrangement employed in the previous airship.

After a series of test flights over Germany the Nazi authorities refused to allow the airship to operate any further passenger services, but the military authorities could see a surveillance use for the giant rigid. The head of the Luftwaffe signals organisation, General Martini, was aware that the British had developed a form of radio location to detect aircraft approaching the British Isles, as indeed the Germans had.

The *Graf Zeppelin II* offered the ideal platform to mount a large quantity of detection equipment, in order to monitor military short-wave wireless and British radar transmissions, to in turn devise counter-measures against them.

On her third flight on 22 September she was thus fitted out as a flying laboratory with banks of electronic detection equipment, together with a group of thirty signals specialists and scientists on board, and made an eleven-hour flight along the Czechoslovakian frontier searching for military wireless intelligence.

This flight was made a week before the Munich agreement, in which the British Prime Minister Neville Chamberlain and his French counterpart Edouard Daladier averted war for another year by agreeing to the ceding of the Sudetenland to Germany. On 14 April 1939 the *LZ-130* made a flight over the North Sea, cruising some 120 miles off the British coast trying to listen for and interpret radar transmissions, but without success. She returned to Germany after a flight of forty-four hours.

The *LZ-130* made several other flights through April to July along the Polish and French frontiers searching for wireless intelligence, again with limited success due to the background clutter of powerful civilian radio traffic.

The *LZ-130*'s final flight made in an attempt to probe the British radar defences took place on 2 August 1939, a month before the outbreak of war. She headed to a point 100 miles off the Norfolk coast and from here she cruised slowly up the English coast to the latitude of the Orkney Islands, turning south to a position level with Newcastle before heading back to base after a flight of forty-eight hours. Throughout the flight the British Chain Home radar tracked the giant airship's every move on their screens, whilst the Germans, in their turn, were unable to detect any indication of the British radar transmissions.

The reason for General Martini's failure to identify British early warning radar was due to the fact that they were looking in the wrong wave bands. The Germans were themselves developing radar but were utilising the ultra short-wave VHF bands of 120–130 MHz, and assumed that the British were using a similar bandwidth. In fact, early radar operated in a much longer wavelength of around 25MHz, an area they failed to search in. Consequently, they received only background static through their speakers and on the cathode ray tube displays during the voyage.

Following the outbreak of war in September 1939 the *LZ-130 Graf Zeppelin II* was laid up with her earlier namesake at Frankfurt am Main, until they were both ordered destroyed along with the huge shed by Hermann Göring in 1940.

20

From the Second World War
to the Present Day

Toward the end of the First World War the US navy operated some twelve or so French-built airships from the Paimboeuf base on the Atlantic coast near Saint-Nazaire. The types included the large Astra-Torres, Chalais Meudon and Zodiac airships, and were employed throughout 1918 on convoy escort, mine spotting and anti-submarine duties. During this period considerable experience was gained by the American crews involved, which was to prove invaluable to the US navy's lighter-than-air programme over the following years.

Following the Armistice, six of these airships were shipped to the United States together with several examples of the large British North Seas class and the smaller SS Zero class. These airships seemed not to have been flown to any extent in the US but they served as examples, where the best features of each were incorporated in their own home-built products. Possibly the best known of these early American naval airships was the C class of 182,000cu ft capacity, these being 192ft in length and powered by two 180hp Union engines.

One of this class, the C5, undertook a journey from her base at Montauk, Long Island, flying 1,300 miles to arrive at St Johns, Newfoundland, to take part in the *Daily Mail*'s transatlantic air race of June 1919. However, whilst being refuelled for the attempt, and fortunately minus its crew, a fierce gale wrested the airship from the hands of the ground crew to be whirled away on the storm and lost in the Atlantic. If not for this accident of fate there is every possibility that this small but highly efficient airship could have accomplished the first direct transatlantic crossing, a month or so before Alcock and Brown in their Vickers *Vimy*.

Throughout the 1920s and 30s the Goodyear Company produced a number of small pressure airships culminating in the *L1*, a 123,000cu ft airship of 1938 that was the standard advertising blimp for the Goodyear Company. The L class was destined to be the prototype for the wartime K class blimps, as well as being used for training purposes. In these airships the gondola was integral with the envelope, being suspended from internal catenary wiring positioned along the upper portion of the envelope

Prior to the outbreak of the Second World War the US navy operated a small selection of assorted airships, some of which, such as the J class, dated from the 1920s, whilst two large TC class airships were inherited from the army after they had abandoned their use in 1936. This motley fleet was augmented during 1939 by the introduction of the experimental K class, with the order for two airships of this modern design.

Recognising the heightened state of international tension during the late 1930s, and the need for the provision of patrol airships for convoy duties in any conflict, Congress authorised a programme to build 10,000 aircraft which include forty-eight non-rigid airships in June 1940. As a first measure an order was placed in October 1940 for six of the improved K class

48. A K class patrol airship over convoy, 1944.

airships numbered *K3–K8* of 404,000cu ft capacity, powered by two Wright R975 440hp radial engines, they had a gross lift of 12.4 tons.

Following the attack on Pearl Harbor in December 1941, an immediate expansion of the navy's airship fleet was implemented, with twenty-one additional K class airships of an improved design being ordered. Numbered *K9–K30*, the capacity of these airships was increased to 416,000cu ft with Pratt & Whitney engines of 425hp replacing the earlier Wright radials.

By November 1942 the US navy had thirty modern airships at their disposal, with initial bases being established at Lakehurst, on the Atlantic coast and Moffett Field, California covering the Pacific coast. This was followed by a further order in January 1943 for twenty-one more airships, numbered *K31–K50*, with an increased capacity of 425,000cu ft, and later in 1943 the final contract was placed for eighty-five K class airships numbered *K51–K136*.

In rapid succession ten additional bases were set up on both coasts of the US mainland, together with others in Trinidad and Jamaica, and two in Brazil. Eventually there were fifteen airship squadrons in commission.

During 1941 over 400 merchant ships were attacked within sight of the US Atlantic coast by German U-boats, but once sufficient airships had became available for convoy escort duties the numbers of attacks on shipping fell to forty-two in 1942 and then to sixteen in

1943. It was even stated that no convoy that was protected by blimps was ever subject to attack by U-boat.

The K class airship had an offensive armament of four Mk10 depth charges, whilst defensive armament comprised of a single .50 calibre machine gun. A crew of ten was carried consisting of commander, two co-pilots, navigator, rigger, two engine mechanics, a radio operator and two electronics specialists.

The 40ft-long control car of the K class was equipped with ASG radar. This had a detection range of 90 miles, sonar buoys and, importantly, Magnetic Anomaly Detection equipment (MAD). This device was originally devised for the exploration of oil and mineral deposits, and measured the natural magnetic variations within the earth's crust, which helped to identify oil traps or mineral lodes.

When installed in an airship flying over the sea, the variation in the magnetic field in a circle several hundred feet in diameter could be recorded as an ink trace on a moving roll of logarithmic paper. Flying at a height of between 50ft to 100ft, a contact would be indicated by a strong magnetic signature from an underwater object such as a U-boat, which once identified could be tracked with relative ease until either attacked by the airship or by surface craft.

Airship Squadron 14 was detailed to operate from Port Lyautey in French Morocco, becoming the first non-rigid airships to cross the Atlantic Ocean in June 1944. Eventually eight K class airships operated with ZP-14, or the Africa Squadron, in the Mediterranean. The primary purpose of the ZP-14 squadron was to fly a 'barrier patrol', employing its MAD equipment to deny the passage of U-boats into and out of the Straits of Gibraltar.

Due to the danger from enemy aircraft these patrols were made by Allied aircraft by day, with the airships taking over the duty during the hours of darkness. The patrols were highly successful, benefiting from the airships' ability to fly slowly and at low altitude, which made

U.S. Naval Airship "K43" - 1943
Goodyear Aircraft Company, Akron Ohio

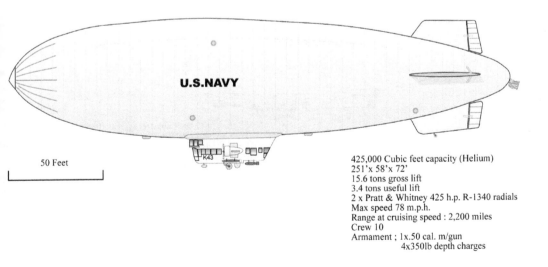

U.S.NAVY

K43

50 Feet

425,000 Cubic feet capacity (Helium)
251'x 58'x 72'
15.6 tons gross lift
3.4 tons useful lift
2 x Pratt & Whitney 425 h.p. R-1340 radials
Max speed 78 m.p.h.
Range at cruising speed : 2,200 miles
Crew 10
Armament ; 1x.50 cal. m/gun
                 4x350lb depth charges

them the ideal craft for the purpose. Once a U-boat was detected the airship would drop depth charges and call up surface craft to continue the attack. Despite the dangers the squadron successfully accomplished the ASW role in this theatre, without loss.

In September 1944 the squadron moved to southern France, stationed at the former French naval airship base at Cures near Toulon – recently vacated by German forces – and were employed in mine spotting and clearance duties. From here the squadron moved to various bases in the Mediterranean, including Oran, Cagliari in Sardinia, Bizerte, Rome, Toulon and Venice. Between May 1944 and September 1945 *ZP-14* successfully operated on convoy duties, mine spotting, harbour clearance and anti-submarine operations, before the squadron was sent back to the United States in early 1946.

In anticipation of a prolonged struggle against the Japanese in the Pacific, a larger class of airship more suitable for operations in the tropics and with longer range was ordered from Goodyear in 1943. The initial contract was for twenty-two of the M class airship of 647,000cu ft capacity, this later being reduced in 1944 to four, the first of which was delivered in February 1944.

The M class blimps were 302ft in length with a diameter of 57ft and possessed a 117ft-long articulated control car, so designed as to spread the loading equally across the internal catenary support in the upper envelope. Power was provided by two Pratt & Whitney R1300–AN-2 radials of 550hp, which gave a maximum speed of 80mph and an endurance of fifty hours carrying a crew of twenty-two. Whilst the offensive ASW armament consisted of eight 350lb Mk47 depth charges.

U.S. Naval Airship "M-4" - 1944
Goodyear Aircraft Company, Akron Ohio

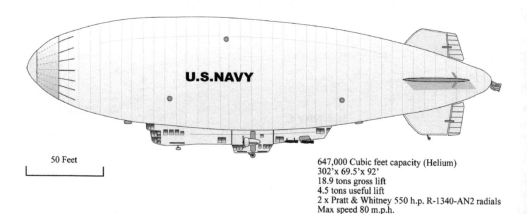

50 Feet

647,000 Cubic feet capacity (Helium)
302'x 69.5'x 92'
18.9 tons gross lift
4.5 tons useful lift
2 x Pratt & Whitney 550 h.p. R-1340-AN2 radials
Max speed 80 m.p.h.
Range at cruising speed : 3,000 miles
Crew 15
Armament ; 1x.50 cal. m/gun
             8 x350lb depth charges

On 27 October 1946 the *M1* left Lakehurst Naval Air Station under the command of Lieutenant H. Walton and headed via Savannah to the Bahamas, Florida and Cuba, eventually landing at Glynco, Georgia after a record-breaking cruise of 170 hours without refuelling. After performing useful service, including many other flights of long duration, the M class ships were finally retired in 1956.

Following the end of the Second World War and the start of the Cold War, a new class of airship was required for surveillance and radar picket duties on the Atlantic coast and the high Arctic, as a defensive measure against the threat of Soviet missile attack. The answer was the N class, or the Nan ship as it was colloquially referred to, of which five were built. They were configured for both anti-submarine warfare and airborne early warning duties.

The first of the class, designated ZPN-1, flew in early 1952, followed by four others during 1954, re-designated as ZPG-2–ZPG-5 with a helium capacity of 1 million cu ft. Three of these airships were converted to carry the AN/APS-20 radar in a streamlined pod below the gondola, together with a height-ranging radar mounted at the top of the envelope, now known as the ZPG-2W class.

The ZPG-2s were employed as a mobile extension to the Distant Early Warning system that covered the US during the Cold War period, operating in both the Atlantic along the 'Contiguous Barrier' and the high Arctic, where conventional radar coverage was sparse. In May 1954 one of these airships, under the command of Commander M. Eppes USN, undertook a flight from Key West, first heading north to Nova Scotia then turning south to Bermuda and Nassau, on over the Caribbean into the Gulf of Mexico and finally landing back at Key West after a continuous cruise of 200 hours.

In 1956 the first airship to utilise a synthetic plastic envelope made from Dacron entered service. The use of this material reduced gas loss substantially and superseded the rubberised fabric materials used. During January 1957 three ZPG-2W operating from South Weymouth NAS maintained continuous patrol coverage, 200 miles out in the Atlantic for eleven days in some of the worst weather conditions ever experienced in that area. In March 1957 Commander J. Hunt USN, aboard a ZPG-2W, left South Weymouth NAS bound for Portugal, from here the airship continued down the African coast to the vicinity of Senegal before re-crossing the Atlantic Ocean back to Key west, covering 9,448 miles in 264 hours without refuelling; a world record for distance and endurance.

Finally, in June 1959, what was to be the last of the US navy's airships, the ZPG-3W, was delivered to Lakehurst NAS. This airship was of 1.5 million cu ft capacity, being 403ft in length and carrying a massive 48ft diameter radar scanner that rotated within the helium-filled envelope. These airships, of which four were built at a cost of $12 million each, were powered by two Curtiss-Wright radial engines of 1,500hp, which gave them a top speed of over 80mph and an endurance in excess of 200 hours carrying a crew of twenty-one. These airships proved to be most successful in undertaking their radar picket duties.

Unfortunately one of the ships, whilst on routine patrol on the east coast, developed a major tear in the Dacron envelope that resulted in a catastrophic loss of gas. This caused the airship to plunge into the Atlantic, drowning all twenty crew members who were trapped inside the gondola.

The three remaining ZPG-3W and the ZPG-2W airships soldiered on for a few more years, until they were replaced by faster and higher flying aircraft such as the Lockheed EC-211 Warning Star (a development of the civil Constellation airliner), which took over their duties between 1959 to 1982 with the US navy. Finally, in November 1962, the ZPG-3Ws, together with all the other remaining airships, were retired from service, ending the US navy's direct involvement with lighter-than-air craft.

During the 1960s and 70s Goodyear continued to operate a few blimps within the continental United States for advertising purposes, and delivered one airship, the *N2A Europa*, to Cardington in England as an initial base for a major advertising programme throughout Europe. Eventually, during 1987 after over seventy years in the business of building airships,

U.S. Naval Airship 'ZPG-3W'-1958

Radar Picket & Surveillance Airship

Goodyear Aircraft Company, Akron Ohio

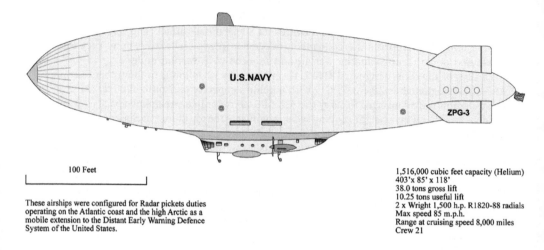

100 Feet

These airships were configured for Radar pickets duties operating on the Atlantic coast and the high Arctic as a mobile extension to the Distant Early Warning Defence System of the United States.

1,516,000 cubic feet capacity (Helium)
403'x 85' x 118'
38.0 tons gross lift
10.25 tons useful lift
2 x Wright 1,500 h.p. R1820-88 radials
Max speed 85 m.p.h.
Range at cruising speed 8,000 miles
Crew 21

the Goodyear Company sold their airship construction arm to Loral Defence Systems. This effectively ended their involvement in the lighter-than-air field, although their small advertising airships continue to fly under the company's control.

At the same time a renewed interest was being shown both in Europe and America in the airship, with various designs being proposed incorporating new materials such as carbon fibre and the use of turboprop engines for propulsion. Cargo Airships, a British company, envisaged a 30 million cu ft helium-filled rigid airship, capable of transporting a large cargo of standard ships containers at 100mph.

This scheme excited some attention at the time and involved the shipping company Manchester Liners, but in truth the whole concept was based on a misconceived business plan, and after a great deal of money had been spent on research and development the scheme was abandoned.

In 1988 the American Blimp Corporation of Orlando, Florida produced a series of small airships equipped with internal lighting for night-time advertising purposes. The Lightship A40 type were of 68,000cu ft capacity, 135ft in length with a diameter of 36ft and powered by two 80hp Limbach motors, giving a top speed of 58mph. Whilst the Lightships did not incorporate any advanced construction methods, they were successfully sold around the world with sixteen being sold to date, some being operated by the Virgin group.

More successful was the British Sky Ship project developed at Cardington in the early 1970s by Airship Industries. The first of these was the Sky Ship 500 of 161,200cu ft capacity, with a length of 104ft and a gross lift of 6 tons. Powered by two Porsche engines, the propellers were mounted in annular shrouds to increase power output and affected thrust vector control that provided excellent manoeuvrability, which allowed it to be operated with a minimum ground crew. Unfortunately, the prototype was destroyed at Cardington in March 1979 when it caught at the mast in a gale, but six others were subsequently built, all of which are still in service around the world.

Roger Munk and his team at Cardington, encouraged by the success of the Sky Ship 500, went on to design a larger version. The Sky Ship 600 was of 235,400cu ft, 193ft long with a maximum diameter of 50ft and capable of 70mph carrying, in one configuration, seventeen passengers. The envelope was made from a polyester fabric, with various composite materials including Kevlar and carbon fibre. The Sky Ship 600 first flew in November 1984 and subsequently seven of this class were built. They were first employed for advertising, as aerial camera platforms and passenger carriers. All these airships are still in operation in various countries.

Sky Ship 600 trials were also conducted with the French navy for coastal surveillance and anti-smuggling activities, whilst the American navy showed interest in possible military applications. An even larger model, the *Sentinel 1000*, followed at the request of the American military for use with the navy for offshore surveillance duties.

First flown in 1994 the *Sentinel 1000* was at the time the largest airship in the world, at 222ft in length with a capacity of 353,500cu ft and a gross lift of 12 tons. This highly promising airship was unfortunately destroyed when fire raged through the No. 2 airdock at Weeksville, South Carolina in August 1995. Prior to this, Airship Industries had gone into liquidation in 1990, with the US company Westinghouse taking over the operation of the existing Sky Ship fleet and acquiring the technological data and the sole rights to build further Sky Ship designs.

Civil Airship "Skyship A600" - 1986
Airship Industries, Cardington, Bedfordshire

50 Feet

235,000 cubic feet capacity (Helium)
195'x50'x70'
Gross lift 6.3 tons
Useful lift 1.5 tons
2x275 h.p.Porsche 930
Max speed 64 m.p.h.
Range 400 miles
Passengers 12
Crew 2

In June 1987 Airship Industries, together with Westinghouse, had signed with the US Defense Department a $170 million contract to build a large, highly sophisticated surveillance craft, the *Sentinel 5000*. This airship was envisaged, in one of its many configurations, as an aerial command post; directing strike aircraft to targets or to intercept incoming attacks. It would mount a large array of radar and electronic countermeasures equipment, and up-to-date avionics and fly-by-wire control systems were to be fitted. The airship was to be 423ft long with a diameter of 105ft and of 2.5 million cu ft helium capacity, which would have made it the largest airship built since the *Hindenburg*.

An initial specification called for endurance of three days at 45 knots and a service ceiling of 10,000ft. The modular three-decked gondola was to have a pressurised cabin carrying a crew of fifteen, and to be constructed, as was the envelope, from composite materials designed to give a low radar signature. Propulsion was effected by two 1,800hp PPB diesel engines to give a speed of 70mph. This was assisted by a single 1,700hp GET-700 turboprop engine that could boost the top speed to 117mph in an emergency.

During 1988 a full-scale mock-up of the gondola was constructed, but delays in the navy acquiring the necessary funding from Congress resulted in Airship Industries going into receivership in September 1990. Westinghouse continued to develop the design for a variety of military uses, including a commercial variant that would have carried 140 passengers,

49. Skyship *A600* with Concorde.

but eventually, without funding being made available by Congress to the navy, the project was abandoned.

Undeterred, Roger Munk and some of the management of the old company formed the Advanced Airship Technology Group, again at Cardington, where they designed the *AT10*, a small airship incorporating new construction features so as to avoid infringing Sky Ship patents. The 135ft-long airship, which had a capacity of 88,000cu ft, was propelled by two 100hp diesel engines and was of a standard design. It was used for pilot and ground crew training prior to the company producing a larger and more advanced design (which they still hope to produce).

During the Second World War the Zeppelin Company wind tunnel facilities at Friedrichshafen had been used in the development of the V2 rocket, which was used during 1944–45 to bombard London and Antwerp where it caused considerable damage and many civilian deaths. The Zeppelin factory also produced components for these rocket weapons, but these facilities and the whole of the Friedrichshafen area was heavily bombed day and night by the RAF and the USAAF, which, fortunately, considerably reduced the number of rockets produced.

In 1994 the Zeppelin Company again entered the airship field with a revolutionary design that employed a unique method of rigid construction. It was known as Zeppelin NT (New Technology). The first of this type, the *SN-01 Friedrichshafen*, flew in 1997, being of 290,000cu ft capacity and 246ft in length with a maximum diameter of 64ft. Powered by three Textron Lycoming motors, the airship has a top speed of 78mph with a duration at cruising speed of twenty-four hours and a useful load of 4,180lb. The comfortably equipped cabin could accommodate a crew of two and twelve passengers. As of 2009 four of these airships had been built, operating in Germany, Japan and the United States.

The prototype model the *SN-01* was leased in 2005 to the De Beers Corporation on a two-year contract to search for diamonds in Namibia. Here, she was equipped with a variety of magnetic and gravity sensors. The *SN-01* successfully accomplished her mission operating continuously from a mast, without the benefit of a shed, in conditions involving great temperature variation. She was wrecked in September 2007 by a storm in Botswana whilst moored to the mast; she was adjudged to be beyond economic repair.

The basis of the design was a rigid structure comprising three longitudinal aluminium girders set at 120°, with the apex of the triangle at the top and thirteen triangular transverse members constructed from fibreglass and carbon fibre. This structure was braced by Aramid carbon fibre wiring, forming an extremely strong framework, which despite its size weighed only 2,204lb. The structure was enclosed within an envelope comprised of a three-layer polyester fabric, the outer surface of which was made from Tedlar (a UV-proof material). This, together with an inner gas-tight layer of a polythene and a sealing membrane, was bonded together forming a robust envelope weighing only 1,200lb.

The envelope was maintained at a positive pressure of 5mb, which, together with two air-filled ballonets, maintained the integrity of the envelope. A three-fin layout was employed, these being attached to the three main longerons at the stern. This arrangement saved weight and reduced the overall height of the craft. When the airship was in operation it was weighed-off in a heavier-than-air condition, typically up to 2,240lb, and was dependant on dynamic lift created by the airflow over the envelope to develop lift to maintain the craft in the air.

As a voyage continued, the consumption of fuel gradually reduced this component and the static lift of the gas was increasingly relied on to carry the load. Power was provided by three 200hp Lycoming engines, giving a top speed of 78mph and range of 600 miles. The two lower longerons mounted two of the motors positioned high above the gondola, which considerably reduced noise in the passenger gondola and allowed the swivelling propellers to be rotated through 120°.

A third motor mounted at the extreme stern of the envelope drove two propellers, one of which could be moved through 180°, whilst the other was arranged to operate in a similar manner to a helicopter tail rotor. The combination of these adjustable propellers, together with the fly-by wire and advanced avionics, enabled the airship to manoeuvre with great accuracy and only needed a ground crew of three to operate under normal conditions.

Currently, one Zeppelin, *SN-03* (registered *D-LZZF* and launched in February 2003), operates pleasure cruises over Lake Constance from the Friedrichshafen base, and at present a fifth airship of similar dimensions is under construction. Plans are in hand to build a larger type, the *NT-14*, which will be a nineteen-seat airship of 500,000cu ft capacity.

Airships like the Zeppelin NT will continue to have a place in aerial transportation for the foreseeable future, filling, as they do, a particular niche. They are also able to perform unique tasks that are beyond the abilities of fixed wing or rotary wing aircraft. The most likely development for its future use will be restricted to pleasure flying, with a number of uses of a scientific nature such as atmospheric research and pollution sampling.

An increasingly important consideration justifying their use is that operating airships offers a low carbon signature, and under certain circumstances an economic and green method of flying compared to heavier-than-air craft.

On a commercial level aerostats are now being employed in forestry management. In particular, large balloons with a lift of up to 15 tons are used in logging operations in areas where ecological considerations and the minimum damage to the environment are considered to be important. Other heavy-lift projects include the *CL160*, proposed by Cargo Lifter AG in the 1990s. Designed for the transport of loads of up to 1,000 tons this airship saw considerable

German Civil Airship
Zeppelin NT-07 "Friedrichshafen" - 2007
Zeppelin Luftschifftecnik A.G. Friedrichshafen

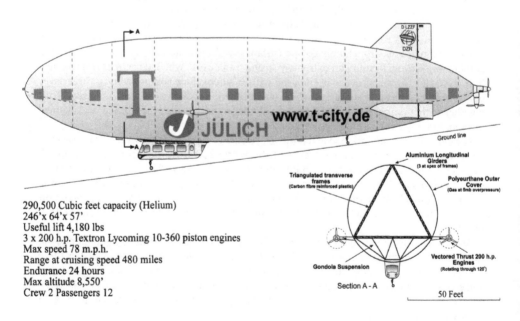

290,500 Cubic feet capacity (Helium)
246'x 64'x 57'
Useful lift 4,180 lbs
3 x 200 h.p. Textron Lycoming 10-360 piston engines
Max speed 78 m.p.h.
Range at cruising speed 480 miles
Endurance 24 hours
Max altitude 8,550'
Crew 2 Passengers 12

development work and the erection of the largest airship shed ever, but failed to come to fruition due to the company going into receivership.

Throughout the history of lighter-than-air development, interest has been shown in studying the composition of the upper atmosphere for scientific purposes. During the nineteenth century many flights of this nature were undertaken in order to learn more about the properties of these regions, such as that conducted by the English balloonists Coxwell and Glaisher, who in September 1862 ascended to a height of 26,200ft plus.

In an open basket balloon the two aeronauts ascended rapidly into the frigid regions above 20,000ft. Here, without oxygen, they both quickly lost consciousness and were only saved by Coxwell briefly recovering at maximum altitude. This enabled him to pull the rip cord with his teeth; his hands were so frozen that he was unable to use them.

On several other such ventures into these icy regions in the nineteenth century the aeronauts paid with their lives in attempting to reach great altitude. They failed to appreciate that as atmospheric pressure falls with altitude, nitrogen bubbles are released into the blood stream leading to a fatal attack of the 'bends'. Not only will lack of oxygen cause narcosis leading to death, but at still higher altitudes of around 62,000ft the low pressure and density of the atmosphere will cause the blood to actually 'boil' if aeronauts are not protected by a pressure suit or contained within a hermetically sealed cabin.

In December 1894, the German balloonist Herr Berson was the first to use oxygen on a successful high-altitude attempt. Aboard the balloon *Phoenix* he reached an altitude of 30,000ft, and later in 1901 Berson and Professor Buring reached 35,000ft in the *Prussen*, launched from Strasburg in Austria.

In 1931 Professor August Pickard, the renowned Swiss physicist, ascended to 54,000ft in a hermetically sealed gondola in a flight of twelve hours, where observations were made on the nature of cosmic rays. Pickard made several other flights of a scientific nature over the next few years, contributing greatly to stratospheric research.

The Russians had a particular military interest in the stratosphere and made several high-altitude flights during the 1930s, accumulating a great deal of information from these regions that aided their development of high-altitude aircraft. Meanwhile, the Americans also entered the high-altitude field with the Explorer series of balloons, culminating in the record flight by Anderson and Stevens in 1935. Their balloon, *Explorer III* of 3,700,000cu ft helium capacity, reached an altitude of 72,395ft (13.7 miles); a record that was to stand for over twenty years.

Following the Second World War and the beginning of the space race, NASA instituted a high-altitude balloon programme of 'Near Space Activity', for the purpose of testing space suit design and escape methods from space capsules. Project Man High began in 1954 and was designed to determine if human beings were capable, physiologically and mentally, of enduring the rigors of prolonged space missions and surviving the effects of the intense radiation that would be experienced during such flights. Newly developed lighter materials were employed for the envelope, which allowed greater altitudes to be reached than in the pre-war attempts.

A truly remarkable flight in the series was that made by Major Joseph Kittinger USAF in August 1960 aboard the balloon *Excelsior III* of 3 million cu ft capacity. Clad in a pressure suit and sitting in an open capsule, he was lifted to an altitude of 102,800ft over a flight of thirty-one minutes, remaining for a further twelve minutes at this altitude. Then, and with 99 per cent of the earth's atmosphere below his feet, Major Kittinger stepped out of the capsule to freefall from almost 20 miles.

Due to the rarefied nature of the atmosphere at this height, Major Kittinger fell at a terminal velocity of 614mph, taking 4.5 minutes to reach 18,000ft where his parachute opened. The outside temperature at the altitude reached was -70°C and due to a malfunction to of one of his gloves Major Kittinger suffered frostbite and a decompression injury to his hand. For this achievement and other flights Kittinger was awarded the Distinguished Flying Cross.

Later, in December 1962, Kittinger took part in Project Stargazer and, accompanied by astronomer William White, spent eighteen hours at an altitude of 82,000ft making astronomical observations from a balloon packed with scientific instrumentation. Here, free from any atmospheric interference, they were able to obtain results that were not to be bettered until the deployment of rocket-launched space telescopes a decade later.

Major Kittinger's 1960 record was exceeded during 1965 by Commander Malcolm Scott USN who, together with Lieutenant Commander Prather, reached a new record altitude of 113,740ft. Unfortunately, during the recovery process when the balloon landed in the sea, Prather lost his life during recovery due to his space suit filling with water and drowning. These flights significantly contributed to the US space effort and extended the science of aviation medicine.

In the regions of the atmosphere above 60–70,000ft, roughly twice the height of the jet stream, weather conditions are relatively quiescent, with a progressive lessening of wind circulatory activity as the air comprising the stratosphere becomes more rarefied. The wind velocity at these heights is typically between 5 to 40 knots with an accompanying lack of turbulence that favoured the operation of High Altitude Airships, or HAVs, in this region.

In order to exploit these regions for commercial and military considerations the advent of the unmanned high-altitude airship is perhaps the most interesting development in recent years. Initial interest in this concept was originally shown by Japanese telecomm firms who sought to develop a more cost-effective system of telecommunication relay than that offered by expensive satellites.

Various research vehicles to exploit this technology have been built both in Japan and European countries, with the aerostats being launched vertically to rapidly achieve these extreme altitudes and perform a varied programme of experimental communication work. In 2005 an American company, Aerostation Near Space Applications, successfully launched a small High Altitude Airship to 74,000ft. It was carrying a 60lb payload and telemetry pod that demonstrated the practicality of such vehicles to relay more powerful and improved broadband and telecommunications functions.

In the 1990s the North American Aerospace Defence Command (NORAD) proposed the provision of eleven high-altitude aerostats to provide overlapping radar coverage, each with a radar footprint of 600 miles diameter, for the protection of the US borders as a major asset in the arsenal homeland defence security. The Lockheed Martin Corporation was awarded an initial $50 million contract to develop an unmanned lighter-than-air vehicle to operate above the jet stream at the limits of the earth's atmosphere. It is in this zone, between 70–100,000ft, where the High Altitude Airship is designed to operate.

The Lockheed Martin HAV is designed to operate autonomously in the upper stratosphere in a geostationary position for sustained, long-endurance operations extending for up to six months at a time, serving as a military command post and communications centre. The envisaged aerostat was 500ft long with a diameter of 150ft with a helium capacity of 5 million cu ft, giving a gross lift of around 130 tons, carrying a scientific payload of between 2 and 20 tons. Powered by an array of photovoltaic (PV) regenerative fuel cells, these provide 15KW of power to the motors, in turn controlling the geostationary positioning propellers. In addition are the onboard electronic detection and telemetric systems transmitting data to the ground stations.

During 2008 the initial contract was cancelled due to budgetary constraints, but was again revived in April 2009 with a new $149.2 million contract – again with Lockheed Martin by the Missile Defence Agency (MDA) of the US government. This contract was for an unmanned radar-carrying surveillance airship to detect and track incoming ballistic missiles, and more specifically for use against low-flying cruise missiles in the light of future perceived threats that might develop from countries such as Iran.

This system was designed to replace the current USAF Boeing E3 early warning aircraft. Again, the continuous radar coverage of the proposed aerostat would be in the order of a 750-mile diameter, against the intermittent footprint of 375 miles of the Awacs E3 aircraft. The employment of these systems offers a comparatively less inexpensive method of placing scientific and military packages in geostationary orbits, compared to rocket-launched satellites for sustained high-altitude missions.

Ultimately, second and third generation HAVs will be operating at still greater heights of up to 100 miles, carrying greater payloads, for missions that could extend into several years. The proposed use of unmanned HAVs could, interestingly, see a return to hydrogen as a lifting agent, both on the grounds of economy compared to the use of expensive and rare helium, but also due to its greater lifting capacity.

Currently, large sounding balloons of up to 4 million cu ft of irreplaceable helium and carrying large scientific payloads are being routinely launched on five-hour missions, with the precious gas being voided to the atmosphere at the end of the mission. At a time when federal authorities have instituted a helium conservation scheme in the United States (so conscious are they of their falling reserves) it goes without saying that hydrogen, the most common element in the universe, should be restored to its former position as the main lifting gas for aerostats.

One further and future use for aerostats, and one that excites the imagination, is the possibility of using them to provide an almost limitless source of energy with the minimal effect to the environment by tapping into the energy contained within the high-altitude jet stream. The jet stream continuously circles the earth in the region between the troposphere and stratosphere, at a height of between 6 to 9 miles, in a stream some 100 miles wide by 3 miles deep with an average velocity of 300mph. Studies are in hand to harness the potential of these high-altitude currents by employing airborne wind turbines, either mounted on large kite-like structures or large aerostats, and introducing them into the air stream to generate electrical power.

One scheme visualises batteries of these turbines tethered to ground stations, with the generated electrical power transmitted down wires rather in the fashion of a wartime balloon barrage. A more elegant solution would be to utilise geostationary aerostats from where the electrical power can be transmitted to the ground station by microwave beams.

Scientists calculate that only 1 per cent of the potential energy contained within the jet stream would be sufficient to meet all the planet's current energy requirements. At the same time, such a widespread system would contribute to the provision of green energy and reduce the reliance on the use of polluting fossil fuels, ameliorating the effects of global warming.

It is a sobering thought that one of most archaic forms of achieving flight, namely the aerostat – first formulated by the Montgolfier brothers and other pioneers in the eighteenth century, and seemingly largely relegated to a minor role in aviation since the advent of heavier-than-air technology – could in the future be a major contributor to solving the planet's demand for energy and reverse the threat of global warming. The vision of fleets of high-flying aerostats supplying unlimited energy to our planet is a stirring concept and a great technological challenge for the years to come. It would represent an enduring legacy to those brave aeronauts who first courageously took to the skies in balloons and airships.

APPENDIX 1

The *R34*'s Transatlantic Flight

*T*he R34's transatlantic flight commenced on 2 July 1919, being the first direct air
communication between Europe and the United States, in a flight of 3,100 miles accomplished
in 108 hours. The return flight of 3,310 miles in seventy-five hours constituted the first double
crossing of the Atlantic Ocean by air.

At the start of the flight, due to a heavy overload of fuel, the *R34* was only just able to maintain
herself in the air by her static lift, but with the engines engaged (providing dynamic lift) she
was able to fly almost 1 ton heavy. By the time she was over Rosyth she was at her pressure
height of 1,500ft flying at 65mph, and as the voyage progressed the consumption of fuel
would further increase her lifting capacity.

As they flew on in the darkness above the Clyde the airship was buffeted by strong
crosswinds coming off the hills, which caused some anxious moments for Major Scott and
General Maitland as the helmsmen struggled to control the pitching craft. Sailing on into
the dawn down the Firth of Clyde, the airship turned westward passing over Rathlin Island
at 4.40 a.m. This was their last sight of land, as before them now lay more than 2,000 miles
of open ocean.

The weather now deteriorated and the *R34* was flying in thick cloud and heavy rain, with
an accumulation of rainwater on the envelope and the contraction of gas reducing lift, giving
cause for concern. Major Scott was obliged to order the airship to fly at an up angle of 12°,
using dynamic lift in order to maintain herself in the air, and continuing in this mode until
such time as the conditions improved and sufficient petrol had been consumed.

The five engines had to be rested at intervals, with usually only four engines running at
any one time, whilst the forward engine in the control car, after running irregularly, had to be
shut down for attention under the watchful eye of the chief engineering officer, J.D. Shotter.
Towards mid-morning, still at an altitude of 1,500ft, the *R34* was flying in clear air between
two cloud layers. As such, the navigator was unable to check either their drift relative to the
unseen surface of the sea, or their position by a sight of the sun.

Around noon the higher cloud began to clear, allowing Major Cooke (the ship's navigator
who had by necessity been using dead reckoning to plot their course) to climb to the upper
platform to take a sextant fix on the sun to more accurately determine their position. With the
ship still relatively heavy with fuel, the effect of the sun on the envelope would cause a rise in
temperature of the gas and, as it expanded, its loss through the automatic valves. This loss of
lift was a situation Major Scott needed to avoid, and he allowed the *R34* to sink into the lower
cloud mass to mitigate the worst effects of superheating.

By 11 a.m. the *R34* was now some 250 miles out in the Atlantic with a following wind.
Major Scott took advantage of this favourable situation by shutting down three of the engines
in order to economise on fuel. There now occurred a unique incident when a stowaway was
discovered in the keel – the miscreant being Aircraftsman Second Class William Ballantyne,
one of the regular crew members who had not been selected for the Atlantic trip but was
anxious not to miss the chance of a lifetime.

When presented to General Maitland, Ballantyne, who was suffering from the effects of breathing hydrogen gas, was put to bed in one of the hammocks until he recovered, whilst the question of punishment had to be put off until a later date. Upon his recovery, Ballantyne was put to work assisting the cook and other duties.

Ballantyne survived his censure and subsequent punishment and continued to serve in the Royal Air Force, rising to the rank of wing commander and retiring after a distinguished career. As it transpired, Ballantyne was not the only stowaway on the trip as a tabby cat, called Wopsie, was also found sleeping in the keel. This little animal had arrived from Inchinnan when the R34 had been delivered to East Fortune and had been adopted by the crew as their mascot, and thus became the first feline to make an Atlantic air crossing.

The R34 was now cruising at 900ft and running on three engines in improved conditions. This allowed the surface of the sea to be seen and there was a chance to check drift by measuring the angle of the ship's shadow against a breaking wave or other rare feature on its surface. Later, the airship climbed to 2,000ft with all engines running into an almost clear blue sky, in which the meteorological officer Lieutenant Harris could discern the signs of a depression approaching from the south-west. This information enabled Scott to alter course in order to take advantage of a favourable wind to help his course westward.

During the first day wireless communications and directional fixes from both East Fortune and Cliveden became erratic and were eventually lost, although at about the same time they picked up faint transmissions from St Johns in Newfoundland. Wireless contact was, however, also made with liners and other vessels on the trade routes; the battlecruisers HMS *Renown* and HMS *Tiger* had earlier been positioned at key points along the route of the flight to offer assistance, should it be required.

At the end of the first day, and with the onset of darkness, the R34 was almost halfway across the ocean at a position of 53°20ft N, 26°00ft W. The airship was flying at 3,000ft; the fall in temperature caused contraction of the hydrogen in the cells, resulting in a subsequent loss of lift. To combat this situation it was necessary for all five engines to be run at flank speed and to fly at a pronounced nose-up angle to produce dynamic lift.

The first night passed without incident (apart from the usual worries over the engines) and after dawn on the morning of 3 July they experienced the clearest weather during the entire voyage. With excellent visibility and cruising at 2,000ft, a further opportunity was used to check the ship's position. Throughout the second day the weather remained benign, even warm, but this happy state of affairs was not to last. After making good progress westward the expected depression made its presence known: it backed to the south-east, forcing the airship off its intended track to the north.

The arrival of the depression also brought with it heavy rain and squally winds, but the R34 rode the storm steadily as she continued to make slow progress westwards. In order to avoid the effects of the gale, Scott climbed R34 to 3,500ft. Here, the wind strength was diminished and she was able to make reasonable progress at the end of their second day after over forty hours in the air.

As the third day dawned to a clearing sky, the tired crew were able to view below on the ocean surfaces their first sight of icebergs floating down from the north. By 9.30 a.m. Newfoundland lay less than 250 miles away, yet it is indicative of the way adverse weather dictates airship operation that from that point it required a further two full days to reach New York.

Once again, as the clouds cleared and the sun's rays beat down on the airship, Major Scott sought the sanctuary of cloud cover to avoid having to valve his precious hydrogen. Also at this time, Scott was becoming anxious about the fuel situation, having already consumed two thirds of the total fuel; fighting the headwinds of the previous day had exceeded estimated consumption. He was also aware that he could expect strong headwinds as the airship cruised down the coast of Nova Scotia and towards New England.

At noon on 4 July, whilst flying at 1,500ft, the outside air temperature abruptly dropped from 42° to 30°F as the ship's track passed over the cold Labrador Current. This differential between the outside temperature and that of the warmer gas in the cell again caused superheating, which Scott countered this time by driving the airship up to the warmer air at 4,000ft. Finally, at 4.30 p.m. on 4 July, the *R 34* sighted the American continent at Trinity Bay on the north-east coast of Newfoundland, after having been in the air for over sixty hours.

The exhausted crew allowed themselves a short congratulation and a chance to reflect on their accomplishment, before applying themselves again to their tasks. The airship's course now took them over Newfoundland, where the weather was sufficiently clear to allow the crew to look down on desolate landscapes of dense dark woods and scattered lakes unrelieved by any sign of human habitation. After traversing this bleak land for some four hours, the *R 34* passed out to sea again over Fortune Harbor on the south-east coast. Here, at last, they saw the small town and harbour beneath them.

The distance from Fortune Harbor to New York was about 900 miles, which, given favourable weather, Scott hoped to cover in the next twenty-four hours at the reasonably attainable speed of 37.5mph. The crossing from Newfoundland to Nova Scotia took five hours, where off the port of Halifax early next morning they sighted a steam ship – the first they had seen on the voyage.

By now strong south-easterly winds that showed no sign of abating were slowing the airship's progress, requiring the engines to be run at increased revolutions and eating in to their dwindling reserves of fuel. The *R 34* was turned westward, flying at a low altitude over a forested landscape towards St John's in order to avoid the worst of the gale over the sea.

Due to the strength of the wind Major Scott realised that he would not be able to reach New York before nightfall, and with only 500 gallons of fuel remaining it was necessary to consider other courses of action to complete the voyage in safety. His options were landing either at Boston or on the northernmost point of Long Island to refuel, or to be taken in tow by a US navy vessel should they be in danger of running out of fuel, which possibly would have been the most humiliating option. Scott, while alerting the American authorities of his plight, was determined to complete the voyage as planned. Now running on three engines to conserve fuel, the airship crept slowly southward against the wind.

Due to the reduction in weight because of the fuel consumed, the *R 34* was now flying in a light condition and possessed of an excess of lift. To remedy this, Scott took the airship above her pressure height to allow the release of gas through the automatic valves to reduce her buoyancy.

At the southern end of the Bay of Fundy, just before midnight, a fierce electrical storm burst upon the ship with driving rain and gale-force winds that tossed the airship violently about, causing it to pitch and roll up and down steeply, suddenly dropping several hundred feet toward the boiling sea. In the control and engine cars, and the keel, the crew held on grimly as men and equipment were thrown around in an alarming manner.

As the storm passed it was succeeded by calmer weather and although the headwind persisted, it gradually began to moderate. During the night, as *R 34* made her way slowly southwards sailing under bright moonlight in clear weather, the airship was again unexpectedly assailed by violent air currents that threw her nose upwards. This was followed by an equally violent nose dive, which called for a quick response from the helmsman. These violent and unexpected manoeuvres were repeated several times, being of a magnitude that caused the hull structure to flex and groan.

This clear air turbulence was attributed by Lieutenant Harris, the meteorologist, to the confluence of the warm Gulf Stream and the cold Labrador Current from the north.

Once the squalls had passed the headwind fell away to be replaced by an increasingly strong tailwind. Despite this latterly favourable turn of events, Major Scott was still worried by the fuel situation and wired the US navy to ask that a landing party be sent to Montauk Point, the northern extremity of Long Island, in order that he could refuel before going on to New York.

As the flight proceeded with steadily improving weather, clear visibility and a strengthening following wind, Major Scott now felt confident that he could reach their goal without an intermediate stop. Scott instructed Lieutenant Shotter, the engineering officer, to scour the 'empty' tanks for the last few gallons of petrol and to transfer it to the gravity tanks above the engines by hand. The airship, now sailing under a clear sky, then passed Cape Cod.

Holding her course for New York, within an hour the giant airship flew over the landing party at Montauk Point and headed straight for Mineola Field (their original destination), which, to the enormous relief of the exhausted crew, they reached at 9 a.m. local time.

Below, an enormous crowd had gathered to witness the arrival of the conqueror of the Atlantic Ocean. As the airship circled the field, with her crew at landing stations, Scott ordered Major Pritchard to parachute to the ground to take charge of the landing party. After four days in the air Lieutenant Pritchard presented a dishevelled appearance. Nevertheless, after smartening his uniform and a quick shave he parachuted from the ship to become the first and only person to enter the United States by this unusual means.

Finally, at 9.45 a.m. local time, the *R34* sank gently into the hands of large ground crew: the first east to west Atlantic air crossing had been completed in 108 hours with less than two hours petrol remaining in her tanks.

Once the great dirigible was secure Major Scott, his officers and crew were welcomed by the vice-admiral, who congratulated them on the success of their epic voyage on behalf of the US navy and the American people. Next, the crew enjoyed well-earned hot baths and a general smartening up, before being swept off to the Garden City Hotel for a civic reception where they received a telegram of congratulation from King George V. Over the course of their three-day stay the crew of the *R34* were treated to a round of receptions and parties. Although pleasurable, this called for a feat of endurance from a crew who had largely been deprived of sleep over the preceding five days.

At Mineola, the *R34* was moored out in the absence of a shed on the three-wire system where a strong mooring point near the bows allowed the airship to fly at anchor some 30ft above the ground. Under this system the airship could be drawn down to ground level by the 400 ground crew to allow for servicing. It goes without saying that the system involved great risk where weather was concerned, and in fact later on the Sunday afternoon a brief storm accompanied by heavy rain swept across the airfield, causing concern to the British officers and some slight damage to the rear gondola.

On Monday engineering officer Lieutenant Shotter and his duty crew of engineers and riggers began the work of servicing the engines, replenishing petrol oil and water ballast, and checking the gas cells and the outer cover for leaks. In the afternoon during these operations under the hot sun and a blustery wind that made the airship roll and pitch at her mooring despite the best efforts of the ground crew, the bow mooring point broke away, fortunately without any structural damage.

Pritchard recalled Scott and Maitland, and with the help of the ground crew repairs were quickly made and the bow mooring point secured again. Tuesday was spent with further maintenance and adding additional gas to the eighteen cells.

On Wednesday, whilst Scott and Maitland were attending another reception, word was brought of an approaching low-pressure system that indicated a storm from the north-east. With the wind increasing in speed, Scott and Maitland prepared for departure. At five minutes to midnight on Wednesday 9 July the *R34* lifted off from Mineola Field to the cheers of the thousands of sightseers and headed towards New York. The great ship cruised over the city bathed in the silvery light of searchlights to the sounds of ship sirens, car horns, factory hooters and the cheers of the multitudes, all drifting up to them from the streets and waterways below.

Turning her bow eastward the *R34* then began the homeward voyage, Scott directing her course to take advantage of the low-pressure area to the north to speed her on her way at 80mph. In contrast to the outward voyage, the return flight was not only quicker – aided by the prevailing wind – but relatively trouble free.

By noon the airship was 600 miles out in the Atlantic, now racing along at 90mph. Later in the afternoon the effect of the wind slowed progress somewhat, but the airship flew on steadily in clear weather. This allowed the navigating officer to climb to the upper platform to take sun sights to fix the ship's position. Towards evening the *R 34* overhauled a large sailing ship, apparently on course for Halifax. This was the only ship they were to sight on the return voyage, demonstrating the vastness of the ocean they were flying over.

Throughout Friday night the dirigible continued her progress without incident, with the watches regularly being changed at four-hourly intervals whilst the engines were rested in turn. Then, at 6.30 a.m. on Friday morning, cruising at 2,500ft with almost a third of the journey completed, the starboard engine in the rear gondola broke down, which was impossible to repair. Cruising on reduced power Major Scott had to rise above cloud, which had formed above and below the airship, eventually breaking into clearer air at 5,000ft. Later in the afternoon Scott attempted to bring the ship below the cloud in order to estimate drift, but abandoned the attempt when he still had cloud below him at 900ft. He had to continue to rely on dead reckoning to guide his course but dropped flares to determine drift.

To add to their woes, additional trouble was also experienced with the forward engine but, fortunately, this was rectified quickly, allowing the airship to continue on four engines. By Saturday morning the wireless messages from Cliveden on the Irish coast were being clearly picked up by the airship. Scott and Maitland had intended to set course to land at East Fortune, where wives and families were waiting to welcome the returning crew. However, at 1 a.m. a wireless message was received from the Air Ministry ordering the airship to divert to Pulham air station in Norfolk.

This action was, naturally, a great disappointment to all the crew and Scott wired a request to be allowed to continue to East Fortune as his engines were defective and he could save five hours' flying time. This was not an unreasonable request considering the condition of the airship and the exhausted crew, but the Air Ministry refused the request and subsequently no explanation was offered for the change of destination.

At dawn the clouds dispersed and in perfect visibility Major Cooke was able to take bearings from the gun platform. From here, in the clear air at 5,000ft with the airship flying at a steady 40mph, Cooke could see for 150 miles in every direction. Finally, towards evening, the Irish coast was sighted near Cliveden. Then, after flying over Belfast, the course was altered to the south-east towards Liverpool. She flew on through the night until, at 6.39 a.m. on the morning of 13 July, the *R 34* dropped her lines over Pulham airfield; her epic voyage of seventy-five hours covering 3,300 miles was at an end.

Airship Development in Great Britain 1900–30

1900   Major J.A. Templer in charge of the Royal Balloon Factory at Aldershot supplies observation balloons for use by the Royal Engineers in the South African War 1899–1902

1902   Stanley Spencer flies airship *Spencer No. 1* from Crystal Palace to Eastcote, Middlesex; the first powered flight in the British Isles

1905   Dr F.A. Barton, after receiving £4,000 from the War Office, fails to successfully fly his 250,000cu ft airship from Alexandra Palace
Royal Balloon Factory produce two experimental envelopes with £2,000 grant from the War Office

1906   E.T. Willows builds a series of successful small airships in Cardiff
Lieutenant Colonel J.E. Capper becomes superintendent of Royal Balloon Factory

1907   Royal Balloon Factory moves to Farnborough
Semi-rigid airship *Nulli Secundus* is completed and flies to London, landing at Crystal Palace

1908   *Nulli Secundus* rebuilt but dismantled after disappointing performance
Plans for rigid airship *Mayfly* commence

1909   First flight of *Baby*, 22,000cu ft, at Farnborough
Messrs Vickers Son & Maxim commence building *Mayfly*

1910   *Baby* rebuilt and lengthened as HMA *Beta* of 50,000cu ft capacity; the first airship to be fitted with wireless telegraphy apparatus. She flies from Farnborough to London and back at night
Large *Clément-Bayard* and *Lebaudy* airships bought from French by public subscription, both making crossing of English Channel
E.T. Willows' airship *City of Cardiff* flies from London to Paris
Non-rigid HMA *Gamma* is launched at Farnborough

1911   Air Battalion of the Royal Engineers formed at Larkhill in April
Rigid airship *Mayfly* is launched: does not fly, breaks back entering shed, becomes total loss
Non-rigid HMA *Delta* is launched at Farnborough

1912   April: Royal Flying Corps formed
         HMA *Beta* and HMA *Gamma* take part in military war games

1913   January: Admiralty takes control of all airships
         Admiralty authorise new rigid, *R9*, from Vickers Son & Maxim (cancelled December
         1914, building reinstated April 1915)
         HMA *Eta* is launched at Farnborough

1914   German Parseval *PL18* and French Astra-Torres *XIV* are purchased by Admiralty
         as HMA *No. 3* and HMA *No. 4*; both make first war patrols over BEF crossing to
         France at outbreak of First World War

1915   Prototype Sea Scout non-rigid airship ordered by Churchill and Lord Fisher – over
         fifty SS airships are built over the war period
         First of thirty-four larger, Astra-Torres tri-lobe Coastal non-rigids built

1916   First of six SSP non-rigids built
         German Zeppelin *L33* brought down in Essex, she becomes the basis of design for
         R33 and *R34* class rigid airships

1917   September: First flight of *R9* rigid airship at Barrow
         First of seventy-seven SS Zero class non-rigid airships enter service
         First of fourteen SS Twin class non-rigid airships enter service
         First of eighteen North Seas class non-rigid airships enter service – these were the
         largest British non-rigids, of 360,000cu ft capacity
         Rigid airships *R23*, *R24* and *R25* launched

1918   Rigid airships *R26*, *R27* and *R29* launched
         June: Requirements are drawn up for a new class of scouting airship of 3 million
         cu ft capacity
         Revised Admiralty A design issued
         5 August: German Zeppelin *L70* is destroyed off Great Yarmouth, large quantities
         of the wreckage are recovered immediately
         August: First flight of the Short Brothers' wooden-framed airship *R31*
         Design work on *R38* commences (dangers inherent of lightened high-altitude
         design are recognised at an early stage)
         September: Order for building *R38* is placed with Short Brothers at Cardington
         October: Design work on *R38* is completed within eight weeks
         November: *R31* is laid up at Howden due to severe girder breakages on delivery flight

1919   February: Construction starts on *R38* by Short Brothers at Cardington
         Work is suspended on *R37* and temporarily on *R38*
         March: Work recommences on *R38*
         First flight of *R33* (Armstrong Whitworth to Barlow)
         First flight of *R34* (Beardmore to Inchinnan)
         July: *R34* completes double crossing of Atlantic
         Non-rigid North Seas class airship *NS11* is struck by lightning off Norfolk coast
         with loss of crew of ten
         September: First flight of Short Brothers' second wooden-framed airship, *R32*
         October: Almost all existing RNAS non-rigid airships are struck off charge

22 October: Air Ministry becomes responsible for all aspects of airship design

5 December: Contract for purchase by US navy of *R 38* completed

1920 March: Training agreement made between Air Ministry and the US navy

20 March: *R 32* is flown from Pulham to Howden, hung up and deflated awaiting USN contingent

20 April: Commander Maxfield and USN contingent arrive at Howden

8 March: Non-rigid *SSE 3* at Howden is placed at disposal of USN for training

May: *R 33* (Pulham) and *R 34* (Howden) are both deflated for reasons of economy and hung in sheds – available to USN for instruction purposes only

8 June: *R 33* is re-inflated – first training flight with USN crew

June: Non-rigid North Seas class airship *NS 7* is based at Howden for training USN crews

19 July: First flight of Vickers airship *R 80*

July: German Zeppelins *L 64* and *L 71* are surrendered at Pulham by German crews

Air Ministry, on instructions from Treasury, must either sell or give away to private consortia, or sell for scrap, all airships

August: *R 34* is re-inflated at Howden

11 August: *R 32* makes first training flight with USN crew

1921 27 January: *R 34* has first training flight with USN crew, collides with hill, is damaged and subsequently destroyed whilst moored out in gale

January: Air Ministry disbands the Airship Service

March: *R 32* performs last training flight with USN crew (203 hours in total)

30 March: *R 80* is placed at disposal of USN crew

April: *R 32* is deleted in strength tests at Howden

R 36 is equipped for passenger carrying and makes first flight (Beardmore to Inchinnan)

May: Government is obliged, due to serious economic situation, to stop work on *R 37*

June: Last flight of *R 36*, after accident at mooring mast, laid up at Pulham, subsequently the damaged bow repaired, scheduled for experimental route proving flight to Egypt in 1926, but cancelled (dismantled June 1926)

Zeppelin prize *L 64* broken up at Pulham in order to re-house damaged *R 36*

23 June: First flight of *R 38/ZR2* (Cardington transferred to Howden)

24 August: *R 38/ZR2* destroyed over Humber on fourth flight prior to proposed Atlantic crossing

September: *R 80* laid up at Pulham (dismantled mid-1925)

1922 May: Imperial Airship Service schemes are under consideration

1923 June: Zeppelin prize *L 71* is broken up at Pulham

1924 August: *R 36* is repaired at a cost of £13,800 for route proving flight to Egypt

Labour government of Ramsay MacDonald implements Imperial Airship Service (*R 100* and *R 101*)

1925 April: *R 33* breaks away from high mast at Pulham, drifts over North Sea, is safely returned to base, repaired/reconditioned and returned to service

1926   June: *R36* broken up at Pulham
            October/November: Aeroplane dropping experiments (Gloster Greebs) at Pulham
            November: *R33* performs last flight and is laid up at Pulham

1927   Contracts for 5 million cu ft capacity airships are awarded to the Airship Guarantee
            Company, Howden, Yorkshire for *R100* and the Royal Airship Works Cardington,
            Bedfordshire for *R101*

1928   Construction commences on *R100* and *R101*
            R33 is dismantled after strength tests at Pulham

1929   October: First flight of *R101* rigid airship at Cardington
            December: First flight of *R100* rigid airship at Howden

1930   July/August: *R100* undertakes a transatlantic flight to Canada and back
            October: *R101* attempts flight to India and crashes near Beauvais, northern France
            Imperial Airship Service scheme ends

APPENDIX 3

The *LZ-127 Graf Zeppelin*

*T*he Graf Zeppelin, *as the most famous airship ever built, has been described in great detail in numerous books over the years by more able authors, so I intend only to present a brief sketch of her career in this section.*

Following the delivery of the *LZ-126 Los Angeles* to the US navy in October 1924, Hugo Eckener turned his attention to what he saw as the next stage in establishing a regular transatlantic airline service to North and South America. His plans were assisted by a clause in the Locarno Treaty of 1925 that finally removed the earlier Allied powers' restrictions on the construction of airships and aircraft in Germany.

Resorting to his pre-war strategy for raising the finance to build a new airship, Eckener organised a public subscription to raise the money. By touring the country and lecturing in the major cities he raised 2 million Marks from the general public, augmented by a further 2 million Marks from the Department of Transportation of the Weimar government, followed by additional government support.

The Construction of the *LZ-127 Graf Zeppelin* commenced in March 1927 at the Friedrichshafen works, making her first flight on 18 September 1928. The *Graf Zeppelin* was constructed of Duralumin and had an overall length of 776ft and a maximum diameter of 100ft. Eckener had wanted to build an airship of larger capacity, but the size of the completed airship was limited by the dimensions of the existing shed.

The total volume of the hull was 4,036,660cu ft, but 1,000,000cu ft of this total was devoted to separate cells in the lower portion of the hull that held a gaseous fuel, *Blaugas* (a propylene/butylene/hydrogen mixture). This fuel, which was only slightly heavier that air, could be used in place of the many tons of petrol that would be required on a long voyage, with this saving being devoted to additional useful lift. The airship was powered by five 540hp Maybach VL-2 twelve-cylinder vee engines, which were capable of (and frequently were) running on either *Blaugas* or petrol, giving a maximum range of 8,400 miles at cruising speed.

The total capacity of the hydrogen-filled lifting gas cells was 3,036,660cu ft, which imparted a gross lift of 105 tons of which 52 tons were available as useful lift. The hull structure was composed of seventeen main longitudinals, interspersed with eleven intermediate longitudinals. The main transverse frames were set at 15m, with two intermediate frames between, enclosing sixteen gas cells.

The airship carried a normal crew of twenty-six, together with twenty to twenty-four passengers. The navigating and passenger accommodation was arranged in a long gondola attached at the bow, comprising control and navigation rooms, wireless room and galley, behind which was a lounge/dining room and twelve double cabins for day and night use. The crew quarters were situated in the keel, with the officers' positioned forward and above the control car, and crew members' to the rear.

In a remarkable and distinguished flying career, filled with incident, over nine years the *Graf Zeppelin* made 590 flights covering 1,059,000 miles in 17,100 hours' flying time, whilst carrying 13,000 passengers. She became the longest lived of the rigid dirigible airships.

The airship was taken out of service following the crash of the *Hindenburg* at Lakehurst, New Jersey in May 1937. She made her last flight from Friedrichshafen to Frankfurt am Main, where, together with the later *LZ-130*, she was demolished on the orders of Herman Goering in May 1940, bringing to an end the era of the rigid airship.

Highlights from history of this remarkable and iconic flying machine include 144 ocean crossings to North and South America. In February 1929 Eckener took the airship on a Mediterranean cruise that included flying over the Vatican, Cairo and over the Dead Sea. Here, due to the depressed geography of the Holy land, the *Graf Zeppelin* found herself flying at almost 1,000ft below sea level!

This was followed in August 1929 by a round-the-world flight. Starting in Lakehurst she went via Paris to Friedrichshafen to refuel, then on non-stop across Russia and Siberia to Kasumigaura in Japan. Here, the great airship was housed in a shed that had been dismantled in Germany at the end of the First World War and awarded to Japan as a war reparation. From here the airship flew across the Pacific Ocean to Los Angeles (the first direct crossing by air) and finally on to Lakehurst, New Jersey in a flight of twelve days' flying time and twenty-one days elapsed time.

During 1931 a scientific exploratory flight was made in conjunction with the Soviet Academy of Sciences to explore the little known regions of the Barents Sea and the Arctic Ocean. This flight commenced from Leningrad and flew over Franz Josef Land, Severnaya Zemlya, the Taymar Peninsular and on to Novaya Zemlya, retuning to Leningrad after a flight of sixty-two hours.

Throughout the 1930s the *Graf Zeppelin* made numerous passenger-carrying ocean crossings to Lakehurst and later to Recife and Rio de Janeiro in Brazil, and other excursions to Iceland and Svarlbard.

In her flying life the airship proved a remarkably reliable and safe form of transport, considering the state of aerial navigation and its technical development of the period. Whilst the *Graf* did initially encounter engine problems and suffered some minor structural damage in storms, she ably fulfilled her primary purpose as a long-distance passenger ship. Over a period of ten years she safely transported her passengers over thousands of miles of ocean and some of the wildest and most dangerous terrain on earth, in a period where the aeroplane range was limited to a few hundred miles.

The First Zeppelin Raid on the British Isles

Before the Great War there was a widely held belief amongst the British public that, in the event of a conflict with Germany, fleets of giant Zeppelin airships would launch bombing attacks on major cities within hours of a declaration of war. Such an attitude had been encouraged over years of mounting international tension, by such writers as Dr Karl Graves in his book *The Secrets of the Red Hohenzollerns*, or through the works of William le Queux, the novelist and self-styled secret agent. Whilst the possibility of actual invasion was described by Erskine Childers in *The Riddle of the Sands*, in which he revealed the potential dangers of a German landing on our unprotected east coast.

Apart from the danger of invasion from the sea, the development of the flying machine and the airship posed the threat of attack from the air; a line of thought that was brilliantly developed by H.G. Wells in his prophetic novel *The War in the Air*, of 1908. In this book Wells describes how a German attack on the United States by a fleet of giant airships destroys the American Atlantic fleet, and then obliterates New York before they are, in turn, overwhelmed by a superior Asiatic air fleet. The ensuing worldwide war leads to the breakdown of organised government and the total disintegration of civilised society.

During the first weeks of the actual war in 1914, as reports of the fierce fighting taking place in Belgium and France against the advancing and seemingly unstoppable German army were received, those at home lay uneasily awake at night expecting, by the hour, to hear the drone of the Zeppelin raiders overhead bringing death and destruction. Their fears were heightened when, on 26 August, two German army Zeppelins attacked Antwerp at night causing widespread damage, killing twelve persons and bombing a hospital. It was an act that outraged British and international public opinion alike.

Count Zeppelin's first airship had flown over Lake Constance in July 1900 and in the intervening fourteen years it had developed into a viable flying machine, first as a commercial passenger carrier, then to be employed for war purposes by the army and navy. At the outbreak of war the German army possessed ten airships, which they used in support of their advancing troops by bombing frontier fortresses and strong points, but in so doing quickly lost three Zeppelins to ground fire. The German navy, on the other hand, only had one Zeppelin in commission: the *L3* based at Fuhlsbüttel near Hamburg. By December 1914 this was joined by five new naval airships numbered from *L4–L8*, whose primary duty was to scout ahead of and protect the High Seas Fleet.

By Christmas 1914 the fighting on the Western Front had settled into the static trench warfare that was to characterise the four-year war; a situation that encouraged the high command to look for alternate ways to carry the war against the enemy. As early as August 1914 Rear Admiral Paul Behncke, second-in-command to Admiral von Ingenohl, commander-in-chief of the High Seas Fleet, had put forward a detailed plan for an attack by airships on the lower Thames, targeting the dock area, Woolwich Arsenal and the Thames forts. Alternative targets included Dover, Portsmouth, Manchester and the port facilities on the Humber. These areas were considered to be legitimate military targets, being 'defended places' as defined in the Hague Convention of 1899 governing the legality of dropping explosives from flying machines.

The authors of this plan were convinced that mass raids of this kind would have an important effect on civilian morale and sap the will of the British populace to continue the war, leading to a cessation of hostilities in Germany's favour. The kaiser initially refused permission for such raids but, under continued pressure from Imperial headquarters, in January he reluctantly agreed to limited attacks on east coast seaports and the lower Thames dock areas east of the Tower of London.

Britain's long preoccupation in viewing France as a potential enemy over the preceding century had allowed them to largely disregard the building of extensive defensive works along the east coast, a region that was now seen to be dangerously unprotected and at risk of invasion from the sea. It was along the flat coasts of Norfolk, Suffolk and Essex that the fear of invasion was most strongly felt; this fear being heightened by a hit-and-run attack on the English coast by ships of the German High Seas Fleet in the early hours of 3 November 1914.

In the early morning light the old gunboat HMS _Halcyon_, patrolling off the Cross Sands light vessel, sighted and challenged a group of unknown vessels emerging from the mist to seaward. _Halcyon's_ challenge was met with a withering blast of fire from the German battlecruisers _Seydlitz_, _Von der Tann_ and _Moltke_, the armoured cruiser _Blucher_ and three accompanying light cruisers.

Against these overwhelming odds the _Halcyon_ pluckily returned fire with her two 4.5in guns, before seeking shelter in a smoke screen laid by the destroyer _Lively_. This allowed the gun boat to turn away with minor damage and escape to the south-west. The German squadron standing off Great Yarmouth then proceeded to bombard the town for thirty minutes, causing extensive damage from their 11in high explosive shells, before retiring unmolested to their base at Wilhelmshaven.

In December the German battlecruisers again dashed across the North Sea to bombard Hartlepool, Whitby and Lowestoft in a daring attack, which once more took British defences by surprise. The shock effect of these raids was to raise the fear of an imminent full-scale landing on the east coast that required the army to permanently station two divisions of troops in the eastern counties to oppose the threat.

Also early in 1915, a squadron of pre-Dreadnoughts were anchored at Sheerness in order to protect the southern approaches and to hopefully counter any further incursions by the German fleet.

An initial attempt to raid the British Isles was then made on 13 January 1915 by the naval airships _L5_ and _L6_ from Nordholz, together with the _L3_ and _L4_ from Fuhlsbüttel near Hamburg. The airships flew westward into increasingly deteriorating weather conditions until, at 3 p.m. flying in heavy rain over the Friesian Islands, the air fleet was ordered to abandon the undertaking and return to base.

Six days later, on 19 January, a second attempt was made with the _L3_ (Kapitänleutnant Fritz) and _L4_ (Kapitänleutnant von Platen Halemund) taking off from Fuhlsbüttel at 10.50 a.m., whilst the _L6_ (Oberleutnant von Buttler-Brandenfels) departed Nordholz at 9.38 a.m.

Korvettenkapitän Strasser (the charismatic leader of Naval Airship Service, who was to develop the Zeppelin airship into a formidable weapon of war in the four-year struggle) was aboard the _L6_, anxious to take part in this epoch-making raid.

The instructions from the commander-in-chief of the High Seas Fleet ordering the raid included the instruction: 'Distant scouting mission to the west only HVB to be carried.' HVB, or _Handelsschiffverkehrsbuch_, was the German merchant service signal code book, which had already been compromised when a copy, captured by the Russians after sinking a German cruiser in the Baltic, had been passed on to the British. British intelligence soon realised that any reference in a wireless message from an airship including 'only HVB on board' indicated its purpose was to attempt a raid on the British Isles.

With the air temperature at 2° below freezing at take-off, the airships were able to carry a maximum bomb load of high explosive and incendiaries, with *L3* and *L4* carrying up to 1,100lb of bombs each, whilst the *L6*, starting from its base nearer the English coast, had a heavier 1,450lb load.

The *L3* represented the latest development in airship design and was 518ft in length with a diameter of 48ft. She had a gas capacity of 794,500cu ft of hydrogen, which imparted a gross lift of 23 tons of which 9.2 tons was available for useful load. Useful load consisted of fuel, water ballast, munitions and crew. Three Maybach CX engines of 200hp were capable of driving the craft through the air at a top speed of 47mph and a theoretical range of action of 1,000 miles.

Strasser, flying aboard the *L6*, detailed her commander to raid the 'England south and the Thames', whilst *L3* and *L4* were ordered to attack the 'Midlands and the Humber area'. The *L6*, taking the more southerly route, encountered severe icing and heavy rain, and after four hours' flying the crankshaft of her port rear engine broke. She was north-east of the Dutch island of Terschelling, 100 miles short of the English coast, and was forced to return the airship to base, much to Strasser's chagrin.

The *L3* and *L4* headed west in clear, icy weather on a bearing of 287° magnetic, which should have brought them to the mouth of the Humber. However, unknown to their commanders, the wind had gone about and a north-easterly wind had set them to the southward over the sea. In the open gondolas in sub-zero temperatures the crews struggled to keep the radiator water from freezing and the motors running, whilst both ships had to drop a third of their water ballast to maintain altitude as ice built up on their envelopes.

At 6.40 p.m. on the Norfolk coast at Ingham, a village a mile and a half from the sea, a man walking down a hill in the dark sighted 'two bright stars low down out to sea apparently one hundred feet apart moving slowly towards him'. He had unknowingly sighted the navigation lights of the *L3* and *L4* as they came in over the Ower Bank light vessel.

At 8.05 p.m. the *L3* crossed the Norfolk coast near Ingham, whilst the *L4* had made landfall ten minutes earlier in the vicinity of Mundesley. Aboard the *L3* Fritz immediately realised that his airship had been driven far south from his intended target and, after dropping parachute flares to orient himself and recognising Happisburgh and the Winterton light vessel, he changed course to head down the coast towards Great Yarmouth, rising to an attack altitude of 5,000ft.

Bearing down on the town the *L3* began her bomb run at 8.20 p.m. Fritz claimed to have been fired on by a battery and responded by dropping nine high-explosive bombs and seven incendiaries, which caused £7,000 worth of damage to buildings over a wide area, killing and wounding a dozen or so residents in the town. Two of the bombs falling in the St Peter Plain area killed a 72-year-old woman, Martha Taylor, who was returning home from shopping and 53-year-old Sam Smith, a shoemaker, who had been working in his shop. They were the first civilians to be killed by aerial bombardment in Great Britain.

After crossing the town Fritz turned the *L3* out to sea again, passing over the Corton light vessel then steering up the coast between the Cockle and Newarp light vessels until he came abeam of Cromer at 10 p.m. Here Fritz stood about to the north-east, returning across the North Sea bringing the *L3* safely back to her base after a flight of twenty-three hours.

On the ground, as the raid progressed, confusion reigned. Even the number of raiders was in doubt; coastguards at Cromer reported seeing six airships, whilst some members of the public said the attackers were aeroplanes. One woman who had seen the raider overhead later said it was 'the biggest sausage I have ever seen', whilst another described it as 'a church steeple travelling sideways'. Additionally, many people commented on the thunderous roar of the motors that added to the sense of panic and helplessness as the aerial visitants cruised, unmolested, across the darkened countryside on that icy cold and wild night.

Meanwhile, von Platen-Hallermund aboard the *L4* initially believed he had reached the southern shore of the Humber in accordance with his orders, and followed the coast northward (passing over Cromer that lay in darkness and was undetected by the airship's crew). Continuing along the coast the airship circled over Sherringham, where the *L4* released a salvo of bombs and incendiaries that caused some slight damage, before turning out to sea, presumably in the belief that they would find the north bank of the Humber and the city of Hull.

Realising his navigation must be in error von Platen-Hallermund again turned south, crossing the Norfolk coast at 9.50 p.m. near Hunstanton and bringing the airship down to 800ft. He dropped flares then aimed two bombs at the Hunstanton wireless station at 10.15 p.m., both of which missed. After circling the town uncertainly for fifteen minutes, the *L4* passed over the village of Heacham where two more bombs were dropped, then flew on to Snettisham where the *L4*'s eighth bomb badly damaged the church.

Continuing southwards the *L4*'s next bombs fell on the Sandringham estate, which caused Queen Mary, who was in London at the time with the king, to later be convinced that the royal couple were the objective of the raid, rather than a chance of fate. At 10.50 p.m., attracted by the lights of King's Lynn to the south-west, the *L4* bore down at speed on the city as the alerted authorities attempted to impose a blackout.

Here, von Platen-Hallermund, after claiming to have been illuminated by searchlights and fired upon, released seven 110lb high-explosive and six incendiary bombs on the streets below in retaliation. The *L4*'s fourth bomb fell in the narrow Bentinick Street and exploded with great violence, causing extensive damage to houses, trapping and injuring many in the rubble and killing a 14-year-old boy.

A further tragedy occurred a few streets away when a 26-year-old widow, Alice Gazley whose husband had been killed in December in the fighting in France, died when struck by shrapnel from a bomb whilst running for shelter in the street. Leaving a total of two dead, thirteen injured and much material damage the *L4* turned eastwards, flying over Norwich at 11.50 p.m., which, being fully blacked out did not attract the Zeppelin's attention.

Finally, at 12.30 p.m., the *L4* stood out to sea above Great Yarmouth, from where her commander reported by wireless that he had 'successfully attacked several fortified places between the Tyne and the Humber'. He then set course for Fuhlsbüttel, where she landed within five minutes of the returning *L3*.

The returning crews were hailed as heroes in their own country and received decorations for their bravery, whilst they were condemned as murderers and 'aerial pirates' in Britain.

In the wake of the raid assertions were made that the Zeppelins had been guided to their targets by a network of spies signalling out to sea, and many innocent motorists driving at night through the eastern counties were stopped and questioned over the coming months as possible enemy agents. These stories were so widespread that the member of parliament for King's Lynn, Holcombe Inglyby, was so convinced that such activities were true that he asked questions in the House and wrote a book to expose the 'spy scandal'.

Although the material damage and casualties incurred were minimal compared to later events, the significance of the raid was that, in spite of being protected by the world's most powerful navy, aerial warfare now exposed the civilian population to new dangers that the authorities seemed powerless to counter.

As a footnote to the raid on 17 February 1915, the *L3* and the *L4*, operating from Fuhlsbüttel again under the command of Fritz and von Platen-Hallermund, were detailed at 4 a.m. to attack British warships reported off the Skagerrak. After reaching the search area successfully without sighting the reported warships, both Zeppelins turned for home against a rising southerly wind.

Fighting against the increasing force of the wind the *L3* suffered engine failure off the Danish island of Lyngvig and while struggling south a second engine broke down, so that at 5.45 p.m. Fritz put the *L3* down on the Danish island of Fanoe. With all the sixteen man crew safe Fritz then set fire to the ship and his secret papers, before surrendering to Danish troops and internment.

Aboard the *L4* von Platen-Hallermund was also in trouble with two engine failures over 200 miles from his base, and was obliged to make a forced landing on the Danish coast at Blaavands Huk. In fierce winds the airship crashed into the raging surf. The crew leapt for their lives, which caused the lightened craft to whirl away on the storm, carrying four of their companions to a lonely death far out in the North Sea. Von Platen-Hallermund and the bedraggled, exhausted survivors then gratefully allowed the Danish police to lead them into internment; their part in the air war was over.

Bibliography

Abbott, Patrick, *Airship* (Adam & Dart, 1973)

Brandenfels, v. Buttlar, *Zeppeline Gegen England* (v. Hasse & Koeler, Leipzig, 1925)

Brooks, Peter W., *Zeppelin Rigid Airships* (Putnam, 1992)

Clarke, Basil, *The History of Airships* (Herbert Jenkins, London, 1961)

Davy, M.J.B., *Aeronautics: Lighter Than Air Craft* (HM Stationary Office, 1950)

Dudley, Ernest, *Monsters of the Purple Twilight* (Geo Harrap & Co., 1960)

Eckener, Hugo, *My Zeppelins* (Putnam, London, 1958)

Forster, and v. Gobel, *Afrika zu Unsern Fussen* (R.F. Koehler, Berlin, 1925)

Gamble, C.F. Snowden, *The Story of a North Sea Air Station* (Neville Spearman, 1928)

Herne, R.P., *Zeppelins and Super Zeppelins* (Bodley Head, London, 1916)

Higham, Robin, *The British Rigid Airship 1908–1931* (Foulis, London, 1961)

James, Admiral Sir William, *The Code Breakers of Room 40* (Saint Martin Press, 1952)

Leasor, James, *The Millionth Chance* (Hamish Hamilton, London, 1957)

Marsh, Lieutenant Colonel Lockwood, 'The Evolution of the Rigid Airship', *Air Annual of the British Empire* (1930)

Marben, Rolf, *Zeppelin Adventures* (John Hamilton, London, 1931)

Meager, Captain G.F, *Leaves from my Log Book* (1960)

Morris, Captain Joseph, *The German Air Raids on Great Britain* (Nonsuch Publishing, 1925)

Mowthorpe, Ces, *Battlebags* (Sutton Publishing, 1995)

Poolman, Kenneth, *Zeppelins over England* (Evans Bros, London, 1960)

Pritchard, S/l J.E.M., 'Rigid Airships and their Development', *Journal of the Royal Aeronautical Society* (1925)

Robinson, D. & Keller, C.L., *Up Ship* (Naval Institute Press, 1982)

Robinson, Douglas, *The Zeppelin in Combat* (Foulis, London, 1962)

Saint George Saunders, Hilary, *Per Ardua* (Oxford University Press, London, 1944)

Shute, Neville, *Slide Rule* (Heinemann Ltd, 1954)

Sinclair, Captain J.A., *Airships in Peace of War* (Rich and Cowan, London, 1934)

Sinclair, Captain J.A., *Famous Airships* (Frederick Muller, London, 1959)

Sprigg, Christopher, *The Air Ship* (Sampson Low, 1930)

Thetford Putnam, Owen, *British Naval Aircraft 1912 to 1958*

Tolland, John, *Ships in the Sky* (Fredrick Muller, 1957)

Vaeth, J. Gordon, *Graf Zeppelin* (Fredrick Muller, London, 1959)

Index

Academie de Sciences 24
Admiralty 57, 61, 63, 65, 67
Ahlhorn 153, 154
Air battalion 52
Air Ministry 172, 179, 181, 195
Aircraft carriers 228, 232
Akron (1911) 34
Akron (1931) 247, 255, 256, 257
Aldershot 41, 42, 43
Alexandra Palace 44
America (1911) 34
Amundsen, Roald 199, 243
Anti-Zeppelin Airship132, 133
Armstrong Whitworth 80, 157, 161, 163
Arnstein, Karl 147
Astra-Torres 70, 77, 78

Baby 47, 48
Bacon, Captain 61
Bacon, Roger 9
Bacton 120
Balfour, Arthur 87, 130
Balloon Company RE 41
Balloon factory 43
Baltic Sea 119, 179
Bannerman, Major Sir Alexander 52, 53
Barrow-in-Furness 60, 62, 97
Barton, Dr F.A. 44
BE2c 137, 186
Beardmore, Wm & Co. 178, 179
Beatty, Admiral Lord 105
Beauvais 255
Berg, Carl 32
Beta 49, 50
Bethmann-Hollweg 120, 229
Bitterfeld 75
Blanchard 20

Blaugas 249
Bocker, Kapitänleutnant A. 139
Bockholt, Kapitänleutnant 150, 151
Boer War 43
Booth, Captain Ralph 183, 251
Branckner, Sir Sefton 255
Brandon, Lieutenant Bath de 176
Breithaupt, Kapitänleutnant J. 120
British Expeditionary Force 77, 118
Buckingham Palace 118
Buttler-Brandenfels, Kapitänleutnant 120, 146, 148, 176

Cadbury, Lieutenant Edgar 177
Campbell, Commander C.I. 161, 221, 224
Capper, Colonel J.E. 43
Cardington 165, 213, 215, 218, 251
Cavendish Dock 61, 66
Cayley, Sir George 25
Churchill, Sir Winston 87, 117, 120, 128
Ciampino 236
Clark Russell 121
Clement-Bayard 51, 52, 53
Coastal class airships 95, 99–101
Coastal Star class airships 111–3
Code-breaking 120–3
Cody, Colonel S.F. 43, 60
Cologne 118
Committee of Imperial Defence 74, 177
Compiegne 187
Cromer 120
Crystal Palace 44
Cuffley 139

Daily Mail 52
Dardanelles 87, 96, 229
Delag 39, 40

Delta 55–7
Dietrich, Kapitän Max 147
Dietrich, Kapitänleutnant Martin 146
Dixmude 235–7
Durr, Dr Ludwig 147
Düsseldorf 119
East Fortune 170, 180
Eckener, Hugo 249
Ellsworth, Lincoln 242
Esperia 238
Eta 56

Farnborough 45, 54, 56, 57
Felixstowe 229
Felixstowe F2a flying boat 231–3
Fisher, Admiral Lord 93
Forlanini 79, 80
Freidrichshafen 381
Fritz, Kapitänleutnant 120

Gamma 48, 49
German High Seas Fleet 136
German Naval Airship Division 70
Gibraltar, Straits of 265
Giffard, Henry 26
Goodyear 263, 265, 266, 268
Graf Zeppelin 249, 250
Graf Zeppelin II 259, 261, 266
Grand Fleet 80, 101, 103, 104, 228
Great Yarmouth 120, 186
Green engine 48
Gross-Basenach airships 34

Haenlein, Paul 26, 27
Haldane, Lord 51, 52
Hall, Sir Reginald 121
HAV 274, 275
Height climbers 152, 153
Heinen, Captain Anton 197, 198
Hendon 254
Hippisley, Colonel Richard 121
Howden 222, 224
Hull 223
Humber River 223
Hunstanton 121

Imperial German Navy 75, 135, 136
Inchinnan 178, 179
India 255
Ismailia 255
Italia 241, 244, 245, 247–9

Jamboli, Bulgaria 150, 151
Jellicoe, Admiral Lord 69, 99
Jutland 104, 105

K class airships 263–6
Kaiser William II 154
Kaltleim glue 168
Karachi 255
Khartoum 151
Kiel 119
King's Bay 246
King's Lynn 120
Kingsnorth 77, 99, 100, 109
Kittinger, Major Joseph 273, 274
Koch, Kapitän 142, 153
Kolle, Kapitän 187
Krebs, Captain A. 28

La France 28, 30,
Lakehurst Naval Air Station 196, 258
Land, Franz Joseph 248
Lansdowne, Lt Cmdr Zachary, USN 198,
   199, 200
Lebaudy 18, 53
Leckie, Major R.W. 230
Leefe-Robinson 17, 136, 176
Lehmann, Kapitän Ernst 137
Lenoir, Étienne 26, 27
Lettow-Vorbeck, General von 150, 151
Little Wigborough 140, 177
Los Angeles 198, 239, 241, 242

M class airships (US) 266, 267
MacDonald, Ramsay 253
Macon 255, 257
Maitland, General E.M. 216, 223
Marix, Flight Lieutenant R.L. 119
Mathy, Kapitän Heinrich 128–30, 136
Maxfield, Commander A.H. 173, 182, 223
Mayfly 59–67
McCary, Captain 197
McMechan airship 71, 133
Mediterranee 236
Meusnier, Jean Baptiste 24
Mineola Field 180, 181
Moffett, Rear Admiral W. 197
Montgolfier Brothers 18
Montreal 251
Morane 124
Morning Post 52, 53
Mullion 243
Mussolini 243

National Physical Laboratory 61, 222, 223
Nobile, General Umberto 242, 243, 245, 246, 248
Nome 245
Nordholz 130, 155
Norge 16, 240, 242, 243
North Pole 246
North Seas class airship 110, 112
Norway 122, 133
Nulli Secundus 45–7

O'Gorman, Mervyn 52, 224
Ohio 200
Ostend 200

Packard engines 197
Parseval 34, 75, 77
Patoka 211, 241
Peizker, Felix 72, 73
Platten-Hallermund, Graf von 120
Pohl, Admiral von 120
Porte, Cdr J. 211, 239
Pratt, Hartley 87
Pritchard, Major J.E. 220–2
Pulham 107, 160, 181–3, 235

R1 (Mayfly) 62
R100 250
R101 253
R23 153
R24 158
R25 158
R26 158
R27 182
R29 163
R31 165
R32 167
R33 178
R34 179
R35 178
R36 192
R37 191
R38 215
R80 204
R85 205
R9 85
Renard Charles 28
Richmond, Colonel V. 253
Roma 196, 239
Rosendahl 200
Royal Air Force 172
Royal Engineers 41, 52

Royal Flying Corps 118, 129
Royal Naval Air Service 119, 128

Santos-Dumont, Alberto 28
Scapa Flow 228
Scheer, Admiral Reinhard 103, 105
Schutt-Lanz 70, 71, 138
Schwann, Cmdr Oliver 61
Schwarz, David 28, 31
Scot, Major G.H. 179, 180, 181, 251
Sea Scout airships 92, 94, 104
Shenandoah 195, 201, 241
Short Brothers 165, 166, 191, 215
Siemens-Shuckert 35
Skyship 269, 270
Slessor, Lt John 131
Spencer, Stanley 43–5
Spiess 81
Spitzbergen 16, 199
SR1 115
SS Twin 16, 113
SS Zero 107, 110
SSP 105, 106
St Paul's 44
Stabbert, Kapitän 133, 186
Strasser, Kapitän Peter 136, 139, 145
Sueter, Admiral Sir Murray 61
Sunbeam engines 192, 218

Tempest, Second Lt Wulfstan 142
Templer, Colonel J.L. 41–3, 142
Thompson, Lord 255
Tirpitz, Admiral von 69
Tissander, Gaston 27
Titania 245
Tondern 133
Treaty of Versailles 154

US Bureau of Aeronautics 216
Usborne, Commander Neville 93

Vadose 246
Vaniman, Calvin 34, 35
Vickers 55, 61, 80, 81, 203–5
Viktoria Luise 70

Wallis, Dr Barnes 87, 203
Wallney Island 97
Wann, Flt Lieutenant A.H. 172, 220, 223
Warneford, Flt Sub Lieutenant 124
Watson, Sir Hugh 69
Wellman, Walter 34, 35

Willows, E.T. 45, 93
Wormwood Scrubs 52
Wright Brothers 43
Württemberg, King of 35

Zeebruge 91, 229, 268
Zeppelin Airships (early)
  Deutschland 39
  Hansa 39
  LZ-1 33
  LZ-2 36
  LZ-3 36
  LZ-4 37
  LZ-5 38
  Sachsen 39, 72
  Schwaben 39
  Viktoria Luise 39
Zeppelin Airships (army)
  LZ-37 124
  LZ-38 123
  LZ-39 124
  LZ-83 231
  LZ-90 137
  LZ-96 139
  LZ-97 139
  LZ-113 213
  LZ-121 213
  Z6 120
  Z7 120
  Z8 120
  Z9 119
Zeppelin Airships (naval)
  L1 74
  L2 73
  L3 120
  L4 120
  L7 133
  L10 132
  L11 146
  L12 131

L13 129
L15 131
L16 138
L20 133
L22 141
L23 141
L24 142
L30 134
L31 136
L32 138
L33 140
L34 147
L36 148
L38 149
L41 186
L42 153
L45 187
L46 186
L47 186
L48 153
L49 189
L50 189
L53 152
L55 186
L59 150
L64 231
L70 152
L71 231
L72 231
LZ-57 150
Zeppelin Airships (civil)
  Bodensee 237
  Nordstern 236
  LZ-126 Los Angeles 241
  Graf Zeppelin 249, 250
  Graf Zeppelin II 259, 261, 266
  Hindenburg 256, 258–60
  Zeppelin NT 271, 272
  Zeppelin, Count Ferdinand 30, 40

The destination for history
www.thehistorypress.co.uk